GREAT EVENTS

1900-2001

GREAT EVENTS

1900-2001
REVISED EDITION

Volume 7
1993-1998

From

The Editors of Salem Press

SALEM PRESS, INC.
Pasadena, California Hackensack, New Jersey

Editor in Chief: Dawn P. Dawson

Managing Editor: R. Kent Rasmussen
Manuscript Editor: Rowena Wildin
Production Editor: Joyce I. Buchea
Photograph Editor: Philip Bader
Assistant Editor: Andrea E. Miller

Research Supervisor: Jeffry Jensen
Acquisitions Editor: Mark Rehn
Page Design and Graphics: James Hutson
Layout: William Zimmerman
Eddie Murillo

Cover Design: Moritz Design, Los Angeles, Calif.

Cover photos: Center image—Corbis, Remaining images—AP/Wide World Photos
Half title photos: Library of Congress, Digital Stock, AP/Wide World Photos

∞ The paper used in these volumes conforms to the American National Standard for Permanence of Paper for Printed Library Materials, Z39.48-1992 (R1997).

Library of Congress Cataloging-in-Publication Data

Great events : 1900-2001 / editors of Salem Press.— Rev. ed.
v. cm.
Includes index.
ISBN 1-58765-053-3 (set : alk. paper) — ISBN 1-58765-060-6 (vol. 7 : alk. paper)
1. History, Modern—20th century—Chronology. 2. Twentieth century. 3. Science—History—20th century—Chronology. 4. Technology—History—20th century—Chronology.
D421 .G627 2002
909.82—dc21

2002002008

First Printing

PRINTED IN THE UNITED STATES OF AMERICA

CONTENTS

COMPLETE LIST OF CONTENTS

VOLUME 1

VOLUME 2

VOLUME 3

VOLUME 4

VOLUME 5

VOLUME 6

VOLUME 7

VOLUME 8

GREAT EVENTS

1900-2001

Algeria Cracks Down on Islamic Militants

Algerian officials reported that security forces had killed dozens of Islamic militants in raids conducted over the past two days.

What: Civil war; Religion
When: November 2, 1993
Where: Algeria
Who:

LIAMINE ZEROUAL (1941-), president of Algeria from 1994 to 1999
CHADLI BENDJEDID (1929-), president of Algeria from 1979 to 1992
KHALED NEZZAR (1937-), minister of defense from 1992 to 1994
MOHAMMED LAMARI, a general charged with quashing Islamic opposition
ALI HUSSAIN KAFI (1928-), chairman of the High Committee of State from 1992 to 1994
BELAID ABDESSALAM (1928-), prime minister of Algeria from 1992 to 1993
DJILALI LIABES (died 1993), education minister of Algeria

Attacks on Islamic Militants

On the first days of November, 1993, Algerian security forces tracked down and killed seventeen Islamic militants in a single raid. Eleven more were killed in five smaller raids, and four suspects were killed in an operation intended to secure the release of three kidnapped French diplomats. The Interior Ministry identified two of the suspects in the kidnapping, who remained at large, as men sentenced to death in absentia for acts of terrorism.

Islamic militants had been blamed for the deaths of seven foreigners in the preceding two months. More than two thousand people had died in politically motivated fighting since January, 1992, when the Algerian government canceled elections that the fundamentalist Islamic Salvation Front (FIS) was predicted to win. Cancellation of the election led to retaliatory bombings and assassinations by Islamic fundamentalists, which the government countered with crackdowns and arrests.

The conflict escalated. Many of the various Islamic groups united in revolt against the government. Women and girls who were not wearing veils, in contravention of Islamic law, were threatened or even killed. Foreigners were warned to leave the country, and the mounting number of assassinations hastened their exit. Newspaper publishers, editors, and writers critical of fundamentalism were targeted for execution.

The government responded by staging raids and arrests, along with employing torture and terror campaigns in neighborhoods and villages known to harbor FIS supporters. The loyalty of the army became a source of concern because the FIS and other militant groups may have made inroads in recruiting soldiers to their cause.

A Clash of Value Systems

Algeria gained independence from France in 1962. The newly freed country was ruled by young, secular socialists who based their government on the European model of social democracy.

The nation faced various economic, political, and moral crises. High birth rates led to a large younger generation by the 1980's and 1990's. Youth had no memory of the revolution against France and thus no true sense of the country's history. In addition, the Algerian government and the educated, French-speaking professional class that had run the country since 1965 had lost touch with the growing ranks of poor urbanites recently uprooted from the countryside.

Cultural shock accompanied urbanization. Crowding, a crumbling school system, and inadequate work for the newly educated created a crisis for millions of Algerians. socialism and Communism provided a belief system and blueprint for the future for many newly decolonized countries, including Algeria. Over the years after independence, however, the socialist dream wore thin, and no other system of values replaced it. The poor grew poorer and the rich obviously richer despite the egalitarian promises of socialism.

The people of Algeria began to look back at their history, to a period of idealized Islam about one thousand years ago. Traditional customs made a resurgence. To many, the Islamic social system represented new hope. Many Muslims had as much hope for an Islamic regime as the preceding generation did for a socialist government.

Disillusionment had set in after three decades of single-party government. Graft and corruption widened the gap between the wealthy and the poor, opening the country to currents of protest similar to those of the 1979 Islamic revolution in Iran. The new generation of revolutionaries, including the FIS, proposed addressing moral corruption by reasserting Islamic values.

AP/Wide World Photos

Liamine Zeroual.

Beginning in 1990, the Algerian government cautiously reacted to public discontent by edging toward democracy. As the promised election campaign of 1992 unfolded, pollsters alerted the government that the FIS, the most vociferous Islamic revival group, appeared as though it would be an overwhelming winner. The agenda of the FIS included replacing French as the language of the educated with Arabic, rescinding many of the laws that had granted women status equal to that of men, and cleansing the society of many foreign ideas. The platform may have been electoral rhetoric, but the government, along with many secular, educated Algerians, reacted with alarm. The government called off the election and began persecution of the FIS and other radical Islamic groups. The militants fought back against police and government officials.

Consequences

Algeria's internal struggles threatened to spill over to the neighboring countries of Tunisia and Morocco. It did spill into France, where Algerian intellectuals and rebels fled. The French government became concerned about the possible influence of the FIS on the approximately three million French Muslims and warned that it would not accept the use of religion as a cover for political movements aimed at causing disorder. In 1994, France expelled known supporters of the FIS.

Algeria's High Committee of State called a national meeting of political parties in January, 1994, to select a new president. All the major parties except Hamas, a moderate Islamic party, boycotted the meeting, so the High Security Council, dominated by the army, appointed Liamine Zeroual, formerly defense minister and a general in the army. Zeroual soon released several FIS leaders from prison following promises from FIS leaders that the militants would respect a pluralistic political system. The FIS, however, refused to yield to demands for an end to violence. Armed opposition to the government included an attack on a prison that led to one thousand prisoners escaping. The Algerian government continued efforts to restructure the economy, reaching an agreement with the International Monetary Fund for standby loans exceeding $1 billion.

Laina Farhat

APEC Nations Endorse Global Free Trade

At a Seattle meeting, the APEC nations took major steps toward worldwide trade liberalization and supported efforts in the Uruguay Round of GATT trade negotiations.

What: Economics; International relations
When: November 18-20, 1993
Where: Seattle, Washington
Who:
BILL CLINTON (1946-1996), president of the United States from 1993 to 2001
RONALD HARMON BROWN (1941-1996), secretary of commerce of the United States from 1993 to 1996
WARREN CHRISTOPHER (1925-), secretary of state of the United States from 1993 to 2001
MICKEY KANTOR (1939-), trade representative of the United States under Clinton

Support for Trade Liberalization

Major achievements of the APEC (Asia-Pacific Economic Cooperation) meeting in Seattle, Washington, on November 18-20, 1993, include the unequivocal endorsement of a trade liberalization program and the decision to achieve free trade in the Pacific region as quickly as possible. The APEC nations agreed to push forward the Uruguay Round of the General Agreement on Tariffs and Trade (GATT) by promising several packages of trade liberalization offers.

The APEC leaders also scheduled a number of ministerial meetings to be held in 1993 and 1994. The environmental ministers had met in the spring of 1993 to discuss global warming and pollution control. The finance ministers were asked to discuss monetary and macroeconomic issues affecting the world. Trade ministers were asked to explore the implications of the successful conclusion of the GATT negotiations and the future role of the World Trade Organization. These meetings were designed to enhance cooperation among APEC members in trade and other economic matters.

The leaders agreed in Seattle to hold annual summits in addition to the ministerial meetings. These summits would provide an opportunity for leaders to become familiar with one another and help economic growth in the Pacific region. These annual leadership summits would also help in resolving some of the bilateral disputes among APEC members. The 1994 summit was scheduled to be held in Bogor, Indonesia, and the 1995 summit in Japan. The agreement on future meetings represented substantial change from 1991, when some of the APEC nations were reluctant to take part in the ministerial meetings.

The APEC powers also decided to impose a three-year moratorium on new members. No new members would be admitted into APEC during 1995, 1996, and 1997. The APEC leaders agreed to start an APEC Education Program that would involve considerable investment in upgrading human resources. The leaders also created two other agencies: the APEC Business Volunteer Program and the APEC Technology Transfer Exchange Center, which would help in the transfer of the latest technology and management skills to less developed countries. The Seattle meeting also produced an agreement to develop an APEC investment code. This code was seen as necessary to promote and protect foreign private investment in APEC nations.

Regional Economic Trade

The APEC forum was established in 1989. It consisted of eighteen nations (as of November, 1994) in the Pacific region: Australia, Brunei, Canada, Chile, the People's Republic of China, Papua New Guinea, Hong Kong, Indonesia, South Korea, Japan, Malaysia, Mexico, New Zealand, the Philippines, Singapore, Taiwan, Thailand, and the United States.

The primary goal of APEC was to restore the credibility of the idea of global free trade. APEC provides a multilateral forum for economic cooperation and investment assistance among its member nations.

The United States conducts more trade with the Pacific nations than with European nations. For example, in 1992, the United States exported $219 billion in goods and services to other APEC members. It is estimated that these export activities created 4 million jobs in the United States. Trade (both imports and exports) between the United States and other APEC nations totaled $533 billion in 1992. Trade between the United States and the European Community totaled only $197 billion in 1992. These numbers demonstrated that APEC nations are major players in world trade. The World Bank estimated that half the growth in world trade and half the growth in global gross national product from 1990 to 2000 would occur in Asia.

APEC has a permanent secretariat to assist in logistical support for APEC meetings and meetings of the various working groups. Major APEC activities during a year are the annual ministerial meetings, where economic ministers meet to discuss APEC's progress and review world economic conditions, and the annual leadership summits. Between these two meetings, ten working groups meet regularly to develop issues for consideration by ministers. Various working groups focus on investment, industrial science and technology, trade promotion, trade and investment data review, regional energy cooperation, human resources development, telecommunications, transportation, tourism, marine resources conservation, and fisheries. In addition, APEC also created, in 1992, the Eminent Persons Group (EPG). The EPG comprises a prominent nongovernment economist from each of the member countries. The EPG is entrusted with the responsibility of developing a long-term vision statement for economic relations in the Pacific region and recommending ways to achieve this vision.

Consequences

A major consequence of the 1993 Seattle meeting was the successful completion of the Uruguay Round of GATT negotiations late in 1993. Other consequences can be summarized by examining the events at the 1994 Bogor summit. The APEC nations signed a major declaration in Bogor, Indonesia, to ease trade restrictions. The highlights of this declaration include an agreement to achieve free trade in the Pacific region by the year 2020, removal of trade barriers by industrialized members of APEC by 2010 and by the developing members by 2020, and establishment of a voluntary dispute mediation organization. At the Bogor meeting, the APEC leaders agreed on a set of investment principles.

A key objective of APEC is to prod full implementation of the GATT agreement. A series of seminars on GATT implementation issues was planned. The United States agreed to host a transportation ministerial meeting in 1995, and South Korea agreed to host a telecommunications ministerial meeting in 1995. The Bogor meeting also set up a permanent business advisory body to APEC. This advisory body will relate private-sector business concerns to the APEC ministers.

Srinivasan Ragothaman

Clinton Signs Brady Gun Control Law

After seven years of debate, Congress passed and the president signed legislation requiring a five-day waiting period for the purchase of a handgun.

What: Law; Crime
When: November 30, 1993
Where: Washington, D.C.
Who:
JAMES S. BRADY (1940-), press secretary to president Ronald Reagan injured in an assassination attempt
BILL CLINTON (1946-), president of the United States from 1993 to 2001
RONALD REAGAN (1911-), president of the United States from 1981 to 1989

New Federal Gun Control Legislation

On November 30, 1993, President Bill Clinton signed PL 103-159, the Brady Law, requiring a five-day waiting period for the purchase of a handgun. The legislation (formally known as the Brady Handgun Violence Protection Act) was named for former presidential press secretary James Brady, who was seriously wounded and disabled in a 1981 assassination attempt on President Ronald Reagan. Brady had become a tireless advocate of gun control legislation.

The law mandated a waiting period of five business days for obtaining a handgun. The waiting period was designed to serve two purposes—to allow an impulsive purchaser an opportunity to "cool off" and to require that sellers run background checks on prospective gun buyers. The final version of the bill provided that the waiting period be phased out in five years. It was expected to be replaced with a national computerized instant check system capable of scanning criminal records to identify felons and prevent the sale of firearms to them. The law also raised licensing fees for gun dealers and required the notification of law enforcement agencies in the case of multiple gun purchases.

In its final form, the bill represented a compromise with respect to when the waiting period provision would be phased out. Gun control opponents argued for a phase-out after twenty-four months. The bill's supporters wanted certification from the attorney general of the United States that the computerized check system was operational before the waiting period provision was allowed to expire. The bill authorized $200 million per year to help states to computerize their records.

Observers believed that the passage of the Brady Law was evidence of strong public support for gun control as a means of fighting crime and also of a decline in the political influence of the National Rifle Association (NRA). A decision by the Senate to separate the waiting period provisions from a larger, more complicated, and controversial anticrime bill accelerated the Brady Law's passage. The Brady Law was the first significant gun control legislation approved by Congress since 1968, when it outlawed the importation of cheap handguns in the aftermath of the assassinations of the Reverend Martin Luther King, Jr., and Senator Robert Kennedy.

The Right to Keep and Bear Arms

Most attempts to pass gun control legislation have run into strong resistance from the NRA. The NRA argues that restrictions on firearms violate the Second Amendment to the Constitution, which reads, "A well regulated Militia, being necessary to the security of a free State, the right of the people to keep and bear Arms, shall not be infringed." The NRA maintains that restrictions on gun sales would harass law-abiding citizens but would have little effect on criminals, who tend to purchase their weapons on the black market.

Until the passage of the Brady Law, the NRA was considered invulnerable. Its membership of

AP/Wide World Photos

Nearly five years after signing the Brady gun control law, President Bill Clinton meets with James Brady to call on Congress to extend the law.

more than three million and its lobbying resources in excess of $18 million annually gave the organization a virtual veto over proposals in Congress to regulate guns. During the debate over the Brady Bill, when it became clear that some form of the legislation would pass, the NRA attempted to substitute an instant-check system for the waiting period.

The 1993 law was the culmination of a seven-year crusade by former White House press secretary Brady and his wife Sarah to persuade members of Congress that a waiting period for handgun purchases would save lives and reduce violence. Brady visited legislators in his wheelchair. The chair and his permanent injuries, the result of shots fired by John W. Hinckley, Jr., in his attempt to assassinate President Reagan, were visible reminders of the effects of gun-related violence. The bill also had the support of law enforcement agencies and mayors of the nation's large cities, who maintained that the bill would save lives and prevent thousands of injuries. Advocates believed that the cooling-off period would force potential killers to reexamine their actions between the time they applied to purchase a gun and the time they actually acquired it. In addition, a gun sale would be refused if the records check revealed that the potential purchaser was a convicted felon or someone with a history of mental illness. Background checks were believed likely to produce the names of fugitives seeking to rearm.

The Bradys' persistent efforts to persuade members of Congress to support their proposal received a boost when Reagan, a longtime NRA member, endorsed it. By 1993, a combination of strong anticrime sentiments among the public and the support of President Clinton made passage possible. Until the final vote, however, supporters feared that the NRA's allies would attach amendments that would weaken the bill fatally.

Consequences

The Brady Law went into effect on February 28, 1994. It applied immediately to thirty states that had no waiting period regulations. In other cases, states moved to introduce computer systems that provided an immediate background check on gun purchasers. Measurements of the law's effectiveness in its first year indicated that forty thousand felons were prevented from buying handguns. It is more difficult to determine whether the law itself has actually reduced crime.

Some state officials have complained about the law's loopholes. The prohibition of sales to persons with a history of mental illness is impossible to enforce because medical records are confidential and do not show up on computer checks. Officials also perceive problems in identifying members of other prohibited groups—illegal aliens and persons with dishonorable discharges from the military. In addition, the background check provision does not apply to "private" gun sales at flea markets or to guns reclaimed from pawnshops. The NRA has encouraged lawsuits by several sheriffs, who have challenged the law both on constitutional grounds and as an unfunded federal mandate.

Mary W. Atwell

General Agreement on Tariffs and Trade Talks Conclude

Successful completion of decades of international negotiations on the reduction of trade barriers opened the way for freer world trade and for more vigorous economic development.

What: Economics; International relations
When: December 15, 1993
Where: Geneva, Switzerland
Who:
Bill Clinton (1946-), president of the United States from 1993 to 2001
Mickey Kantor (1939-), American special trade representative who concluded negotiations
Peter Sutherland (1946-), international director of GATT negotiations
Renato Ruggiero (1928-), first head of the World Trade Organization
Carla Anderson Hills (1934-), American special trade representative

Concluding Negotiations

Years of delicate and complex international negotiations over the fate and final contents of the General Agreement on Tariffs and Trade (GATT) hung in the balance late in 1993. Marathon talks between American special trade representative Mickey Kantor and the European Community's (EC) exterior minister, Sir Leon Brittan, had been conducted over the previous weeks. These talks, involving the future economic relationships of 117 nations, had moved from Toronto, Canada, to New York City, then to Geneva, Switzerland. The special "fast track" negotiating authority given to President Bill Clinton by the U.S. Congress was about to run out. That authority allowed Clinton to reach an agreement without facing the usual congressional power to renegotiate it in detail. Clinton's special authority was due to expire in March, 1994. If it did, decisions about the U.S. role in GATT would pass to Congress. Congressional leaders already were signaling their opposition to the president's policy favoring freer international trade. He had committed the United States to the successful completion of the seventh round of talks on GATT, the Uruguay Round, begun in 1989. Clinton also had pushed for approval of the North American Free Trade Agreement (NAFTA), which passed in the House of Representatives in November, 1993.

Last-minute compromises made possible the signing of the Uruguay Round talks in Geneva on December 15, 1993. The text of the agreement was nearly five hundred pages and contained forty separate agreements embodying the efforts of fifteen working groups. Although numerous details remained to be worked out, the Geneva agreement provided a number of broad guidelines for GATT members. Industrial tariffs were to be reduced by one-third, and some were to be eliminated. Comprehensive protection was to be afforded to intellectual property, copyrights, and trademarks. Over a six-year period, agricultural subsidies were to be lowered, and outlines were established for the global regulation of financial services. Subsidies to basic and applied research were to be curtailed, and clearer standards and regulations were to be drafted to strengthen measures against unfair trade practices known as "dumping." Finally, to replace GATT, a World Trade Organization was to be established as a permanent oversight body.

Years of Conflict

GATT was launched as a specialized agency of the United Nations in 1947. The agency was supposed to have been placed under an International Trade Organization, but the United Nations failed to establish that agency. GATT functioned as a regular negotiating forum for multilateral trade agreements aimed primarily at reducing trade barriers. GATT sought to achieve this reduction by encouraging the removal of import quotas (restrictions on the quantity of a good that could be imported into a country) and the reduction of tariffs (import taxes) through general application of most-favored-nation arrangements. GATT negotiations also helped resolve certain problems created by countries' international balance of payments, or the difference between the values of imports and exports.

Of the numerous problems characterizing GATT talks, none was more persistent than the stalemate on agricultural subsidies. Industrialization throughout the developing world during the twentieth century had dramatically reduced the number of farmers. The few who remained included only 1.1 million American farmers, of whom fewer than 20,000 (primarily giant agribusinesses) exercised significant effects on markets. While demanding that other countries abandon subsidies in the interest of free trade, Congress, bowing to political pressure from states with farming economies, insisted on maintaining American agricultural subsidies. Similarly, Japanese governments fought tenaciously to retain subsidies on native-grown rice; French governments stubbornly defended price supports for French wheat, cheese, wine, and other products; and the EC struggled to preserve its members' protective arrangements for such commodities as oilseed, cereals, and bananas.

Added to these agricultural problems were those involving tariffs and hidden subsidies that shielded from foreign competition such items as industrial goods including cars, electronics, and textiles; airlines; the shipbuilding industry; and shipping. Battles over these and other barriers to freer trade had stalled GATT talks badly in 1979. Another decade passed before they were resumed in Uruguay in 1989. Carla Hills began that round of negotiations as the U.S. representative.

Consequences

Global trade regulations were signed in Marrakech, Morocco, in April, 1994. On July 1, 1995, GATT member nations were to begin lowering tariffs on a wide range of goods. After heated battles, the United States and most other GATT participants agreed to reduce agricultural subsidies, the United States by 43 percent and the EC by 50 percent. In December, 1994, both houses of Congress approved the Geneva agreements by significant majorities. Dividing along hemispheric lines, GATT nations then jostled to select the leader of the new World Trade Organization. The United States and Western Hemisphere nations backed Mexico's president, Carlos Salinas, but lost to the European Community's choice, Italian diplomat Renato Ruggiero.

The successful conclusion of GATT negotiations was applauded by President Clinton as "a vision of renewal," but French premier Eduoard Balladur, along with many others, was less enthusiastic. Economists around the world were generally optimistic, predicting that the dropping of agricultural subsidies alone would save the developed world's consumers $300 billion annually in taxes and higher prices. They estimated that the new world trading arrangements would add at least $230 billion per year to worldwide production.

Clifton K. Yearley

Astronomers Detect Black Hole

Astronomers used the Hubble Space Telescope to find evidence for the existence of a black hole in the center of galaxy M87.

What: Astronomy
When: 1994
Where: Space Telescope Science Institute, Baltimore, Maryland
Who:
Holland C. Ford (1940-   ) and
Richard Harms, American astronomers who investigated M87
Karl Schwarzschild (1873-1916), a German physicist and astronomer
Albert Einstein (1879-1955), a German American physicist whose work provided the theoretical background for understanding black holes

Stars' End

A black hole is a region of space where the force of gravity is so strong that no matter, light, or communication of any kind can escape. Details of the structure of a black hole are calculated from Albert Einstein's general theory of relativity and from other more recent theories of gravity. The existence of black holes was predicted in 1916 by German astronomer Karl Schwarzschild, who used Einstein's theory to estimate what the radius of a black hole would be.

Astronomers believe that a black hole occurs when a sufficiently massive star runs out of nuclear fuel and is crushed by its own gravitational force. If the star's mass is more than three times the mass of the sun, a black hole will be the final fate of the star. Stars of smaller mass will evolve into less compressed bodies, either white dwarf stars or neutron stars.

Although the existence of black holes had long been theoretically predicted, proving their existence was not easy. Black holes are impossible to observe directly, since they emit no light. Yet researchers believed that it would be possible to find black holes indirectly, by observing the effects of their enormous gravitational fields on nearby matter. Early searches concentrated on observations of binary star systems that emitted X rays. A binary star system is a pair of stars that revolve about each other. If a black hole were a member of a binary star system, researchers reasoned, matter would flow into it from its companion star; this matter would become intensely heated and would radiate observable X rays before spiralling into the black hole.

Another scenario that might indicate the existence of a black hole would occur if a very massive black hole were located at the center of a star cluster or galaxy. The tremendous force of gravitational attraction provided by the black hole would result in observable increases in the densities and velocities of stars near the center. Thus, large volumes of interstellar gas and dust would collect and collapse into these black holes, giving off enormous amounts of energy.

Finding Something Invisible

Within the Virgo cluster of galaxies, approximately 50 million light-years from Earth, lies a giant elliptical galaxy known as M87. Since 1917, it has been known that M87 has a bright, optical jet of light emanating from its central region. In 1991, the Faint Object Camera of the Hubble Space Telescope took an image of M87 that revealed unprecedented details. Around the central region were found numerous globular clusters, each containing hundreds of thousands of stars. This central cluster of stars is thousands of times denser than the spread of stars in the neighborhood of our sun, and at least hundreds of times denser than expected for a galaxy like M87. The jet was found to be approximately 40,000 light-years long, with features as small as ten light-years across. Astronomers suspected that the jet was associated with a massive black hole.

Stronger evidence than the mere concentration of starlight near the core of M87 was needed to confirm the existence of a black hole; however, astronomers insisted on seeing the actual movement of matter orbiting the black hole. In 1994, the Hubble Space Telescope provided what many astronomers considered to be conclusive evidence of a black hole at the heart of M87. Holland C. Ford, an astronomer and physicist with the Space Telescope Science Institute, and Richard Harms, an astronomer with the Applied Research Corporation, were investigating the photographs from the Hubble Space Telescope. Their analysis revealed a disk of gas, only sixty light-years out from the center of M87, that was moving at more than 1.6 million kilometers/hour (about one million miles/hour). As Ford and Harms pointed out, the only known explanation for such a massive disk of gas moving at such high velocity so close to the galaxy's center would be the presence of a black hole with a mass equal to two to three billion suns, with a size no larger than our solar system.

Consequences

Do black holes really exist, and are they accessible targets for study by astronomers? Most astronomers now answer in the affirmative. The discovery of an orbiting disk of gas about the center of M87 in 1994 represents overwhelming evidence for the existence of a black hole. In fact, most astronomers now conclude that the enormous energy associated with galactic systems, such as M87, can only be attributed to the collapse of millions or billions of solar masses of interstellar gas and dust into a supermassive black hole. It is very exciting for astronomers to find an object that was once considered to exist only in the realm of science fiction. The discovery of the black hole in M87 gives astronomers hope that they may find other black holes embedded in the centers of other distant galaxies.

In the very early universe, the density of matter was extremely high, and it is believed that some regions of enhanced density most likely collapsed to form black holes. Consequently, many astronomers believe that by observing and studying black holes, they may unlock the secrets of the origin of the universe itself. The detection of a black hole in galaxy M87 in 1994 provided renewed optimism for finding some ultimate answers that will transcend the present understanding of the laws that govern physical science.

Alvin K. Benson

Hubble Space Telescope Transmits Images to Earth

The Hubble Space Telescope has provided astronomers with clear images of distant objects in the universe.

What: Astronomy
When: January, 1994
Where: Earth orbit
Who:
HERMANN OBERTH (1894-1989), a German rocket scientist
STEVEN A. HAWLEY (1951-), an American astronaut
RICHARD O. COVEY (1946-), one of two pilots of the *Endeavour* repair mission
KENNETH D. BOWERSOX (1956-), the other pilot
CLAUDE NICOLLIER (1944-),
STORY MUSGRAVE (1935-),
JEFFREY HOFFMAN (1944-),
KATHRYN C. THORNTON (1952-), and
TOM AKERS (1951-), *Endeavour* astronauts

An Orbiting Telescope

Before the space age, distant objects could be studied only through telescopes on Earth's surface, but Earth's weather creates problems for astronomers making the observations. Winds blur the images, and astronomers cannot observe when the skies are cloudy. Information contained in some parts of the electromagnetic spectrum cannot penetrate the atmosphere. In the 1920's, Hermann Oberth suggested that telescopes be placed in orbit above Earth to avoid weather-related problems. The National Aeronautics and Space Administration (NASA) began a study of Oberth's idea in 1962. Fifteen years later, Congress approved the plan, and a space telescope became an official NASA project. It later became known as the Hubble Space Telescope (HST) in honor of Edwin Hubble's important contributions to astronomy. The HST was designed to function somewhat like an Earth-based telescope. As the telescope orbited Earth, two solar panels pointed toward the sun to provide energy for scientific instruments. A door at one end of the telescope opened and light struck the larger (primary) mirror, which was 94.5 inches in diameter. This mirror reflected light toward the smaller (secondary) mirror, which was 12.2 inches in diameter. From there, the light was again reflected and passed through a hole in the primary mirror. The focused light was converted to an electrical signal, which was transmitted by satellite to White Sands, New Mexico, and then on to the Goddard Space Flight Center and the Space Telescope Science Institute, both of which are in Maryland.

The scientific instruments on the HST helped scientists analyze the light that it gathered. The instruments included the wide-field/planetary camera, the faint-object spectrograph, the high-resolution spectrograph, the high-speed photometer, the faint-object camera, and fine-guidance sensors. All these instruments were modular in design so that they could be replaced in case of a system failure.

The HST was designed as part of a series of orbiting observatories dedicated to measuring many parts of the electromagnetic spectrum. The instruments on the HST make possible measurements of infrared and ultraviolet radiation as well as visible light. Other orbiting observatories in the series are the Gamma Ray Observatory, the Advanced X-Ray Astrophysics Facility, and the Space Infrared Telescope Facility.

The HST, which weighed approximately 25,000 pounds, was loaded into the cargo bay of the space shuttle *Discovery* and launched into orbit from Cape Canaveral, Florida, on April 24, 1990. On the following day, the telescope was deployed by astronaut Steven A. Hawley, using a 50-foot mechanical arm.

The HST's primary mirror was designed at the Perkin-Elmer Corporation in Danbury, Connecticut. Two months after deployment, astronomers discovered that technicians had made it slightly too flat, causing the starlight to be out of focus and producing fuzzy images. Once the defect was identified, scientists began planning a mission in which astronauts would repair the HST, allowing light to be focused as originally planned.

The "Rescue" Mission

On December 2, 1993, seven astronauts aboard the space shuttle *Endeavour* were launched into a 357-mile-high Earth orbit with the mission of repairing the HST. Aboard the shuttle they carried a number of instrument packages to repair and improve the telescope, including new gyroscopes, solar-power panels, and two packages containing the corrective mirrors that would bring into focus the starlight that had been unfocusable as a result of the faulty primary mirror. The astronauts planned to make at least five spacewalks during the eleven-day mission in order to install the new and corrective devices.

The most important of these devices were the corrective mirrors, which NASA had commissioned to a small firm, Tinsley Laboratories, located behind a shopping mall in the San Francisco suburb of Richmond. Despite its low profile, Tinsley was the right choice: Larger corporations such as Kodak, Hughes, and United Technologies had been unable or unwilling to meet NASA's exacting requirements for grinding the mirrors, and Tinsley not only ground thc mirrors to within a few atoms' length of their specified dimensions, but also accomplished this amazing feat well within the twelve-month schedule and at half the cost that NASA had anticipated.

The shuttle was piloted by Colonel Richard Covey of the Air Force and Commander Kenneth Bowersox of the Navy. Claude Nicollier, a Swiss astrophysicist with the European Space Agency, would be the one to operate the shuttle's mechanical arm, which would reach out to grab the telescope. The remaining four astronauts—Story Musgrave, a physician; Jeffrey A. Hoffman, an astrophysicist; Kathryn C. Thornton, a physicist; and Lieutenant Colonel Tom Akers of the Air Force—would perform the spacewalks necessary to repair the telescope: installing two new gyroscopes, replacing its tracking system (to enable it to point at appropriate celestial objects), re-

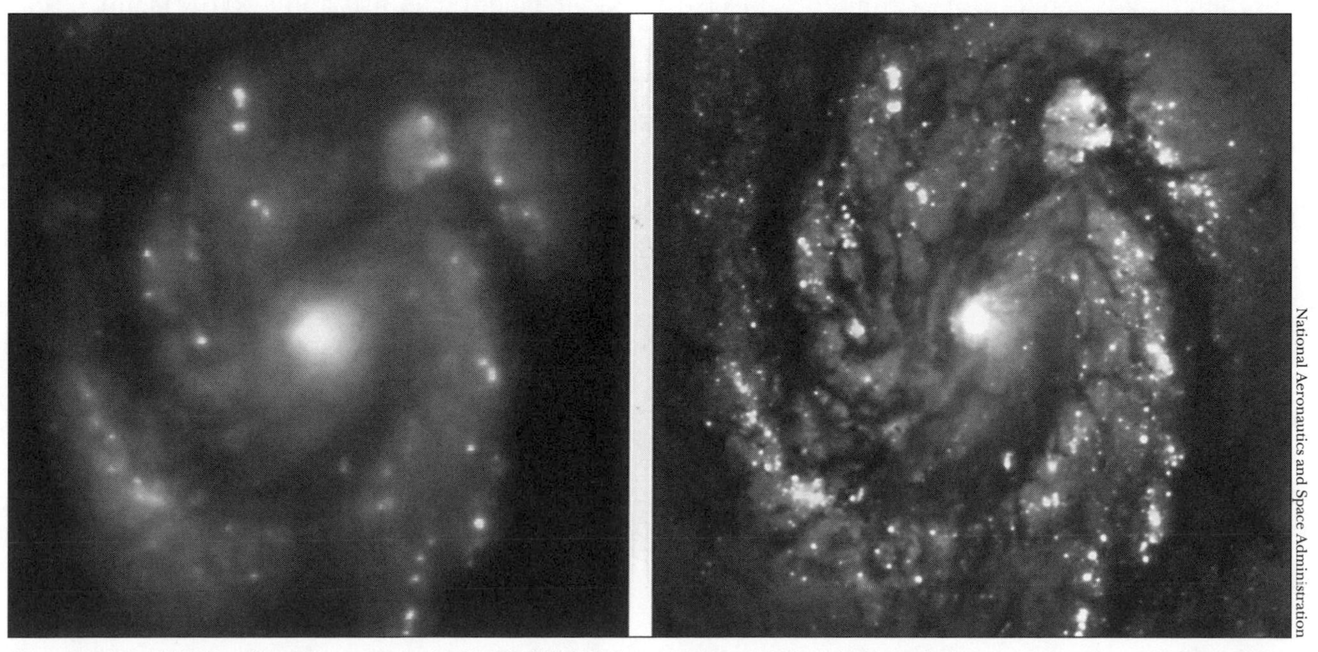

National Aeronautics and Space Administration

The core of the M100 galaxy as photographed by the Hubble Space Telescope before (left) and after the November, 1993, servicing mission.

placing wobbly solar arrays, and, most important, installing the wide-field planetary camera and a set of new instruments—the faint-object camera and two spectrographs.

As the world watched, the spacewalkers performed the intricate maneuvers to place the instruments and make the necessary repairs over the following days. The operations went smoothly, and when the shuttle returned safely to Earth, scientists and astronauts were ecstatic; the great scientific promise of the telescope, which had been delayed for nearly four years since the launching of the Hubble, now seemed on the verge of being fulfilled. When the telescope began to transmit its first images in January of 1994, the incredibly clear detail astonished astronomers as well as the general public. The Hubble, according to the mission's chief scientist Edward J. Weiler, had been "fixed beyond our wildest expectations," and James Crocker, who had overseen part of the corrective mirror project, stated of the test pictures: "These images are as perfect as engineering can achieve and the laws of physics allow."

Consequences

The original HST had cost $1.6 billion; the repair mission would cost $629 million. In comparison with the superconducting supercollider, another basic science project funded with taxpayers' money, these costs are reasonable: The supercollider cost $2 billion before it was completed and its funding was canceled in October of 1993; the Hubble Space Telescope, for about the same cost complete, promises to deliver answers to some of the most fundamental questions about the creation and workings of the universe until the year 2010. Astronomers will be able to see objects 10 to 12 billion light-years away—almost as old as the universe itself. Such sharp vision is comparable to seeing a firefly eight thousand miles in the distance.

With the data and images provided by the HST, scientists will probably be able to verify the existence of black holes, objects with such great gravity that not even light can escape from them. These objects have been theorized to exist for some time, but proof of their existence has not yet been found. In addition, the HST will take pictures of benchmark stars that will help scientists to calculate the universe's size and shape. Already, the HST has provided valuable pictures: The globular cluster 47 at Tucanae, 16,000 light-years away, for example, has revealed some previously unknown white dwarf stars; another picture, of the remnants of Supernova 1987A, is extremely sharp and clear.

Some have criticized the spending of government funds on basic science in view of the United States' overwhelming budget deficit and the need for reform of health care and other basic social systems. When compared with the projected costs of such programs, however, the funds directed at the HST are minimal, and the promised returns unimaginable. The repair mission alone taught scientists and engineers much that will be used on future missions that may pave the way toward long-term inhabitation of space. Moreover, basic research, far from being useless, has resulted in countless useful technologies, from the lightbulb to computers.

Indian Peasants Revolt in Mexico's Chiapas State

The Zapatista National Liberation Army, a guerrilla army of Indian peasants, seized several towns in Chiapas to demand that Mexican government officials end long-standing injustices.

What: Civil rights and liberties; Civil strife
When: January 1, 1994
Where: Chiapas, Mexico
Who:
Rafael Sebastián Guillén Vicente (1958-), spokesman of the Zapatista National Liberation Army
Carlos Salinas de Gortari (1948-), president of Mexico from 1988 to 1994

Indian Peasants Occupy Towns in Chiapas

On January 1, 1994, armed Indian peasants occupied several towns in the highlands of Chiapas. Chiapas is the southernmost state of Mexico, on the border with Guatemala. The Indian peasants, calling themselves the Zapatista National Liberation Army, declared war on the Mexican government. They claimed, through spokesman Rafael Sebastián Guillén Vicente (known as Comandante Marcos), that the government had neglected the Indian people of Chiapas and allowed their exploitation.

More than one hundred people died in four days of gun battles between the Mexican army and the Zapatistas. As many as two thousand guerrillas participated in the uprising and occupied at least seven towns in Chiapas before they disappeared into the mountains, pursued by the Mexican army.

The uprising occurred at an embarrassing time for Mexico's president Carlos Salinas de Gortari. January 1, 1994, the day the uprising began, was also the day that the North American Free Trade Agreement (NAFTA) took effect. That treaty, signed by Mexico, the United States, and Canada, removed barriers to trade among the three countries. Salinas declared that Mexico's participation in NAFTA meant that Mexico was well on its way to becoming a modern, prosperous nation. The uprising in Chiapas suggested otherwise.

Mexico still faced serious economic and social problems. In its drive to modernize the country, the Mexican government had failed to address the needs of large segments of its population who lived in dire poverty and were exploited by a few powerful and wealthy people.

Although the Mexican government tried to dismiss the uprising in Chiapas as the work of outside agitators from Guatemala and El Salvador, nearly all the rebels were Mexicans. They were motivated by their pent-up frustration with political corruption and by economic and social conditions that made it nearly impossible for them to improve the desperate condition of their lives.

The Struggle for a Decent Life

The Zapatista uprising occurred not only because Chiapas was one of the poorest states of Mexico but also because the Indian peasants of Chiapas could not prevent their communal lands from falling into the hands of wealthy rural bosses and ranchers, the caciques.

For decades the government of the state of Chiapas had failed to prosecute the caciques for illegally seizing Indian lands. Instead, the government attacked the Indians for resisting confiscation of their land. The caciques stole the land to enlarge their farms and ranches and to increase their production of coffee and cattle for export.

With only 3 percent of Mexico's people, Chiapas produced one-fourth of its land disputes. By 1990, fifteen thousand Indians were in prison on charges related to land conflicts. As friction in-

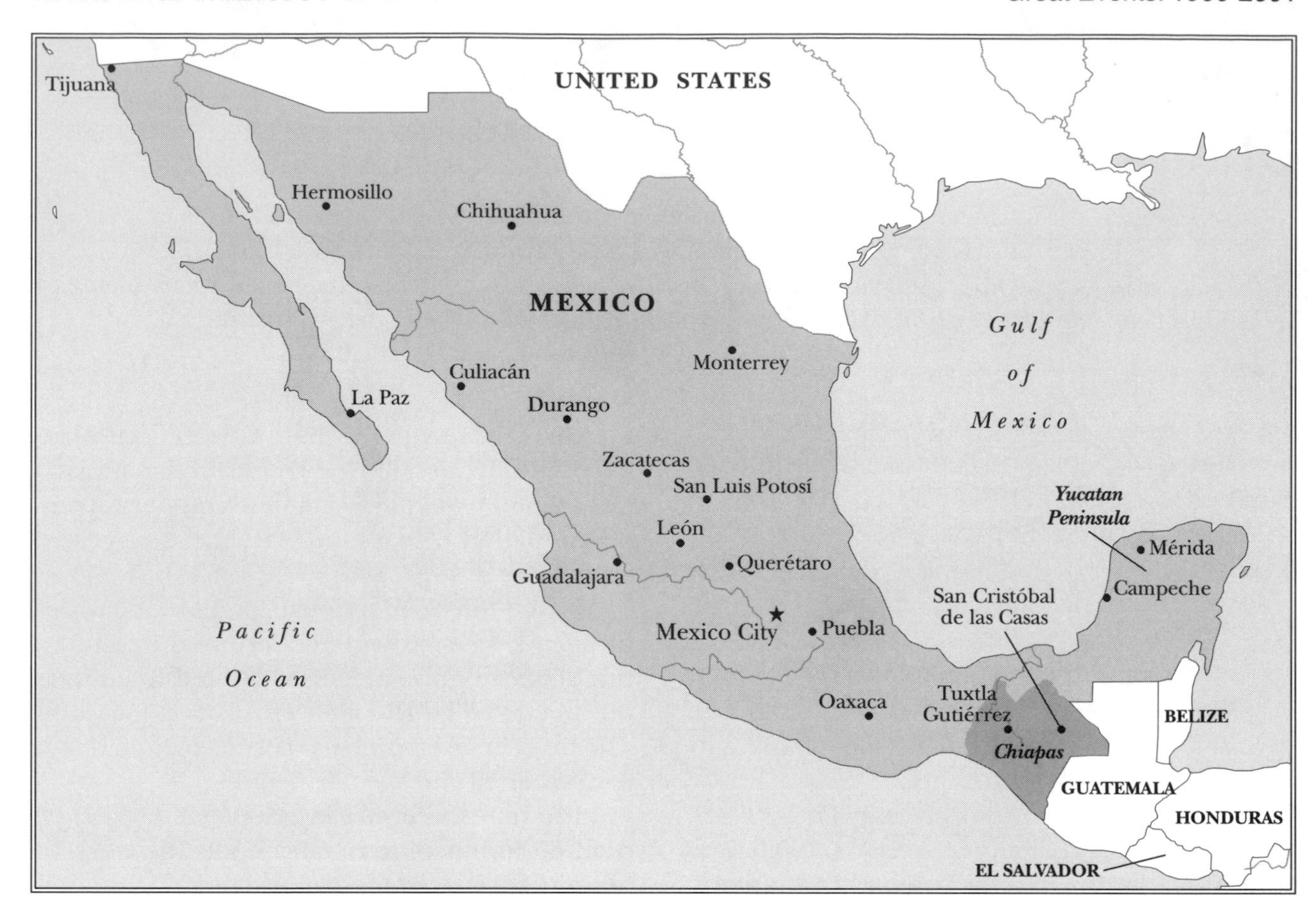

creased between the Indians and the caciques, the Mexican army began occupying Indian villages in 1993.

According to Human Rights Watch, the Mexican army had troops in Chiapas for several years before the Zapatista uprising. These troops intervened on the side of the caciques in disputes with the Indian peasants over land and natural resources. Several leaders of Indian groups who worked peacefully for the restitution of Indian lands were killed in the decade before the uprising.

The administration of President Salinas destroyed Indian peasants' hopes to reclaim their lands when it passed amendments to the Mexican constitution in 1992. These amendments repealed the right of every Mexican to own land and removed the prohibition on the private purchase of communal lands belonging to Indian communities. This made it possible for wealthy investors to set land prices and buy Mexican land. It effectively denied Indian peasants land ownership.

In many ways, Chiapas is Mexico. The Zapatista uprising underscored problems faced by the entire Mexican nation. In Chiapas, for example, 30 percent of the population is illiterate, 72 percent of the children do not complete first grade, and 32 percent of the people speak only an Indian language. Although Chiapas produces 55 percent of Mexico's hydroelectric power, 34 percent of the homes there have no electricity. The bitter poverty that fueled the Zapatista uprising in Chiapas affected Indian groups in almost every state of Mexico and could spark unrest anywhere.

Although the Salinas administration lowered inflation, sold off state-owned enterprises, and ended federal budget deficits in the years leading up to the implementation of NAFTA, it failed to stimulate enough economic growth to eliminate poverty. Economic policies that created a handful of billionaires left almost half the population in deep poverty. During the 1980's, Mexico's poor focused on improving their lot through the electoral process, but the Institutional Revolutionary Party (PRI), which has dominated Mexican politics for more than sixty years, stole elections repeatedly and continued to ignore the Indians' demands for justice.

Many factors contributed to the Zapatistas' decision to take up arms against the Mexican government. They were unable to stop the seizure of their land or elect a government that would protect their rights through legal channels. Their living conditions were so desperate that they concluded they had nothing to lose by waging war against the government.

Consequences

The Mexican government negotiated with the Zapatistas to bring an end to the violence in Chiapas. It agreed to almost all the demands made by the Zapatistas. The government promised to speed up rural electrification projects; to provide land for Indian peasants; and to provide more housing, health clinics, schools, and bilingual education (Spanish and Indian languages). It also agreed to increase Indian representation in the Chiapas state legislature, to reform the repressive justice system, to strengthen laws prohibiting discrimination against Indians, and to help Indian communities successfully deal with economic changes brought about by Mexico's participation in NAFTA.

Because Indian grievances were widely publicized in the wake of the Zapatista uprising, hundreds of Indian groups, primarily in Chiapas, occupied communal lands to reclaim them in the first months of 1994. Despite attempted negotiation and government reforms, violence continued to plague Chiapas throughout the 1990's.

Evelyn Toft

North American Free Trade Agreement Goes into Effect

The North American Free Trade Agreement reduced barriers to the flow of goods and services among Mexico, Canada, and the United States.

What: Economics; International relations
When: January 1, 1994
Where: Washington, D.C.
Who:
George Herbert Walker Bush (1924-), president of the United States from 1989 to 1993
H. Ross Perot (1930-), businessman and unsuccessful presidential candidate in 1992
Bill Clinton (1946-), president of the United States from 1993 to 2001

Adopting Free Trade for North America

Negotiators representing the United States, Mexico, and Canada formulated the North American Free Trade Agreement (NAFTA), but congressional approval was needed to carry out its many provisions. As finally approved, it was a long and complex document. The U.S. House of Representatives approved NAFTA on November 17, 1993, and it became effective January 1, 1994. It had four major types of provisions, as follows.

First, NAFTA reduced and ultimately would eliminate all tariffs (taxes on imports) and most nontariff trade barriers (such as quantitative quotas on imports) among the three countries. These liberalizations were spread over fifteen years, but two-thirds of Mexican exports to the United States and half of United States exports to Mexico were or became duty free immediately. Second, NAFTA provided rules to protect investment and intellectual property rights. NAFTA expanded Canadian and United States companies' ability to set up or buy businesses in Mexico and made it easier for them to sell out. Restrictions on bringing profits back home were removed. Protection of intellectual property rights involved patents, copyrights, trademarks, and computer software. United States firms strongly desired protection against unauthorized copying of books, records, videotapes, and software. This had been more of a problem in Mexico than in Canada. Third, NAFTA reduced barriers to trade in services, such as banking and finance, transportation, and telecommunications. Mexico extended temporary work permits to service providers from Canada and the United States. Fourth, NAFTA provided administrative procedures to settle disputes concerning how each country applied the rules.

United States Trade Policy

In 1934, when the world economy was in a deep depression, the United States adopted a Reciprocal Trade Agreements policy. Agreements were negotiated with many countries in which the United States reduced tariffs on its products if trading partners agreed to do the same for United States products. This helped trade to expand and gave each country an opportunity to sell more exports as well as buying more imports. At the end of World War II, this policy was extended by formation of the General Agreement on Tariffs and Trade (GATT), which involved many countries negotiating at once. Now U.S. tariffs might be reduced on German automobiles, for example, if Germany reduced tariffs on British woolens and Great Britain reduced tariffs on U.S. machinery. GATT negotiations occurred in a number of rounds, with the Uruguay round ending in new agreements in 1994.

Within each country, some industries benefit from liberalizing trade because they can sell profitably to other countries. Other industries, however, believe that they cannot compete with imports. U.S. companies producing clothing and shoes, for example, complained that they could not compete with imports from low-wage countries such as China. Wages are low in China, however, because labor productivity is also low. Most economists argue that relatively free international trade encourages each country to specialize in the products it can produce most efficiently and that consumers will benefit from lower prices and higher productivity. There are, however, short-run harms during the period of adjustment to free trade.

Many firms and many labor unions have opposed free trade, fearing that competition from imports will lower prices and decrease job opportunities. These issues were strongly argued in the election of 1992. President George Bush had encouraged the formulation of NAFTA and initialed it in 1992. Candidate Bill Clinton also supported NAFTA, but Ross Perot strongly argued against it. He claimed there would be a "giant sucking sound" as U.S. jobs were transferred to Mexico. Many environmentalists also opposed NAFTA, arguing that Mexican firms had an unfair advantage because they were not required to practice environmental protection to the same extent as U.S. firms. In contrast, some libertarian groups opposed NAFTA because it did not truly provide free trade. This claim was based on the substantial bureaucracy required to carry out its many complex provisions.

Supporters of NAFTA argued that many types of U.S. businesses would gain as a result of improved market conditions in Mexico. For example, privatization of the Mexican telephone system in 1991 created profit opportunities for U.S. firms, which were among the world leaders in this sector. United States firms producing motion pictures, recorded music, television programs, and computer software derived a large amount of revenue from sales to other countries and were often damaged by intellectual piracy. Pro-NAFTA forces also argued that NAFTA would increase the prosperity of the Mexican

AP/Wide World Photos

President Bill Clinton signs legislation implementing the North American Free Trade Agreement on December 8, 1993, as (from left to right) Vice President Al Gore, House Minority Leader Bob Michel, and House Speaker Thomas Foley watch.

economy, thus increasing Mexican wage levels and decreasing the large flow of Mexican immigrants across the southern border of the United States. They also noted that the economies of Mexico and Canada were far smaller than that of the United States and thus unable to flood United States markets with goods. President Clinton signed NAFTA in 1993, but it had to go through Congress before it could be implemented.

Consequences

Perot's opposition to NAFTA did not prevent his running far behind Bush and Clinton in the 1992 elections; voters apparently did not agree with his anti-NAFTA message. A few months after NAFTA was approved, Congress also approved the creation of a World Trade Organization to extend and administer GATT. NAFTA did not have a large impact on economic relations between the United States and Canada, because their trading relationship already had been close to that specified by the agreement.

For the first year after NAFTA, both the United States and Mexico appeared to benefit. Both United States export sales to Mexico and imports from Mexico increased substantially. Mexico benefited from substantial capital inflow, increasing production capacity and improving technology. In December, 1994, however, Mexico was hit by a financial crisis that resulted in a devaluation of the Mexican peso by about half. Foreign investors were largely responsible, as they sold Mexican securities and used the proceeds to buy dollars and other foreign currency. The crisis had many causes, however, and might have occurred without NAFTA in place.

Paul B. Trescott

Guatemala Takes Steps Toward Peace

Hope rose for an end to Central America's longest-running war when Guatemala's government and guerrillas agreed on steps toward a final peace agreement in 1994.

What: Civil war
When: January 10, 1994, and March 29, 1994
Where: Mexico City, Mexico
Who:
Ramiro de León Carpio (1942-), president of Guatemala from 1993 to 1996
Hector Rosada Granados, chief peace negotiator for the Guatemalan government
Rodrigo Asturias Amado, guerrilla leader known as Commander Gaspar Ilom
Jorge Soto Garcia, guerrilla leader known as Commander Pablo Monsanto
Ricardo Ramirez de León, guerrilla leader known as Commander Rolando Moran

A Negotiating Timetable

After thirty-four years of war against the government of Guatemala, guerrillas of the Guatemalan National Revolutionary Unity (URNG) signed two agreements with that government early in 1994. These agreements aimed toward negotiating a final peace agreement. On January 10, 1994, the basic framework for further negotiations was signed. A second, more substantive agreement was reached on March 29, 1994. In Annex I of the second agreement, the Guatemalan government of President Ramiro de León Carpio and the URNG made a human rights agreement. Annex II set specific dates for negotiations on each remaining barrier to a final peace treaty. These two elements, a human rights agreement and a timetable for a final peace agreement, were seen as breakthroughs that might end Central America's longest-running war.

This framework, timetable, and Comprehensive Agreement on Human Rights of 1994 all reflected the growing role of the United Nations (U.N.) as a peacemaker in the years after the end of the Cold War. After November, 1993, U.N. negotiator Jean Arnault facilitated discussions between the de León government and URNG representatives in Mexico City. Both the framework agreement (January 10) and the Comprehensive Agreement on Human Rights (March 29) called for verification by the United Nations. Both the government and the guerrillas requested that the U.N. secretary general appoint an impartial non-Guatemalan to establish offices inside Guatemala to verify that progress toward peace was occurring. The United Nations also was called on to convene each negotiating session agreed upon in the timetable.

A Vicious "Dirty War"

Many social and political problems contributed both to the outbreak of internal war in Guatemala in November, 1960, and to its long persistence. The 1994 agreements sought to address these underlying causes along with ending the fighting.

Poverty and landlessness had left millions of Guatemalans hopeless, especially after elected reform governments were removed from power violently by the Guatemalan army in 1954. Widespread social discrimination against indigenous Mayan Indians, the poorest Guatemalans, long divided the opposition during the era of military rule (1954-1986). Thousands of people who voiced grievances were menaced by death squads, abductions (known in Guatemala as "disappearances"), assassinations, and torture. Teachers, university students, Roman Catholic priests,

rural base of support. Between 1981 and 1983, more than four hundred rural villages, most in Indian areas north and west of the capital, were burned. Death squad killings of reform leaders continued even after the army transferred power in 1986 to a civilian administration.

Civilian administrations of Vinicio Cerezo (1986-1991) and Jorge Serrano (1991-1993) initiated several fruitless rounds of talks with the URNG after 1987. Guatemala made no serious efforts at land reform as part of this democratization process, unlike neighboring El Salvador. This failure occurred in part because the Guatemalan URNG never came as close to victory as the Salvadoran guerrillas. In this context, few in the Guatemalan army favored basic reforms along the lines that brought about the Salvadoran peace, brokered by the United Nations, in 1992.

The few tentative steps taken toward mending the torn social fabric were threatened when President Serrano attempted to overturn the country's constitution and reestablish dictatorship in May and June of 1993. After days of crisis and pressure from the Bill Clinton administration in the United States, the Guatemalan army sided with Serrano's opponents. Human rights ombudsman Ramiro de León Carpio was selected by the Guatemalan Congress to complete Serrano's term of office, lasting until January, 1996. De León thus had a unique opportunity to move the nation beyond internal war.

The 1994 agreements demanded an end to human rights violations by all sides. Further negotiations were to focus on the rights of indigenous Indians, agrarian reform, constitutional reform, and the strengthening of democratic institutions. The agreements envisioned a day when the URNG would be integrated into a

trade union members, and political party leaders all were victimized, especially after 1978. At least 35,000 perished in these types of human rights violations, and more than 100,000 people were killed between 1960 and 1995.

Prevented from opposing this system by way of meaningful elections and barred from any other effective peaceful protest routes, many Guatemalans joined or supported guerrilla opponents to military rule. In 1982, the rebel groups Guerrilla Army of the Poor, Rebel Armed Forces, and Revolutionary Organization of the People in Arms joined with the Communist Party to form the URNG. The Guatemalan army responded to the broadening guerrilla movement by attacking its peaceful Guatemalan society in which all would be protected from violence.

Consequences

Only some aspects of the 1994 agreements were realized. The U.N. monitoring team, headed by Argentine Leonardo Franco, arrived in November, 1994, and issued its first report in March, 1995. It found continuing human rights violations, with most of the 288 cases occurring after November, 1994, the fault of the army. In March, 1995, agreement was reached in Mexico City regarding the rights of indigenous Indians. This was nine months after the date specified in the 1994 timetable.

The 1994 agreements also set timetables for talks about land reform (July, 1994) and the role of the army (August, 1994). These issues were not resolved. More fundamentally, the planned cease-fire (September, 1994) and the overall peace settlement (December, 1994) did not occur until a general peace accord was signed at the end of 1996.

Gordon L. Bowen

United States and Vietnam Improve Relations

The United States ended a nineteen-year trade embargo against Vietnam, paving the way for restored diplomatic and trade relations between one-time bitter foes.

What: International relations
When: February 3, 1994
Where: Washington, D.C.
Who:
Bill Clinton (1946-), president of the United States from 1993 to 2001
Warren Christopher (1925-), U.S. secretary of state from 1993 to 2001
Le Duc Anh (1920-), president of Vietnam from 1992 to 1997

Ending the Embargo

The U.S. administration of President Bill Clinton risked domestic criticism when it decided to end a nineteen-year trade embargo against Vietnam. The embargo, which banned trade with Vietnam, had been in place since the end of the Vietnam War. The decision to end it paved the way for a restoration of full diplomatic and trade relations in 1995. The decision closed the chapter on the Vietnam War for many Americans and reflected the emergence of Vietnam as a normal country in the eyes of its neighbors and former foes.

Domestic opinion over the merits of the embargo was divided. Most Americans agreed on the undesirability of the communist regime running Vietnam. It had a poor human rights record and, until late 1992, kept an army of occupation in neighboring Cambodia. Although its human rights record improved only marginally in the late 1980's, Vietnam finally withdrew its troops from Cambodia under United Nations supervision. Through reforms of its trade and investment policies, it signaled a new willingness to end several decades of diplomatic and economic isolation as well as a desire to improve relations with neighbors and old enemies.

American veterans were particularly involved in, and divided on, the embargo issue. Some called for retention of the embargo. Others, including several prominent veterans from both parties in Congress, said it was time to end the embargo and bring the quarrel between Vietnam and the United States to a close. Congressional support from well-known veterans such as Senator John McCain was an important factor influencing Clinton's decision. A significant segment of the veteran community regarded Clinton's war record negatively. He had avoided the draft, then obtained a college deferment while studying at Oxford University.

Strong support for lifting the embargo came from the U.S. business community, which viewed positively Hanoi's decision to abandon its inefficient, planned economy in favor of market reforms. Many important American businesses and even entire industries lobbied for an end to the embargo, fearing that they would be left behind by foreign competitors in the scramble to profit from the opening and growth of Vietnam's economy.

A difficult and emotional issue faced the United States, that of unaccounted-for prisoners of war (POWs) and soldiers missing in action (MIAs) from the Vietnam War. Suspicion about mistreatment of unreturned POWs and MIAs had, for nearly two decades, contributed to cool relations between the United States and Vietnam. Misgivings were fed by occasional stories of sightings of MIAs and POWs and the widespread belief that information about MIAs was being concealed deliberately by the Vietnamese. That belief reflected deep distrust on the part of many

Americans toward the Vietnamese regime. At a more emotional level, it represented a naïve, though understandable, unwillingness to accept that a tropical jungle does not easily give up its dead. The MIA issue thus was a major obstacle to improving trade and diplomatic relations. Clinton was able to lift the embargo only after Vietnam agreed to a joint investigation of outstanding reports of sightings and to cooperate in the search for and return of MIA remains.

War and Retribution

The Geneva Accords of 1955 divided Vietnam at the 17th parallel into North Vietnam and South Vietnam. These states existed in suspended hostility until 1959, when the climactic phase of the nationalist/communist struggle, known as the Vietnam War, broke out. President John F. Kennedy raised the number of U.S. advisers to South Vietnam from 600 to more than 16,500 before he was assassinated in November, 1963. Under President Lyndon B. Johnson, the number of U.S. advisers rose even more. The United States became directly involved in large-scale combat in 1964. The American phase of the war raged until 1973. After the pullout of American troops, South Vietnam fought on alone. It succumbed to a full-scale northern invasion in 1975.

After the war, a united, communist Vietnam bent to the task of reconstruction. The U.S. imposed its trade embargo partly to punish the north for its conquest of the south and partly to punish it for defeating the United States. In Vietnam, rigid ideology and economic mismanagement, the embargo, and continuing war with several neighbors resulted in nearly two decades of dramatic decline.

Meanwhile, elsewhere in Southeast Asia various states emerged as prosperous modern economies. In 1978, Vietnam invaded Cambodia. It did not vacate that country for fourteen years that included a clash with China in 1979. The end of the Cold War saw Vietnam's economy in shambles and its Soviet ally gone. In 1990, the inflation rate in Vietnam surpassed 1,100 percent per year.

The United States and Vietnam still had not agreed on full diplomatic relations, largely as a result of American insistence on a satisfactory accounting by Vietnam for all American POWs and MIAs from the Vietnam War. Responding primarily to growing corporate pressure, in mid-1993 Clinton lifted both a ban on joint business ventures and a longstanding U.S. veto of World Bank loans to Vietnam. At the end of the year, he approved U.S. business participation in internationally approved and funded trade ventures and other joint ventures with Vietnam. In February of 1994, he ended the embargo, over objections from families of MIAs and others but to wide support from the business community and the general public.

Consequences

The decision to lift the embargo broke a trade and diplomatic logjam and led to new levels of cooperation on the MIA issue. U.S. businesses joined the international hunt for trade and investment opportunities in Vietnam, and Vietnam hoped to improve its economy through trading with the United States. By the end of July, 1995, the United States and Vietnam had formally established diplomatic relations. Vietnam opened its embassy in Washington, D.C., at the beginning of August. Vietnam continued to shed the worst features of communist economics, joining the world financial and trading community. In November, 2000, relations between the United States and Vietnam reached a new level when President Clinton made a formal state visit to Vietnam.

Cathal J. Nolan

Church of England Ordains Women

Following one of its most divisive debates, the Church of England agreed to a fully ordained ministry for women.

What: Religion
When: March 12, 1994
Where: Great Britain
Who:
George Carey (1935-), archbishop of Canterbury from 1991
Barry Rogerson (1936-), bishop of Bristol from 1985
Graham Leonard (1921-), bishop of London from 1981 to 1991
Angela Berners-Wilson (1954-), first woman to be ordained within the Church of England

Women Become Anglican Priests

On November 11, 1992, the General Synod of the Church of England, the state church and the largest and most influential of the English churches, voted by the required two-thirds majority to allow women into the full Anglican ministry. Voting procedures were complicated in that the Synod was divided into three houses of bishops, clergy, and laity. Each house had to produce the required two-thirds majority. The bishops and clergy each achieved clear two-thirds majorities, but the laity barely reached that mark, at 67.3 percent.

The debate over women in the ministry had been divisive, and there were threats of clergy and churches leaving the church to form their own "Continuing Church of England." A similar situation had occurred in the United States when the Episcopal Church had taken a similar vote in 1976. The leading opposition speaker, the Right Reverend Graham Leonard, cautioned against leaving the church, however, asking opponents to continue their opposition within their own parishes.

The Synod vote was only an enabling vote, to allow church (or "canon") law to be altered. This had to be done through Parliament, then the queen's consent gained. This process took a further eighteen months, and it was not until March, 1994, that the appropriate legislation and liturgy were in place.

The first ordinations were performed on March 12 in Bristol Cathedral by the Right Reverend Barry Rogerson, the diocesan bishop, together with his assistant bishop. Thirty-two female deacons of the diocese were ordained, with 170 male priests participating. Protests were muted. The first woman to be ordained was Angela Berners-Wilson, a university chaplain. The first woman to celebrate Holy Communion was the Reverend Glenys Mills, at the 8 A.M. service the next morning at Christ Church, Clifton, Bristol.

Is the Priesthood Male?

Women had first been allowed to minister within the Church of England in a recognized official capacity with the revival of the order of deaconesses in 1862. Subsequently, women were also appointed as full-time parish workers, Bible women, nuns, and workers in the Church Army, the Anglican equivalent of the Salvation Army.

In the first half of the twentieth century, the scope of women's ministry was subject to a number of reports and debates, beginning with a 1919 report, titled *Women's Ministry*, by a committee set up by the Archbishop of Canterbury. Debate tended to center on whether the order of deaconesses was theologically equivalent to the order of deacons. Deacons in the Church of England may perform most offices and services but may not consecrate the bread and wine at Communion or give absolution.

The general theme of the debate was that the two orders were not the same and that women could not be priests. Their training deteriorated compared with men's, as did their pay and conditions of service. The tide did not turn until the 1960's, perhaps as part of a much wider feminist movement. For the first time, women were allowed entrance into Anglican theological colleges. Pay and conditions improved, and deaconesses became *ex officio* members of parish church councils. They were allowed to preach and conduct liturgical services, rather than being restricted to nonstatutory services, as they had been previously.

In 1967, the Church Assembly debated the report *Women and Holy Orders*, produced by an archbishops' commission. The assembly decided it needed more time to consider this report but did not reject its positive findings. The Church Assembly was then reorganized into the General Synod. In 1975 this new Synod took up the debate, declaring that there were "no fundamental objections to the ordination of women to the priesthood." This proved to be a major turning point and opened the door in 1982 for legislation allowing women to be ordained as deacons. About one thousand women, some previously deaconesses and others who had graduated from theological colleges, were ordained as deacons between 1987 and 1989. They began serving as such, usually under male vicars and rectors as curates, or as chaplains.

In 1984 and 1988, attempts were made in the General Synod to pass legislation for full ordination for women, but there were insufficiently large majorities. Opponents tended to be Anglo-Catholics and Evangelicals. Arguments against full ordination included biblical teaching concerning gender, the tradition of a male priesthood, a loss of fellowship with Roman Catholic and Orthodox churches, and the fear of division and the loss of membership. Various pressure groups emerged, the most powerful being the Movement for the Ordination of Women.

In 1978 and more decisively in 1988, the Lambeth Conference (a meeting of Anglican bishops worldwide, held every ten years) accepted the different practices of the various provinces. By 1988 there were at least one thousand Anglican women priests in Hong Kong (beginning in 1971), the United States (beginning in 1974), and elsewhere. By 1994, half the worldwide Anglican community ordained women.

Consequences

During the following three months, a further twelve hundred women were ordained in forty-two dioceses. The leader of the Church of England, the Most Reverend George Carey, the archbishop of Canterbury, had always been in favor but sought to retain the loyalty of dissenting Anglicans. He appointed two roving bishops to minister to those who opposed ordination of women, and financial arrangements were made to cover resignations by dissenters. In fact, there was no formal split, although some clergy and laity moved to the Roman Catholic Church and, in a few cases, to Orthodox churches. In certain parishes, female priests remained unacceptable, but generally there was widespread supportive acceptance.

David Barratt

Mexican Presidential Candidate Colosio Is Assassinated

The favorite in the 1994 Mexican presidential elections, Luis Colosio, was shot and killed at a Tijuana political rally, raising questions about the prospects for meaningful political reform.

What: Assassination; National politics
When: March 23, 1994
Where: Tijuana, Mexico
Who:
Luis Donaldo Colosio (1950-1994), presidential candidate of the Partido Revolucionario Institucional (PRI)
Carlos Salinas de Gortari (1948-), president of Mexico from 1988 to 1994
Ernesto Zedillo Ponce de León (1951-), president of Mexico from 1994 to 2000

Colosio's Assassination

Five months before the August 21, 1994, elections, Luis Donaldo Colosio, the presidential candidate of the Partido Revolucionario Institucional (PRI, or Institutional Revolutionary Party), was brutally shot and killed at a campaign rally in Tijuana. The charismatic Colosio, who during the campaign had promised that if elected he would reform and democratize the nation's authoritarian political system, was considered likely to win the upcoming elections. The PRI had not lost a presidential election since its inception in 1929.

Although political violence occurs regularly in Mexico, Colosio's assassination shocked many Mexicans. It marked the first time that such a prominent political figure had been killed since presidential candidate Álvaro Obregón was assassinated in 1928. Outgoing president Carlos Salinas de Gortari began an investigation of the killing. He quickly named Ernesto Zedillo, a young, inexperienced bureaucrat, to step in as the PRI's presidential candidate for the election. Zedillo went on to win the election convincingly, though there were claims of electoral fraud by opposition candidates.

Colosio's death did little to convince either the international community or Mexicans that the PRI was serious about democratic reform. The assassination came at a time of economic ferment, only two months after the implementation of the North American Free Trade Agreement (NAFTA), a set of economic agreements among the United States, Canada, and Mexico aimed at reducing tariffs and protectionist practices and forging a regional common market. The murder of Colosio raised questions in Washington and Ottawa about Mexico's political stability as that country was becoming more closely integrated with the North American economies.

Political analysts have advanced a number of elaborate theories of who killed Colosio and why. The government probe was criticized by many, principally because the official version of the events of March 23, 1994, changed several times. The initial version—that a deranged Tijuanan, Mario Aburto Martínez, was the lone assailant—gave way to more complex theories. As evidence was uncovered and made public, it became increasingly clear that infighting within the PRI was largely responsible for the slaying.

Mexico's Authoritarian Past

To better understand the assassination of Colosio, one needs to trace the evolution of the PRI's stranglehold on national politics. The PRI took power in Mexico in the wake of the catastrophic Mexican Revolution (1910-1921) that cost more than a million lives and left the country in ruins. It set up an authoritarian system that promised to provide political stability while it paid lip service to democratic freedoms and institutions. The party deftly employed both rewards and punishments as means of coopting enemies and crushing dissent.

The president wields almost absolute power for a six-year term. An elaborate set of institutions such as the legislature, the judiciary, and a seemingly outspoken opposition hides the fact of authoritarian rule. In an effort to legitimate the PRI's rule, the system sacrifices meaningful political participation for stability and continuity.

Mexico's political culture also ensures that every presidential change of power follows a similar script. The outgoing president chooses a successor, and the electoral campaign becomes part political theater and part repression of opposition political parties, with an unhealthy dose of political infighting. Colosio's campaign was no exception to this script. The presidential successor goes on an exhaustive political campaign, not to ensure victory—the result is inevitable—but rather to reconstitute the coalitions within the PRI for the next six years and to begin the process of legitimization of the new ruler. Once the new president is sworn in, he builds his clout by lashing out against his predecessor's inefficiency and corruption while in office.

Mexican presidents, who rule with virtual impunity for their six-year terms before turning over the reins of government to a successor, increasingly have used political office for their personal gain or for the benefit of their cronies. The problem was exacerbated by the growing political clout of the drug cartels involved in the transport of drugs from Colombia to the United States.

AP/Wide World Photos

Luis Donaldo Colosio standing beside a portrait of President Carlos Salinas de Gortari.

Several murders, including Colosio's, have been linked to the Guadalajara and Culiacán drug organizations.

The 1988 presidential elections showed the first crack in the PRI's sixty-year monopoly of political power. The election came on the heels of one of the worst economic recessions in Mexican history. Many observers were convinced that Salinas actually lost the election to moderate-left candidate Cúauhtemoc Cárdenas. Official results showed Salinas receiving 50 percent of the vote and stealing the election. The 1988 electoral scare convinced Salinas and his political ally Colosio that political liberalization was necessary, but that it had to come slowly and only after economic reforms were implemented.

Consequences

Aburto Martínez, who confessed to the killing of Colosio, was convicted on October 31, 1994, and sentenced to a forty-two-year jail term. Mexico has no death penalty. Colosio apparently was killed by shots to the head and the abdomen, but Aburto Martínez fired only once, and the shell casings found at the scene were not the same caliber used by the convicted killer. The investigations conducted by the Salinas administration did not inspire public confidence because the official explanations continued to change as new evidence was unearthed.

When Zedillo became president on December 1, 1994, he assigned the case to his new attorney general, Antonio Lozano García. He concluded that there were numerous irregularities in the previous investigation and that evidence had been manipulated. He also announced that new tests proved that the two bullets that had taken Colosio's life could not have been fired by the same person.

On February 25, 1995, authorities arrested Othon Cortes Vazquez, alleged to be the second gunman. He was working as a member of Colosio's security detail the day of the assassination. Although a number of theories continue to be debated in Mexico, the most plausible explanation argues that Colosio represented interests who sought to reform the PRI and gradually democratize the political system. Hardliners resistant to change reputedly conspired with the drug cartels to eliminate him.

Allen Wells

Ugandans Prepare to Draft Constitution

After decades of political instability, Ugandans elected representatives for the first time in fourteen years, anticipating presidential elections and the drafting of a new constitution.

What: National politics; Politics
When: March 28, 1994
Where: Uganda
Who:
IDI AMIN (1925-), military dictator of Uganda from 1971 to 1979
YOWERI KAGUTA MUSEVENI (1945-), president of Uganda from 1986
APOLLO MILTON OBOTE (1924-), president of Uganda from 1966 to 1971 and 1980 to 1985

A Constitutional Assembly Is Elected

On March 28, 1994, Ugandan voters cast ballots in their first election in fourteen years. They elected a Constituent Assembly of 214 members from a field of 1,100 candidates. The new assembly, distinct from Uganda's parliament of 279 members, was charged with writing a new constitution that would define the political relationship among Uganda's four historic kingdoms. The election was the first legitimate balloting since the country's independence in 1962.

The most important issue in the campaign was whether candidates supported an immediate transition to multiparty government or an additional period under President Yoweri Kaguta Museveni's no-party movement system. Although existing parties were prohibited from publicly supporting candidates, in most cases party allegiances were well known. Those favoring the multiparty system were often supporters of the Uganda Peoples Congress (UPC), which was identified with former dictator Milton Obote and various northern tribal groups.

During campaigning, Museveni was criticized for allowing foreigners to dominate his government, particularly Banyarwandans and Rwandan refugees, most of whom had lived in Uganda for decades. Three weeks before the election, three UPC leaders were arrested and charged with sedition stemming from publication of a manifesto that repeated such charges. Museveni was also engaged in a heavy-handed counterinsurgency campaign against a relatively disorganized band of northern rebels. This campaign led to charges of authoritarianism and military abuse.

Although all constituents were elected as purported independents, known supporters of President Museveni were elected to 114 of the 214 seats, and two-thirds of the electorate voted for an amendment to extend nonparty politics for five years. Multipartyists won nearly all the seats in the north. It was estimated that about 70 percent of Uganda's seven million potential voters participated.

In November, 1994, multipartyists began a boycott of the constitutional assembly when the amendment extending nonparty politics was passed. The Baganda, the most powerful ethnic group in Uganda, wanted their ceremonial leader, the kabaka, to enjoy real power in a federal system. Museveni responded by shutting down their rallies and firing several profederalist ministers. Of the forty-eight ministers who held cabinet posts at the end of 1994, forty-two firmly supported nonpartyism.

An Authoritarian Past

Although medieval Uganda boasted some of the most sophisticated states in Africa, the country had little preparation for democracy and was beset by the interrelated effects of colonialism and tribal rivalries. Some forty distinct ethnic groups were gradually consolidated under the rule of four Bantu-speaking kingdoms: Buganda, Bunyoro, Ankole, and Toro. By 1850, when Arab traders arrived, the kingdom of Buganda dominated all others, with a large army and a highly developed but autocratic political system.

AP/Wide World Photos

Yoweri Kaguta Museveni.

British explorers and missionaries arrived in the 1860's. Buganda was made a protectorate by Great Britain in 1894. Although the other kingdoms were incorporated into the protectorate later in the decade, the more highly developed Buganda retained special privileges. This established a foundation for its preeminence in the postcolonial era.

After World War II, Africans played a larger role in governing Uganda. The independence movement was slower to develop than in neighboring Belgian Congo and Tanganyika. Movements for independence were hindered by the unwillingness of Buganda to surrender special privileges. When the Democratic Party began to challenge the British, prominent members of the Baganda tribe (associated with Buganda) organized the Kabaka Yekka Party in opposition. The deadlock was finally broken when, in 1960, Milton Obote founded a third party, the Uganda Peoples Congress (UPC), based in the underdeveloped north. There supporters of the Kabaka Yekka Party entered into a coalition with the UPC against the Democratic Party and thus created a government strong enough to achieve independence, finally granted by Great Britain on October 9, 1962.

Obote, a member of the northern Langi tribe, became the country's first prime minister. Kabaka Mutesa, a Baganda, was elected president in October, 1963. In 1966, Obote charged him with plotting to overthrow the government. Obote seized control of the government, abolished separate local kingdoms, and established himself as dictator. Extension of autocratic rule led to his overthrow in 1971 by the Ugandan army under Major General Idi Amin Dada, who established himself as military dictator.

In 1972, Amin gained temporary popularity by forcing between forty thousand and fifty thousand Asians out of the country. This action quickened the pace of economic decline, which had begun under Obote. As an increasing number of groups began to speak out against Amin, his brutality increased. It is estimated that he executed 300,000 Ugandans. In order to distract an increasingly restive army, in 1978 he used the pretext of a border dispute to invade Tanzania. Tanzanian troops counterattacked, and within five months they occupied most of Uganda, driving Amin into exile in 1979. A new civilian government was elected in December, 1980, with Obote again becoming president.

Obote continued to use violence as a political tool. He ordered execution of an estimated 300,000 Ugandans. By the mid-1980's, Uganda reached a state of general social and economic collapse. Obote was again overthrown, this time by the military National Resistance Movement (NRM). Yoweri Museveni, former defense minister of Uganda, finally stabilized the government in 1986.

Museveni's political strategy was to invite opponents into the coalition NRM, then ban public demonstrations by other political groups. Although he wanted to move away from the brutal authoritarianism of the Amin and Obote regimes, he feared that a multiparty state would simply heighten the ethnic divisions that historically had plagued Uganda.

Consequences

By returning an assembly that supported Museveni's no-party system, Ugandans supported the almost decade-long move away from ethnic parochialism toward the creation of a middle class that cut across ethnic lines. Museveni continued to rebuild the judiciary and civil service, and to diversify and professionalize the army. These moves made Uganda, in the mid-1990's, once again the most stable country in East Africa. Many feared, however, that order was tied too closely to Museveni. With the stability of a new constitution, slated for completion in 1995, Museveni promised to hold new presidential elections.

Uganda's new constitution provided that political party activity was not to resume until elections in 2001. In the meantime, the popular elections for president held in May, 1996, demonstrated that most Ugandans supported Museveni, who won 74 percent of the vote. This indicated popular approval of Museveni's go-slow policy with regard to the establishment of multiparty elections. In March, 2001, Museveni was again reelected, this time with 69.3 percent of the vote.

John Powell

Commonwealth of Independent States Strengthens Economic Ties

During 1994, the Commonwealth of Independent States, as a result of several meetings, consolidated economic cooperation by adopting an Interstate Economic Committee.

What: Economics
When: April-October, 1994
Where: Moscow, Russia; Minsk, Belarus; and other CIS capitals
Who:
Boris Yeltsin (1931-), president of the Russian Federation from 1991 to 1999
Nursultan A. Nazarbayev (1940-), president of Kazakhstan from 1990
Leonid Kuchma (1938-), president of Ukraine from 1994
Alexander Lukashenko (1955-), president of Belarus from 1994

Creation of an Economic Committee

On October 21, 1994, the twelve member nations of the Commonwealth of Independent States (CIS)—the former republics of the Union of Soviet Socialist Republics—agreed to form the Interstate Economic Committee (IEC) as a permanent body coordinating their economic relations. This was one of several developments in 1993 and 1994 designed to bring the former republics into a closer union. The new economic relationship, worked out over several months, would facilitate trade and economic cooperation among the republics, enabling them to trade on favorable conditions without tariffs or multiple bilateral agreements. It would also involve some transnational matters, such as transportation, gas and oil pipelines, communications, and energy.

The process of forming a closer economic union began in 1993 but escalated in 1994. On April 15, 1994, a summit conference was held among the CIS leaders. The meeting addressed about thirty issues, ranging from the disposition of the Black Sea fleet to creation of a single economic entity. Leaders were reluctant to make binding agreements but did plan to set up a customs-free zone among the CIS republics, to go into effect in stages. Discussion and negotiations continued through the summer and fall. The resulting IEC was ratified unanimously at the summit conference among the CIS states in October, 1994. The headquarters of the IEC was placed in Moscow.

The future role of Ukraine was ambivalent because Ukraine appeared to want only associate member status in the new agreement. Ukraine would apparently participate only in areas of its choosing. Nevertheless, the IEC was regarded as an important step toward economic integration of the CIS.

Simultaneous with their involvement in organizing the IEC, in July, 1994, the republics of Kazakhstan, Kyrgyzstan, and Uzbekistan, under the leadership of President Nursultan Nazarbayev of Kazakhstan, formed a closer alliance of the Central Asian states. The alliance would provide closer cooperation in the areas of economics, politics, and armed forces. In September, a Eurasian Integration Fund was created to further explore processes of closer union. Nazarbayev was elected honorary president of the fund. These developments suggest that the IEC, while important, would not be the only forum for economic cooperation among the republics. The IEC would strengthen the ties of the other republics with the most powerful CIS state, Russia.

The Path to Cooperation

During the years of the Soviet Union (1922-1991), the constituent republics, although nominally separate, were joined in a system with centralized control in Moscow. Joseph Stalin instituted policies during his long reign (1924-1953) to ensure that the republics were intertwined and dependent on the central government. The economies were linked so that independent existence would be virtually impossible. When the Soviet Union dissolved at the end of 1991, the republics were determined to be independent, but the reality of their economic interdependence remained. They had to seek ways to continue working together.

The Russian Federation was the largest single state, occupying approximately three-fourths of the territory of the former Soviet Union. Cooperation often took the form of bilateral arrangements between Russia and the newly independent states. Bilateral agreements among republics solved some immediate problems but were not a permanent solution. It was necessary to find ways to cooperate more broadly. In September, 1993, nine of the republics signed an agreement on an economic union. Georgia joined the pact a few months later.

Separation of the republics also had military and security ramifications. The military establishments of the republics were interconnected because the Soviet army had forces in all the republics. Russia inherited the major military, security, and foreign policy instruments of the Soviet Union. Each independent republic formed an army, often by nationalizing the Soviet forces and equipment in its territory Nuclear weapons were deployed in several republics. About 70 percent of the weapons were in Russia, but 14 percent were in Ukraine and about 13 percent in Kazakhstan and Belarus. Russia tried to consolidate control of nuclear weapons but found Ukraine, in particular, reluctant to surrender its nuclear capability.

Periodic meetings were held from 1992 to 1994 to plan closer economic cooperation. Although there was general agreement on its desirability, numerous obstacles existed, including rapid turnover of leadership in some republics, such as Ukraine and Belarus; the weakness of some republics' currencies; and various military conflicts between republics. Some republics endeavored to adopt their own currencies while others kept the ruble. The maintenance of a ruble zone of currencies contributed to the further decline of the ruble and suggested that dependence on Russia would continue and perhaps even increase. The process of negotiations was long and characterized by a series of agreements in principle that would be hard to implement.

Consequences

As linkages among the CIS states are solidified, the consequences may include greater dependence of the smaller, less powerful republics on Russia. The ruble zone, regional and local military conflicts, and economic arrangements will all likely contribute to solidifying Russian influence in the CIS republics. The IEC, as adopted in October, 1994, was a vague body, and adherence to its policies was optional for the republics. The basic dilemmas remained. Cooperation was desirable, but there was reluctance to adopt mandatory measures. Ukraine, in particular, wanted to be able to avoid participation in projects that might be detrimental to its national interest. The formation of the separate Central Asian alliance suggested that those states could become a voting bloc in the larger IEC.

It appeared that closer integration of the CIS economies would probably take place eventually, despite some republics' fears of Russian domination. That fear was offset by recognition of dependence on the resources and power of Russia.

Norma C. Noonan

U.S. Military Enters Bosnian War

By United Nations request and under NATO auspices, American air forces became involved in the protracted war in Bosnia between the Serbs and Bosnian Muslims (Bosniaks).

What: Civil war; International relations
When: April 10, 1994
Where: Gorazde, Bosnia
Who:
Slobodan Milošević (1941-), president of Serbia from 1997 to 2000
Franjo Tudjman (1922-1999), president of Croatia from 1990 to 1999
Bill Clinton (1946-), president of the United States from 1993 to 2001
Radovan Karadzic (1947-), leader of the Bosnian Serbs from 1992 to 1996
Alija Izetbegovic (1925-), president of Bosnia from 1990 to 1996

Widening Armed Conflict

At dusk on the evening of April 10, 1994, two American F-16 jets from the American air base in Aviano, Italy, flew over Gorazde, Bosnia. The planes dropped three five-hundred-pound bombs on Serbian positions used to bombard the city, which had been designated by the United Nations as a safe zone for Bosnian Muslims (Bosniaks). The strike, under North Atlantic Treaty Organization (NATO) auspices, brought the United States military further into the civil war being waged in the former Yugoslav republic of Bosnia. In February, 1994, U.S. airplanes had shot down Serbian aircraft violating the "no fly" zone established by the United Nations. The April 10 bombing was the first time in forty-five years that NATO air forces attacked ground targets in Europe.

The attack occurred after repeated warnings from the United Nations, NATO, and American president Bill Clinton to the Serbs that continued bombardment of Gorazde would bring retaliation. Gorazde was one of several regions the United Nations had designated as havens for Muslim refugees and was garrisoned by U.N. peacekeeping troops.

When the Serbs refused to stop their bombardment, Yasushi Akashi, the U.N. representative in Bosnia, advised U.N. secretary general Boutros Boutros-Ghali to call NATO strikes. NATO charged the American command to carry out the task, and Admiral Leighton Smith, Jr., commander of the NATO southern forces, ordered the strikes.

The bombing at first appeared to stop the Serbian shelling, but the Serb leader in Bosnia, Radovan Karadzic, claimed that the strikes brought the United Nations into the war on the Bosniak side, and he threatened retaliation against U.N. forces. President Clinton responded that the air attack was a clear expression of the will of NATO and the will of the United Nations. When Serb shelling resumed, a second attack was carried out on April 11.

Centuries of Conflict

The collapse of communism in Eastern Europe in 1989 revealed long-standing national and communal hostilities that dictatorship had hidden for forty-five years. Nowhere was this more evident than in the Federation of Yugoslavia, a collection of six republics that was home to eight major nationalities. Created after World War I, Yugoslavia never came to grips with its nationality problem, a major reason the country fell prey to fascist aggression in World War II. After that conflict, the powerful and charismatic communist leader Tito was able to maintain stability in the country.

After Tito died in 1980, Yugoslavia began to unravel because of national differences. Hostilities that extended back to the Middle Ages appeared in full bloom once more, and the republics declared their independence. The first serious war between the republics erupted in

1991, pitting Serbia, led by Slobodan Milošević, against Croatia, led by Franjo Tudjman. Croats and Serbs are members of the same nationality but differ in religion (the Croats are Roman Catholic, the Serbs Eastern Orthodox) and historical experience. Both republics claimed territory in Bosnia.

Bosnia itself was a republic Tito had created in 1945 for the Bosniaks, descendants of Serbo-Croatian Christians who converted to Islam in the fourteenth and fifteenth centuries. Milosevic and Tudjman considered dividing Bosnia between them. In response to this possibility, Bosniak leaders headed by Alija Izetbegovic declared Bosnia's independence from Yugoslavia and appealed to the nations of the world for recognition. This was slow in coming because of the problems the former Yugoslavia presented on the international stage at that time. European nations approached the situation carefully because traditional alliances divided them between the Serbs and the Croats. These peoples, especially the Serbs, reacted to Bosnian independence by launching a war against the Bosniaks. Serbs living in Bosnia soon formed an independent force under Karadzic. Milosevic came to his aid with all the resources of the Yugoslav state and military. As violence increased, especially toward civilian populations, the United Nations placed an arms embargo on the area. This worked to the advantage of the Serbs, and attacks against the Bosniaks increased. The nations of the West finally recognized Bosnia, but the civil war continued.

Western nations were forced to act by the pressure of public outrage, fanned by detailed media reports of atrocities, particularly the systematic raping of Bosnian women by Serbian soldiers and the deaths and maiming of hundreds of children. NATO voted to take military action but not to send in ground forces against the Serbs. Pressure was applied on Serbian leaders in Belgrade to stop the support of their conationals in Bosnia.

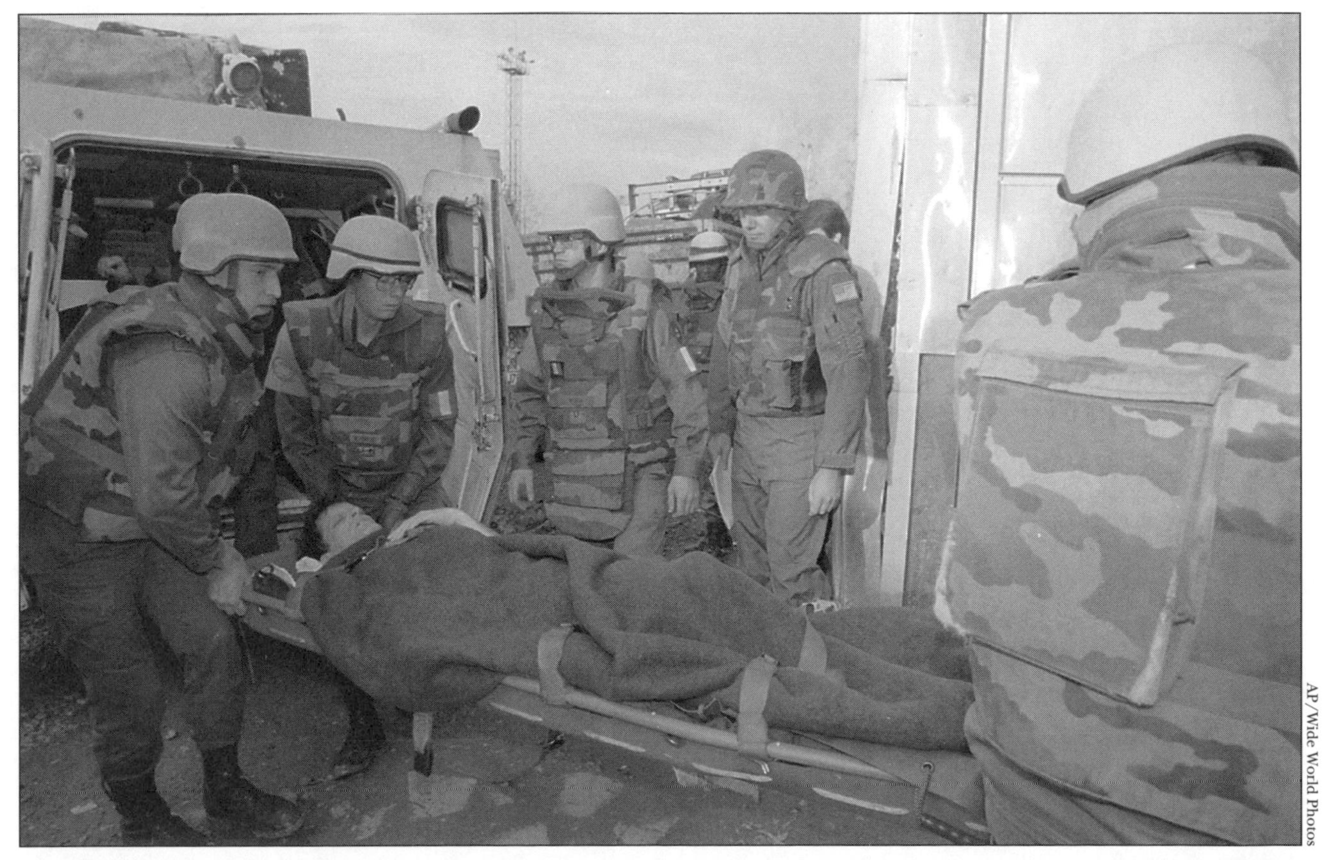

AP/Wide World Photos

Wounded soldiers are transferred from U.N. vehicles to a U.S. aircraft at Sarajevo airport under the supervision of U.S. military personnel.

Numerous plans for a negotiated settlement failed. Serbian military units surrounded Bosniak cities, including the safe zones the United Nations had established, and bombarded them from the surrounding highlands. NATO then decided to attack these Serbian outposts from the air. President Clinton agreed and, when asked, ordered American planes to begin raids.

Consequences

The action had little effect in lessening the war. Serbs continued their onslaught with success, with ground forces increasing the territory they controlled. In December, 1994, former U.S. president Jimmy Carter visited Bosnia to mediate a peace and worked out a cease-fire with Karadzic, only to see it fail within a few weeks. An airplane piloted by American Scott F. O'Grady was shot down in June, 1995, but he was rescued after a six-day ordeal hiding in the mountains. In July, the Serbian forces captured Gorazde and then moved on to the other safe areas. Within the United States, Congress, administration officials, military leaders, and numerous presidential candidates debated the course of action to take—lift the arms embargo, continue existing policies, commit ground troops, or simply ignore the war. NATO forces, with a large U.S. contingent, entered Bosnia in December, 1995, to enforce a peace agreement.

Frederick B. Chary

Florida Sues for Reimbursement of Costs of Illegal Immigrants

Lawton Chiles, governor of Florida, sued the federal government for nearly $1 billion to reimburse the state for the expenses of handling illegal immigrants.

What: Law; National politics
When: April 11, 1994
Where: Miami, Florida
Who:
Lawton M. Chiles, Jr. (1930-1998), governor of Florida from 1991 to 1998
Doris Meissner (1941-), commissioner of the INS from 1993 to 2000
Janet Reno (1938-), U.S. attorney general from 1993 to 2001
Edward Bertrand Davis (1933-), federal district judge who dismissed the lawsuit

Florida Sues for Expenses

After claiming that illegal immigration to Florida is out of control, Governor Lawton M. Chiles, Jr., filed a lawsuit against the federal government demanding more than $884 million reimbursement for education, health, and prison expenditures incurred by the state of Florida. These expenses were related to illegal aliens who entered the state during 1993. To emphasize the serious economic pressures on local governments caused by illegal immigration, both the Dade County Public Health Trust and the county's school board joined Governor Chiles as plaintiffs in the suit. The lawsuit named as defendants commissioner of the Immigration and Naturalization Service (INS) Doris Meissner and U.S. attorney general Janet Reno, along with regional and district officials of the INS.

The complaint was filed on April 11, 1994, as *Chiles v. United States*, 94-0676 (S.D. Fla.). It alleged that the federal government, through the defendants and their predecessors, had since before 1980 "abdicated its responsibility under the immigration laws and the Constitution to enforce and administer rational immigration policies." The suit also pointed out that, according to the INS's own figures, at least 345,000 undocumented aliens resided in Florida in 1991. That number constituted at least 36 percent of all immigrants in the state. The lawsuit stated that Florida is a victim of an ongoing immigration emergency that endangers the lives, property, safety, and economic welfare of the residents of the state.

On December 20, 1994, Federal District Judge Edward B. Davis threw out the lawsuit. In his decision, Davis said that he recognized the financial burden to Florida resulting from "the methods in which the Federal Government has chosen to enforce the immigration laws" but ruled that this did not give the state the right to recover money. He concurred instead with the federal government's argument that the case represented a political rather than a legal question over "the proper allocation of Federal resources and the execution of discretionary policies" dealing with immigration.

A National Immigration Crisis?

In November, 1994, Governor Chiles faced a reelection campaign in a state in which much of the electorate was tired of successive waves of refugees from Cuba and Haiti. Political considerations were important in his decision to file the lawsuit early in the campaign year.

Judge Davis, in denying the suit, admitted that Florida faced "a Hobson's choice." His written decision stated that if Florida chose not to provide services to illegal aliens, including closing schools, emergency rooms, and other service providers, the impact on the health, safety, and welfare of its citizenry could be devastating. If, on the other hand, Florida chose to provide the services, the costs could cripple the state.

In the meantime, Governor Chiles's lawsuit and other pressures pushed the Bill Clinton administration to ask Congress later that year for $350 million to help governors pay the costs of imprisoning illegal aliens convicted of felonies. In November, 1994, Attorney General Reno and INS Commissioner Meissner promised a 40 percent increase in border patrol resources along the Mexico-U.S. border within the next six months.

Such moves were clearly in response to the popular perception—in the other politically powerful border states of Texas and California as well as in Florida—that the federal government was not meeting its responsibility for controlling immigration. Many citizens were concerned that illegal immigration diminished the number of jobs available in an already tight job market. California's Proposition 187, a ballot initiative passed by a large majority in November, 1994, threatened to cut off most public services to illegal immigrants. These services, according to Governor Pete Wilson of California amounted to more than $3 billion annually in a state with increasingly serious budget problems.

More than simple economics was at work. Both the Florida lawsuit and California's Proposition 187 reflected a growing public backlash against large numbers of illegal immigrants and perhaps even against the relatively high rate of legal immigration. Not only was Chiles reelected in Florida, but Wilson also was re-elected, by a substantial margin. Most observers attributed Wilson's margin of victory in large part to his tough stance on illegal immigration. Some immigrant-rights advocates accused Wilson of pandering to racial and ethnic divisions in a state with the highest unemployment rate in the country.

Consequences

Illegal immigration clearly remains a politically volatile issue. During the year following Florida's unsuccessful lawsuit, a new Republican-controlled Congress gave evidence of responding to growing public concern over illegal immigration. In March, 1995, House Speaker Newt Gingrich formed a bipartisan task force to consider strategies to crack down on illegal immigration. That same month, Senator Dianne Feinstein of California, a Democrat, assailed the Clinton administration for lax enforcement of immigration laws. Political pressures indicated the likelihood of much larger appropriations for INS enforcement of immigration laws.

Increased funding for the INS eventually materialized. The service's 2001 budget was set at $4.8 billion—a figure roughly triple its 1994 budget.

Anthony D. Branch

Calderón Sol Is Elected President of El Salvador

After a decade of civil war in El Salvador, the candidate of the conservative ARENA party won the presidency in an open election, ushering in an era of political reconciliation.

What: National politics
When: April 24, 1994
Where: El Salvador
Who:
Armando Calderón Sol (1949-), president of El Salvador from 1994 to 1999
Ruben Zamora (1942-), former guerrilla leader and candidate of the Democratic Convergence Party
Roberto D'Aubuisson (1944-1992), founder of the ARENA party and suspected organizer of death squad activities during the civil war

Return to Democracy

On April 24, 1994, Armando Calderón Sol was elected president of El Salvador in the first free elections in Salvadoran history. The elections came after twelve years of brutal civil war in which seventy-five thousand people died and another million were forced into exile.

The fact that all political groups in the country, from the far left to the far right, were willing to lay down their arms to participate signaled a clear change from the violence of the recent past. The election of Calderón Sol, who won with 68 percent of the popular vote, was not regarded as a ringing endorsement of his probusiness platform and his political party, the Nationalist Republican Alliance (ARENA). Foreign observers instead suggested that ARENA's victory reflected the electorate's desire for peace and stability above all other factors.

At the same time, although the peaceful acceptance of the results of the April elections gave Salvadorans some reason to feel optimistic, it was generally agreed that democracy in the country was still a long way off. A full 30 percent of the voters had been ruled ineligible because of irregularities in the electoral registers. In some rural districts, known supporters of former guerrilla Ruben Zamora and his Democratic Convergence Party discovered that their names had been entered incorrectly into the registers in order to prevent them from casting ballots. Even in the cities, fear remained a factor in determining the direction of voting.

In his victory speech, Calderón Sol gave full recognition to the gains made in consolidating the democratic system. At the same time, the president-elect could not resist the temptation of eulogizing Roberto D'Aubuisson, the late founder of ARENA and a controversial figure who was frequently linked to the activities of death squads during the civil war. Calderón Sol's speech concluded with the singing of the ARENA anthem, the words of which proclaim that "El Salvador will be the graveyard of the Reds." This was hardly the note of reconciliation that moderates might have desired.

A Bloody Civil Conflict

El Salvador is a small country of about 8,200 square miles. Much of the land is devoted to the cultivation of coffee. The coffee industry had for generations been under the control of a tiny elite of only fourteen families. These families, who ruled in conjunction with high-ranking military officers, had never shown an inclination toward economic or political reform. This left the great majority of the country's six million peasants with little hope for the future and only about one dollar a week in income from their work on the coffee plantations.

In October, 1980, several communist-oriented guerrilla organizations united in founding the

AP/Wide World Photos

Armando Calderón Sol (left, with ribbon) and his wife, Elizabeth de Calderón.

Farabundo Martí National Liberation Front (FMLN). The FMLN immediately launched an all-out campaign to overthrow the status quo and establish a revolutionary Marxist regime. The government, for its part, refused to implement land reforms that had been announced and instead focused exclusively on a military solution. The Ronald Reagan administration in the United States, fearful that a Communist takeover of El Salvador would destabilize the rest of Central America, opened a floodgate of arms and economic aid to the Salvadoran government. About $12 billion in assistance reached the Salvadoran government in less than a decade.

Neither the government nor the guerrillas could win an outright victory. The struggle degenerated into protracted bloodletting with little meaningful direction. The collapse of the Soviet Union and the election of a new democratic government in Nicaragua left the Salvadoran guerrillas with few options but to negotiate. The Soviet Union and Nicaragua had provided support to the forces opposing the government.

In January, 1992, representatives of the two contending sides met to sign a peace accord. The government agreed to reform the military, accept the participation of former guerrillas in the new civilian police force, and open the electoral process. The guerrillas, for their part, agreed to disarm and to accept a market economy based on private enterprise. The all-important question of land reform, although vaguely alluded to, was purposely left unsettled. Since the signing of the 1992 accord, El Salvador has enjoyed an uneasy peace.

Consequences

The election of Calderón Sol, though significant in political terms, did not bring anything resembling prosperity to the country. Coffee production remained in bad shape, and the destruction of roads and bridges made it impossible to compete with the world's other coffee producers. Charges of corruption in the army, the government, and financial institutions continued to dash hopes for a rapid recovery from the war. In addition, although the FMLN turned in its weapons, many unemployed and homeless people did not do likewise. This fact represents a major challenge for the new depoliticized police force and for society as a whole.

Despite these problems, there was room for hope. In the election the Salvadoran people showed their commitment to peace and to the democratic process. They voted with patience, sometimes waiting in line for more than an hour to vote, and the elections proceeded with no noteworthy disturbances. The question remained, however, how long Salvadorans would retain their patience and abide by the democratic process.

Thomas Whigham

African National Congress Wins South African Elections

In South Africa's first election in which blacks as well as whites were allowed to vote, the African National Congress swept to power and Nelson Mandela was elected president.

What: Politics; Political reform
When: April 26-May 9, 1994
Where: South Africa
Who:
NELSON MANDELA (1918-), leader of the African National Congress; president of South Africa from 1994 to 1999
FREDERIK WILLEM DE KLERK (1936-), president of South Africa from 1989 to 1994
MANGOSUTHU GATSHA BUTHELEZI (1928-), head of the Inkatha Freedom Party

Victory for Mandela and the ANC

April 26, 1994, was a historic day for South Africa. For the first time, black South Africans, who accounted for 74 percent of the population, were allowed to vote in an election. Up to that day, South Africa had been dominated by whites, who composed only about 14 percent of the population. For more than forty years, the whites had pursued a policy of apartheid, a system of strict segregation by race that denied blacks any political power.

The election was peaceful, despite the fears of many that there would be violence. When the counting of votes was completed a week later, the main black party, the African National Congress (ANC), led by Nelson Mandela, was victorious. The ANC won 63 percent of the vote, which gave it 262 seats in the 400-member National Assembly. The mostly white National Party, led by President F. W. de Klerk, came in second, with 23 percent of the vote. The National Party had ruled South Africa since 1948. In third place was the ANC's rival, the Inkatha Freedom Party, led by Mangosuthu G. Buthelezi and representing the Zulus. Inkatha received slightly less than 7 percent of the vote. The white Freedom Front, which sought a separate homeland for whites, received nearly 3 percent of the vote.

Black South Africans poured onto the streets of their cities to celebrate the victory. Mandela called the occasion "a joyous night for the human spirit." De Klerk said, "After so many centuries, all South Africans are now free."

On May 9, 1994, Mandela was elected by the 400-member National Assembly as the first black president of South Africa. He was unopposed. A day later, in a pageant in Pretoria, he was sworn in. In a speech, he emphasized the need for reconciliation between former enemies and for a new social order based on justice for all. Mandela headed a five-year government of National Unity, dominated by the ANC but with de Klerk's National Party playing a junior role. De Klerk became one of two vice presidents, and a number of white ministers in the former government remained in cabinet posts.

The ANC's program for removing the effects of past discrimination included building a million houses in five years and providing free education for everyone. The National Assembly, assisted by the 90-member Senate, also faced the task of drawing up a permanent constitution for the country within two years.

The Crumbling of Apartheid

Whites had first settled in South Africa in the mid-seventeenth century and solidified their hold on the country during the twentieth century. Apartheid became law in 1948.

In the 1950's, the ANC became a mass protest movement, but it was banned by the government in 1960. In 1964, Nelson Mandela was sentenced to life in prison. In 1976, an uprising began in the black township of Soweto, near Johannesburg. It soon spread to other townships. Unrest continued to grow in the 1980's as the government tried to counter the protests with increased repression.

The South African government became isolated internationally. Many countries, including the United States, imposed economic sanctions. There was also a sports boycott. South Africa's economy began to slide amid industrial strife. The ANC stepped up its campaign of sabotage from outside South Africa's borders. Although the government made token attempts to include blacks in the political process and to create a black middle class, many blacks came to believe that apartheid could be overthrown only by violence. It looked as if South Africa might be heading toward a full-scale civil war.

De Klerk became president in 1989. He announced that he wanted to bring a gradual end to apartheid. He was convinced that only radical change could avert political, social, and economic disaster. He legalized the banned ANC and released black political leaders from prison. Mandela was released in February, 1990. A long and difficult period of negotiation began between de Klerk's government and the ANC. De Klerk and Mandela formed a sometimes turbulent partnership that helped pave the way for black majority rule. De Klerk had to convince white South Africans that it was in their own interests to give up power, and Mandela had to promise his supporters quick and fundamental change. At the same time, whites had to be reassured that they would have a future in an ANC-ruled country.

The next four years were marked by much bloodshed. More than thirteen thousand people

AP/Wide World Photos

Nelson Mandela casts his vote at Ohlange High School in Inanda, South Africa.

died in political violence. Most of the violence was the result of a feud between Zulus and ANC supporters, but the white government was also implicated when the ANC accused it of secretly fomenting violence in a bid to divide and weaken the black majority. Against this troubled background, the peaceful election and the smooth transfer of power to the ANC were remarkable achievements.

Consequences

President Mandela promised a dynamic first hundred days. On August 18, 1994, he made a televised address to parliament to review the progress that had been made. He said that a national consensus had been established and soon the benefits of democracy would come.

Mandela's speech was intended to answer criticism that the new government had no tangible achievements to its credit. His supporters pointed out that political violence had subsided, whites had not fled the country, and the mostly white military and police were continuing to obey civilian authority, even though that authority was now wielded by blacks. A newspaper poll showed that blacks and whites felt better about each other than they had before the election.

Many challenges remained, particularly the need to dismantle the segregated educational system and to boost the economy by attracting foreign investment. The major achievement, however, was clear: South Africa, after many years of isolation, was once more a respected member of the international community.

After completing his elected term of office in June, 1999, Mandela retired from politics as one of the most respected statesmen in the world.

Bryan Aubrey

Channel Tunnel Links Great Britain and France

Completion of the first uninterrupted link between the British Isles and the European continent since the end of the last Ice Age represented one of history's greatest accomplishments in engineering.

What: Technology; Transportation
When: May 6, 1994
Where: English Channel between Dover in England and Calais in France
Who:
Margaret Thatcher (1925-), prime minister of the United Kingdom from 1979 to 1990
François Mitterrand (1916-1996), president of France from 1981 to 1995
Elizabeth II (1926-), queen of Great Britain from 1952

The Opening of the Channel Tunnel

On Friday, May 6, 1994, Queen Elizabeth II of Great Britain and President François Mitterrand of France officially opened the Channel Tunnel linking their countries. Together they rode on a special train through the 30.7 miles of the tunnel, 24 miles of it under the seabed. The Channel Tunnel took six years to build and cost almost $16 billion. It is the world's longest undersea tunnel and connects the British Isles with the European Community via a fixed link, that is, a station-to-station roadbed route.

Response to the official Channel Tunnel opening was mixed. The French celebrated, with Calais offering nine days of festivities, replete with bands, dances, carnival delights, and the National Orchestra of Lille playing a hymn of peace. The British were far more subdued. British public opinion was decidedly lukewarm concerning the entire project, and the nationalistic fears of the British made the opening less than grand. The project was sometimes called the Chunnel by supporters and the Funnel by detractors. Cost overruns and gloomy predictions regarding the Channel Tunnel's ability to turn a profit made British politicians leery of appearing to be too keen on supporting the project.

The Channel Tunnel is an astounding feat of engineering. At its deepest point, it lies 148 feet beneath the seafloor. Drilled through solid, watertight chalk marl rock, it has natural protection against leaks. The tunnel is actually three parallel tunnels. One tunnel takes cars, trucks, buses, freight, and passengers by rail from France to England; a second tunnel does the same from England to France; and a third tunnel is a service tunnel sandwiched between the other two. Safety features range from fire doors to protect against terrorist attacks to "stun mats" to prevent rabid animals from invading Britain.

Following the official opening, the Channel Tunnel was closed for further safety testing. Passenger service between Waterloo Station in London and Gare du Nord Station in Paris began in November, 1994. Eurostar, an express train capable of reaching speeds of 186 miles per hour, traveled between the two capitals in about three hours and was roughly comparable to air travel in time and cost. "Le Shuttle," a no-frills automobile train between Folkestone (near Dover) and Calais, was ready for operation in 1995. At peak times, trains departed every fifteen minutes, with passengers driving their vehicles into eighteen-foot-tall, double-decker boxes, with ten autos per box. Thirty-five minutes later, cars detrained at the other end. The cost was roughly comparable to a ferry trip but with neither the frills nor the seasickness.

Attempts to Create a Link

For more than ten thousand years the English Channel has separated Great Britain from the continent. It is considered the first line of de-

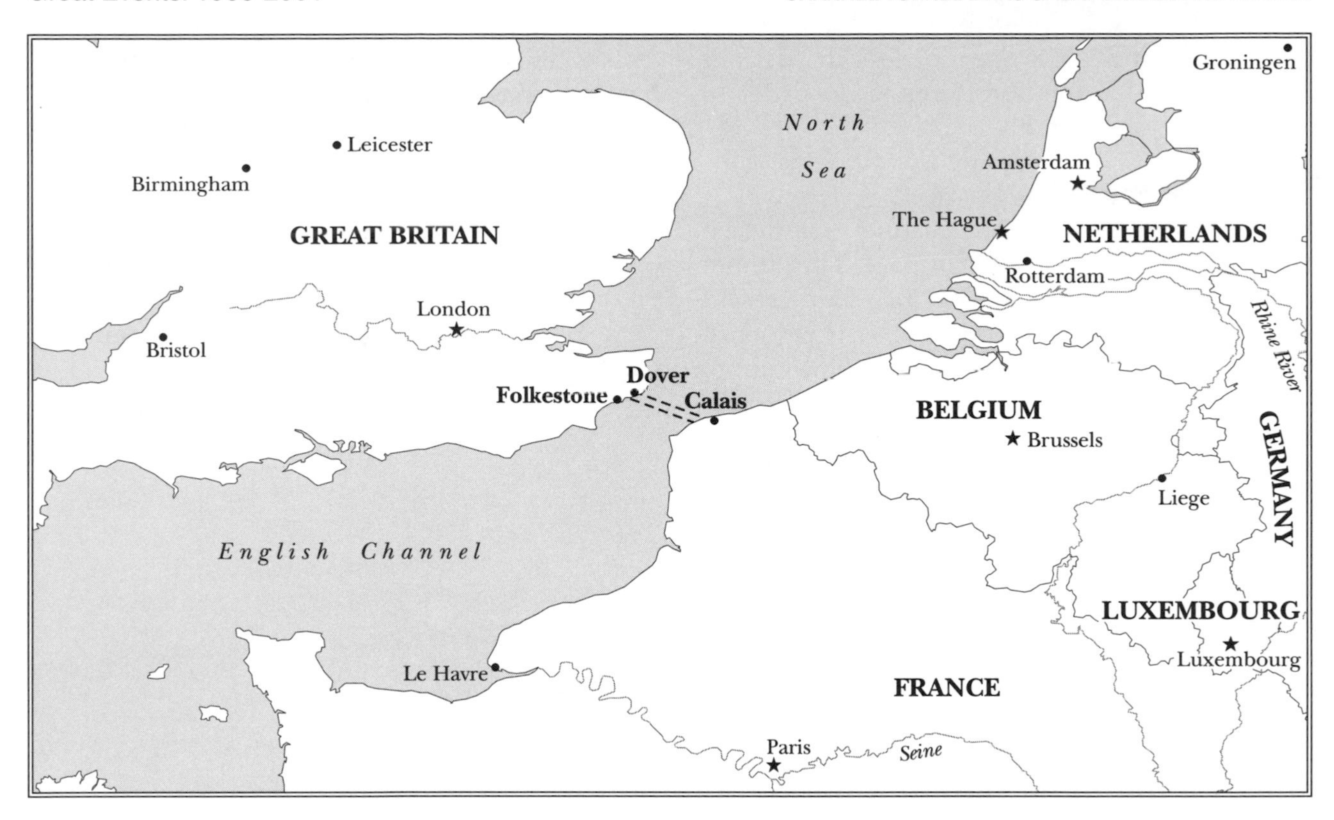

fense by Britons; no invader has mounted a successful attack on the island since 1066. For most of the last thousand years, Great Britain and France were each other's primary enemies. Efforts to bridge the twenty-three-mile water link between the two countries foundered not only because of nationalist concerns but also because of the complexity of the engineering and financial commitment required.

The first significant channel tunnel plan was proposed in 1802 by Albert Mathieu-Favier. Coming as it did during the era of the French Revolution, Great Britain viewed the tunnel proposal as a plan to negate Great Britain's naval power and to allow France's superior army access to Britain. Baimé Thomé de Gamond proposed various plans for a link, including a submerged tube (1834), a bridge (1830's), and a tunnel (1856). Although political and commercial support existed in France, only commercial support was available in Britain. Tunnel boring actually began in the 1880's, but the British War Office, citing strategic and military concerns, forced a halt to the work.

British strategic opposition to the tunnel continued during the world wars of the twentieth century. Following World War II, perceptions changed regarding the efficacy of the country's moat. In 1957, Great Britain chose not to join the European Economic Community (EEC). As the country prepared to join the EEC officially in 1972, Prime Minister Edward Heath and French president Georges Pompidou agreed that a joint channel tunnel project would be a symbol of commitment by Great Britain to Europe. After a period of neglect by the Labor government of 1974-1979, the idea of a channel tunnel was brought forward by Prime Minister Margaret Thatcher's Conservative government.

On February 12, 1986, a three-tunnel scheme was agreed upon by the two countries. It was to be financed and managed completely within the private sector. Eurotunnel, a consortium composed of the Channel Tunnel Group of Great Britain and France Manche of France, was to design and construct the Channel Tunnel, then operate it for fifty-five years. Work was begun in 1988, and on December 1, 1990, a breakthrough was made in one of the tunnels. By 1994, the Channel Tunnel had cost more than double the original estimates and had required two extra years to build. Completion of the tunnel is a significant case study of how countries can work out political and economic differences in order to achieve a common goal.

Consequences

It may be years before the Channel Tunnel exerts a significant economic impact. The tunnel connected Folkestone with Calais, but beyond those points rail connections were the responsibility of the national governments. The tunnel linked a modern rail transport system to one typical of the nineteenth century. The Eurostar travels at 186 miles per hour in France, 100 miles per hour through the Channel Tunnel, and less than 60 miles per hour in England. Moreover, British Rail was not integrated into the European rail network, a situation that prevented full utilization of the railway freight shipping potential of the Channel Tunnel. A high-speed link between London and the tunnel was begun after completion of the tunnel and was expected to begin partial operation in late 2003.

William S. Brockington, Jr.

Colombia Legalizes Some Private Use of Drugs

Colombia, a major exporter of cocaine to the United States, legalized the possession and private use of small amounts of cocaine, marijuana, and some other drugs for its citizens, provoking anger among U.S. drug enforcement officials.

What: Law; Social reform
When: May 6, 1994
Where: Bogotá, Colombia
Who:
Gustavo de Greiff (1929-), prosecutor general of Colombia from 1991 to 1994
César Gaviria Trujillo (1947-), president of Colombia from 1990 to 1994, also at the time of decriminalization
Alfonso Valdivieso, prosecutor general of Colombia from 1994 to 1997

Drug Decriminalization Decreed

The Constitutional Court of Colombia, that nation's highest court, on May 6, 1994, unexpectedly overturned two provisions of a 1986 narcotics law. That action effectively decriminalized the possession and private use of small amounts of certain drugs. In a five-to-four vote, the court ruled that the law in question violated the right of citizens, guaranteed by the Colombian constitution, to develop their personalities freely. Individuals became free to possess twenty grams (about two-thirds of an ounce) of marijuana, one gram of cocaine, five grams of hashish, and small amounts of certain other drugs.

The resulting celebration among marijuana-smoking youths was quickly overshadowed by the furious denunciations by U.S. politicians, the U.S. State Department, and drug enforcement officials. Criticism also came from many Colombians, including the four dissenting members of the court, who angrily labeled the decision as deplorable. Also opposing the ruling were all the candidates for president in the upcoming election and many government officials and business leaders. Leaders of the Roman Catholic Church were among those who asserted that the decision would have a negative effect on young people and families.

Polls indicated that a majority of Colombian citizens was opposed to the decision. Colombians at all levels expressed concern about possible international repercussions, particularly from the United States, and many believed that the change would intensify the perception of Colombia as a lawless nation. Some were concerned that lessening sanctions against personal drug use would make it even more difficult to prosecute Colombia's powerful drug lords effectively. Many Colombian advocates of drug legalization also were dissatisfied with the ruling, which they believed was not broad enough either to address the problems of drug-related corruption and violence or to decrease the economic power and influence of drug traffickers.

The U.S.-Colombian Drug War

In 1994, Colombia was the world's largest producer of cocaine. Large quantities of that drug were smuggled into the United States. Colombia also was a major producer of marijuana and had begun to increase its production of heroin. The export of illegal drugs was a mainstay of the Colombian economy. Sophisticated organizations of drug traffickers, known as "cartels," have had a growing influence on Colombian society and were suspected of having influence over many politicians and military leaders, in addition to their considerable economic power. Both human rights violations and common crime have been rampant in the country, and political corruption has grown steadily.

Much of the war on drugs in the United States has focused on Colombia. For many years, Colombia's government has received more U.S. funds for narcotics eradication and enforcement than any other country in the world. Colombia also has been the largest recipient in Latin America of U.S. military aid. Although it has been estimated that one in sixty Colombian citizens uses drugs and that 300,000 of the country's 23 million citizens are addicted to drugs, the country's vast drug trade is fueled primarily by demand from the United States. Despite continually escalating military efforts to wipe out the cultivation of drugs, the amount of land in Colombia devoted to such production increased almost fourfold in four years. Less than 10 percent of the crop was being confiscated, and the street price of cocaine in the United States was lower in the 1990's than it was in the 1980's.

Some months before the court's decision, Prosecutor General Gustavo de Greiff (the equivalent of the U.S. attorney general) had begun a vocal public campaign advocating that the personal use of drugs be legalized. Many prominent Colombian intellectuals, including Nobel Prize-winning novelist Gabriel García Márquez, also had joined in the call for decriminalization or legalization of drugs for personal use. Proponents of decriminalization asserted that efforts to control personal drug use through criminalization, which consistently had failed, were a waste of resources that could better be used for education and treatment.

In addition, by keeping the price of drugs high, these efforts allowed drug traffickers to earn large profits, thus promoting corruption. De Greiff, like many other proponents of legalization, asserted that the so-called war on drugs was futile when consumer demand, particularly in the United States, continued unabated. The drug trade was enormously profitable, and attempts to prosecute major drug lords generally had been less than successful. De Greiff advocated an open discussion of options, ranging from an uncontrolled, free market to one that would be centralized and heavily regulated. He also crusaded for relatively lenient plea bargains for drug traffickers who were willing to surrender and provide useful information. His calls for some form of controlled legalization, not only in Colombia but also in the United States and other countries, were harshly rebuked by drug enforcement officials and political leaders in the United States.

Consequences

Within days after the Constitutional Court's ruling, César Gaviria Trujillo, the outgoing president and a strong opponent of drug legalization, banned the use of drugs in public places and in the presence of children. Automobile drivers found to be under the influence of drugs could be fined the equivalent of $2,500, and nursing mothers who used drugs could lose custody of their children. The president also announced his intention to try to gather enough signatures for a national referendum on the recriminalization of drug use. If such a measure were to pass, the constitution could be changed to prohibit drug use.

In August, 1994, de Greiff was forced from his post as Colombia's first prosecutor general. He and many others attributed the ouster to political maneuvering. He was replaced by Alfonso Valdivieso, who strongly opposed the legalization of drugs.

Irene Struthers Rush

Pope John Paul II Forbids Ordination of Women

In a significant setback to women in the Roman Catholic Church, Pope John Paul II issued an apostolic letter to bishops that ruled out any possibility of the ordination of women as priests.

What: Religion
When: May 30, 1994
Where: Vatican City
Who:
John Paul II (Karol Wojtyła, 1920-), pope of the Roman Catholic Church from 1978
Edward M. Kennedy (1932-), U.S. senator from Massachusetts from 1962
William H. Keeler (1931-), archbishop of Baltimore; president of the National Conference of Catholic Bishops in 1994

The Decision Against Women as Priests

In his apostolic letter "On Reserving Priestly Ordination to Men Alone," issued on May 30, 1994, Pope John Paul II definitively proclaimed that Roman Catholic women could not be priests. Although he was prepared to allow altar girls as well as altar boys, the pope saw women as inherently unqualified for priestly ordination because women had not been among the twelve individuals chosen as apostles by Jesus Christ. He stressed that he was restating views he shared with Pope Paul VI and that the church historically had been consistent in ordaining only men.

Despite his realization that many Catholic feminists would object to his teaching in this episcopal letter, Pope John Paul II did not believe that he was discriminating against women in denying them priestly status. He did describe the presence of women in the church as "absolutely necessary and irreplaceable."

In an obvious attempt at damage control with potentially disaffected liberal Catholics, Archbishop William H. Keeler of Baltimore, president of the National Conference of Catholic Bishops, asked those Roman Catholics having difficulty accepting this episcopal letter to pray for understanding in accepting it. He also said that the letter did not change the Roman Catholic Church's basic belief in the fundamental equality of men and women, although it did assign men and women different leadership roles.

Roman Catholics troubled by this episcopal letter included Italian theologian Inos Biffi and American psychologist Sidney Callahan. Biffi claimed that the letter's significance lay in its form rather than its content, which merely restated established Catholic teaching. Callahan explained that she found the letter difficult to accept because she believed that the heart of the Gospel emphasized equality and inclusiveness.

Reaction to the letter included assertions that Pope John Paul II may have wished to encourage Anglicans who were opposed to the ordination of women by their denomination to consider conversion to the Roman Catholic faith. Other observers believed that the pope wished to preempt any discussion of female ordination at an October synod on religious life.

Barriers to Female Clergy Fall

As the Roman Catholic Church increasingly became short of priests in North America and Western Europe, many liberal Roman Catholic clergy and laity gave increasing consideration to the possibility of priestly roles for women. Senator Edward M. Kennedy issued a statement supporting the ordination of women priests during the course of his 1994 reelection campaign in order to increase his support among female voters.

Interest in ordaining women has been stimulated by the fact that nuns have assumed increasingly significant roles in parish and institutional administration in order to allow the dwindling supply of male priests to concentrate on the sacramental role now reserved for men. In addition, growing numbers of lay Catholic women have assumed professional roles in Roman Catholic religious education and charitable programs.

Increases in religious responsibilities for Roman Catholic women since the 1960's have taken place as North American and Western European women have begun to challenge traditional concepts of gender roles that limit women to the roles of wife, mother, and homemaker. Modern women in these regions have become increasingly unwilling to accept anything less than full participation in the economic, political, religious, and social lives of their communities.

In the sphere of religion, women have assumed a sacramental or priestly role in most Protestant denominations. Protestant seminaries have admitted growing numbers of female students despite substantial grassroots resistance to ordained women in the pulpit in most traditional Protestant denominations. Successful female clergy in mainline Protestant denominations now provide potential examples for those Roman Catholics interested in allowing women to serve as priests in an era of acute clergy shortage and rising role expectations for women.

Even the tradition-oriented Protestant Episcopal Church has ordained women since 1974. The Church of England first ordained women in 1994, and most other members of the worldwide Anglican communion to which these two churches belong have ordained women or will probably do so eventually.

In the higher ranks of the ordained clergy, there are now female bishops in hierarchical or semihierarchical churches such as the leading Episcopal, Lutheran, and Methodist bodies in the United States. Women have assumed respected leadership roles in these denominations.

Of the non-Roman Catholic Churches in North America and Western Europe, only the Eastern Orthodox churches now maintain the traditional practice of having a completely male clergy. The Orthodox vision of the clerical role as sacramental service in a hierarchically based church governance structure very closely resembles the Roman concept of priesthood.

Consequences

Pope John Paul II's strong stand against female priests may make it difficult for his eventual successors to reconsider this important issue. Although his episcopal letter on the ordination of women to the priesthood stands short of infallible doctrine in Roman Catholic teaching, women are unlikely to obtain priesthood in the Roman Catholic Church in the near future.

Because both groups have refused to ordain women, the Roman Catholic Church may find more common bonds with the various Eastern Orthodox bodies, with which they already have close doctrinal ties. The Roman Catholic Church will probably move further away from the Protestant Episcopal church in the United States and other Anglican bodies, which have decisively broken historical barriers to the ordination of women and are unlikely to return to a traditional stance on this issue.

Susan A. Stussy

Veterans Observe Fiftieth Anniversary of D-Day

Allied veterans gathered in Normandy, France, to honor fallen comrades and their own achievements of June 6, 1944, when they successfully opened a new front in the European theater.

What: War; International relations
When: June 6, 1994
Where: Normandy, France
Who:
BILL CLINTON (1946-), president of the United States from 1993 to 2001
DWIGHT D. EISENHOWER (1890-1969), supreme commander of the Allied Expeditionary Force
FRANÇOIS MITTERRAND (1916-1996), president of France from 1981 to 1995
ERWIN ROMMEL (1891-1944), German field marshal in charge of German forces from the Netherlands to the Loire River

D-Day Veterans Return to France

On June 6, 1994, half a century after Allied forces landed in Normandy as the first stage in their return to northwestern Europe, survivors of that day gathered with world leaders to honor the invasion code-named Operation Overlord. Government and military leaders representing the nations whose soldiers debarked in France celebrated the successful invasion known as D-Day. British veterans formed the largest contingent, but veterans from the United States, France, Canada, Poland, Norway, the Netherlands, Belgium, Luxembourg, the Czech Republic, Slovakia, Australia, and New Zealand gathered and were honored, perhaps for a last time.

All the towns and villages along the coast where the fifty-nine-mile front once stretched held celebrations honoring the returning few. Americans were addressed at Utah Beach and Omaha Beach by U.S. president Bill Clinton and French president François Mitterrand. The Canadians gathered at Juno Beach, and the British at Arromanches. French officials participated in tributes to wartime allies and in a ceremony honoring the French commandos who landed on D-Day.

For some, the fiftieth anniversary was a pilgrimage into the past. For many, it was a time to honor and learn from the sacrifices of others.

Thirty-eight American veterans representing the thirteen thousand airborne soldiers of June 6, 1944, reenacted their parachute drop that heralded the invasion. Veterans also came by sea, as in the case of the *Jeremiah O'Brien*, a Liberty ship that was chartered and sailed to France by retired seamen and officers. Many veterans returned with their families. It was an opportunity to educate the young about D-Day and the battle of more than two months that broke the German encirclement. Many Normans joined the anniversary, remembering the civilian dead and the eventual end of German occupation. As President Mitterrand remarked during his speech, "June 6 sent a signal. It meant that though nothing was yet won, everything was possible."

Allied Forces Invade France

When the United States entered World War II, the German eastern front stretched for more than twelve hundred miles, and in the west a line ran from Greece to France and north into Scandinavia. Between 1941 and 1944, German aggression led to conquest on the European continent. By 1944, the situation had changed. Allied bombers decimated Germany's heartland from the air while Allied soldiers fought northward up the Italian peninsula.

Hitler and his staff believed that an Allied attack in northwestern Europe would occur in the Pas de Calais, where the English Channel was narrowest. This belief was reinforced by an Allied "paper army" code-named Fortitude South, cre-

AP/Wide World Photos

Standing in front of headstones at the American cemetery in Cambridge, England, President Bill Clinton addresses World War II veterans during events to mark the fiftieth anniversary of D-Day.

ated by the Allies to fool the Axis Powers. Hitler's Atlantic Wall, an incomplete and unevenly supported series of defenses, guarded the western approaches to Europe.

General Dwight D. Eisenhower, Supreme Allied Commander, had planned to invade on June 5, 1944, but stormy weather stalled operations for twenty-four hours. The next morning, the Allied force, divided between the British Second Army in the east and the U.S. First Army in the west, moved to secure their objectives. While Allied forces arrived from the air and sea, the French Resistance immobilized German communications and transport networks. When news of the landing reached the headquarters of Field Marshal Erwin Rommel, commander of the Atlantic Wall defenses, it was discounted as a minor diversionary movement to distract German forces from the Calais region. The Germans had intercepted deliberately planted messages indicating the presence of Allied Forces.

On beaches named Utah and Omaha for the Americans, Gold and Sword for the British, and Juno for the Canadians, soldiers hammered away at the Atlantic Wall defenses and fought inland. French commandos liberated Ouistreham on its 1,435th day of occupation, making it the first French town liberated by French forces. On D-Day more than 35,000 men and 3,000 vehicles landed on Omaha Beach alone, where more than 2,000 Americans lost their lives. The civilian toll was also high. The Normans suffered nearly 4,000 civilian dead on June 6 and 7, with another 10,000 deaths in the following two months of intense fighting.

Sixteen hours after the Normandy landings began, 132,715 Americans, Britons, Canadians, French, Poles, Czechs, and others were ashore. By June 7, the Allies had secured their toehold in Norman France. During the months preceding the end of the war in 1945, that toehold was extended to the Elbe River in Germany. Throughout this time, the first electronic media coverage of any war sent reports of the Allies' battlefield progress directly to the various home fronts.

Consequences

Operation Overlord set the stage for the end of the war in Europe and remains the greatest single military expedition ever mounted. In subsequent conflicts, electronic journalism became the standard means of conveying battlefield events. It was employed for coverage of the fiftieth anniversary.

The celebrations resulted in some controversies. The major contributions of the Soviet Union received little notice. Germany was not invited to participate despite the fact that the previous month the Italians, whose wartime leader was a staunch fascist, were included in ceremonies marking the Allied invasion of Italy.

Two opportunities were overlooked. The first was to underline, with Germany, that the past is past. Second, an opportunity was bypassed to formally thank veterans of the former Soviet Union, a critical wartime ally. Without the Soviet sacrifice, many more German forces could have been thrown at the Allied beachhead and ultimately at England. Beyond the heroics, the invasion is understood as an example of the power of collective efforts directed toward a common purpose. It was recalled as such during the 1991 Persian Gulf War.

William B. Folkestad

Vatican and Israel Establish Full Diplomatic Ties

Ushering in a new era of Roman Catholic-Jewish relations and strengthening hopes for Middle East peace, Israel and the Vatican announced full diplomatic ties.

What: International relations
When: June 15, 1994
Where: Jerusalem, Israel
Who:
YOSSI BEILIN (1948-), Israeli deputy foreign minister from 1992
ANDREA CORDERO LANZA DI MONTEZEMOLO (1925-), the Vatican's first official envoy to Israel
JOHN PAUL II (KAROL WOJTYŁA, 1920-), pope of the Roman Catholic Church from 1978

Shaking Hands

On June 15, 1994, Israel and the Vatican established full diplomatic ties. The day before, Israeli deputy foreign minister Yossi Beilin and Vatican envoy Monsignor Andrea Cordero Lanza di Montezemolo had signed an accord formalizing the relationship.

Six months earlier, on December 30, 1993, the two sides had officially recognized each other by adopting a fourteen-point Fundamental Agreement. Israel ratified the agreement on February 20, 1994, and the Vatican on March 7.

The agreement provided for the exchange of special representatives, to be raised to the status of full ambassadors. This exchange took place on January 19, 1994. On August 16, the Vatican's first official diplomatic emissary to Israel, Monsignor Lanza di Montezemolo, presented his letter of credence to Israeli president Ezer Weizman. Israel's first ambassador to the Holy See, Shmuel Hadas, presented his credentials on September 29 to Pope John Paul II. The event marked formal completion of the establishment of diplomatic ties.

Although questions of taxation remained to be decided by a special committee, the agreement did ensure the Roman Catholic Church the right to own property in Israel. It also affirmed the church's freedom of expression and right to pursue charitable and educational activities there. In addition, it guaranteed safe and free access to Christian holy places.

For its part, Israel expanded its network of international relations through the accord. By signing the agreement, the Vatican became the forty-first state to establish, renew, or upgrade ties with Israel since the start of Arab-Israeli peace negotiations in 1991.

The Vatican said it would open its embassy in Jaffa. Jaffa, which is part of the city of Tel Aviv, has a large Christian Arab population. The Holy See had long demanded international control over Jerusalem, which Israel claims as its capital, because the city is sacred to the faiths of Jews, Christians, and Muslims. In the years leading up to the Fundamental Agreement, the Vatican softened its stand. The Fundamental Agreement sidestepped the issue, leaving it to the Middle East peace talks.

Praise for the announcement of diplomatic ties was broadcast around the globe. Nevertheless, the agreement was condemned by the Arab press for not coinciding with the establishment of a state for Palestinians. A group of ultrareligious Jews also protested the move. They insisted that the agreement's pledge to combat anti-Semitism fell short of a formal apology by the Vatican for centuries of Catholic teaching that, they charged, had inflamed hatred against Jews.

Past Hatreds, Present Hopes

"A stroke of the pen possibly has opened a new epoch," declared *L'Osservatore Romano*, a Vatican publication, after the signing of the Fundamental Agreement. The establishment of diplomatic relations between Israel and the Vatican appeared to mark a turning point in the long and often bitter history of Judaism and Christianity. Episodes of persecution included the Spanish Inquisition and the expulsion of Jews from Catholic Spain in 1492.

Attempts to heal the rift began in 1965, when the Second Vatican Council issued a document rejecting the idea of collective Jewish guilt for the death of Jesus. The document, "Nostra Aetate," also called for an end to the "teaching of contempt" for Jews and affirmed the lasting importance of Judaism.

"Nostra Aetate" also rejected the belief that the Jewish people had been cast out of their land for their sins and could, therefore, never return. It thus denied religious objections to Israel's right to exist. In 1985, another document, "Notes on Preaching and Teaching," recognized Israel's importance to modern Jewish identity. Still, the Vatican balked at full diplomatic recognition. Israel's occupation of the West Bank and Gaza Strip, it said, was a barrier. The Vatican had concerns for the fate of Palestinian Christians. It also feared that ties with Israel might jeopardize Christians living as minorities in Arab countries.

On September 13, 1993, a pact between Israel and the Palestine Liberation Organization (PLO) on mutual recognition and Palestinian self-rule was reached. This pact, known as the Declaration of Principles, caused a breakthrough in diplomatic talks between the Vatican and Israel. These talks began secretly in 1991, following the start of talks among Israel, the Arab states, and the Palestinians that same year. With the collapse of the Soviet Union and the Gulf War in 1991 came an Arab opening to the West as well as progress in the Middle East peace process. The Vatican-Israeli talks went public in July, 1992, when the two sides appointed a joint commission to negotiate a settlement.

Eight days after the signing of the Declaration of Principles, Yisrael Meir Lau, chief rabbi of Israel's European Jews, met with Pope John Paul II. The meeting, the first between a pope and a chief rabbi since the establishment of the state of Israel in 1948, was interpreted as a possible step toward establishment of full Vatican-Israeli relations.

Lau later credited the pope with furthering the cause of diplomatic ties. The pope had known Lau's grandfather, also a rabbi, in Crakow, Poland, and had personal experience of the Holocaust.

Consequences

During his audience with Pope John Paul II, Ambassador Hadas renewed Israel's invitation to the pope to make his first visit to Jerusalem. The pope later outlined a plan to usher in the year 2000 with a trip to the Holy Land. During the summer of 1994, the first Israeli exhibit at the Vatican opened with a display of the Dead Sea Scrolls. These events symbolized a new spirit of cooperation between Catholic and Jewish leaders, sparked by the normalization of relations between the Vatican and Israel.

The rise of post-Cold War conflicts in which religion mixes with politics requires a balancing of religious and political diplomacy. The Vatican-Israeli accord may provide a model for such peacemaking. Following the massacre of Muslim worshipers by a Jewish militant in Hebron, a meeting between Pope John Paul II and Israel's prime minister Yitzhak Rabin was said to have helped keep the Israeli-Palestinian peace talks on track.

Amy Adelstein

Indonesian Government Suppresses Three Magazines

The Indonesian government closed three news magazines during a period of oppression brought on by charges of political corruption.

What: National politics; Civil rights and liberties
When: June 21, 1994
Where: Indonesia
Who:
SUHARTO (1921-), president of Indonesia from 1967 to 1998
HARMOKO (1939-), minister of information and chairman of Golkar from 1993
FIKRI JUFRI, editor of *Tempo*
BACHARUDDIN JUSUF HABIBIE (1936-), minister of Research and Technology

The Closure of Three News Magazines

On June 21, 1994, Minister of Information Harmoko ordered the closure of Indonesia's three top magazines, *Tempo, Editor,* and *Detik.* This action violated the rights of a free press guaranteed by the 1982 Indonesian Press Laws. The Press Laws prohibit the banning of newspapers and provide for months of warning before the revoking of licenses for publications.

This sudden ban of *Tempo, Editor,* and *Detik* affected an important class of Indonesians. Their closures deprived 700,000 middle-class readers of knowledge about important government pronouncements and the ideas of opinion makers.

Detik, with a readership of 450,000, was well known for its tough political reporting. The charge against it was covering politics without a proper license. *Editor* was similarly charged. *Tempo,* with a circulation of 190,000, had a valid permit for political reporting and was surprised by the government's action.

The government was angered by *Tempo*'s reporting of a disagreement between Minister of Research and Technology Bacharuddin Jusuf Habibie and Finance Minister Mar'ie Muhammad over the purchase and refitting of thirty-nine formerly East German warships. Defense Minister Edi Sudrajat sided with the finance minister.

President Suharto considered such reporting as undermining the national security and social stability of Indonesia. He interpreted such liberal reporting as influencing who would succeed him as the next president in 1998.

Many peaceful rallies against press censorship were held in Jakarta, Bandung, Jogjakarta, Surabaya, and other cities and university centers in Indonesia. The military was used to stop boisterous demonstrations in the capital, Jakarta, on June 27. This antigovernment rally had the backing of eighty artists and intellectuals, who signed a petition supporting the freedom of the press and rejecting a government system of oppression and deception. The editor of *Tempo,* Fikri Jufri, signed the petition because he believed that Indonesia could not advocate economic openness without press freedom to expose social inequality and corruption.

The June 27 protests were significant because they coincided with the arrival of Australian prime minister Paul Keating in Jakarta to inaugurate Australia's largest-ever overseas trade exhibition. This promotional event, "Australia Today—Indonesia 1994," was intended to establish stronger economic ties and improve relationships that were strained because of Australian criticisms of Indonesia's human rights record. Keating was under political pressure at home to address the Indonesian ban on the three news publications. The ban conflicted with the economic openness supported by Suharto.

From Guided Democracy to New Order

Suharto, as a general in the Indonesian army, brought law and order to Indonesia after the failed communist coup and political bloodletting of 1965. Before the coup, President Sukarno had ruled Indonesia since its independence from the Dutch in 1949 and had monopolized political power with the principle of guided democracy. This principle emphasized authoritarian rule by a strong president who would lead the nation toward harmony and unity under the national ideology of *Pancasila*—the five principles of nationalism, humanitarianism, democracy, social welfare, and belief in God.

When Suharto succeeded Sukarno as president of Indonesia in 1967, Suharto replaced guided democracy with his creation of a new order. This change gave more power to the military, which was called on to play the dual roles of providing security and leadership in the government bureaucracy. Furthermore, Suharto created Golkar, a form of national party made up of several hundred interest and functional groups, to replace competing political parties. The new order used Golkar to pack the thousand-member People's Consultative Assembly with electors who meet briefly once every five years to choose the president. Suharto was chosen in each election beginning in 1968 and made efforts to control the choice of president of Indonesia in 1998.

Suharto had decided to promote a fast-growing, export-oriented economy. To achieve this, he opted for economic openness to remove trade restrictions and attract foreign investments. Indonesia's economy grew at an annual average rate of about 7 percent in the late 1980's and early 1990's. As the economy prospered and the country became more open, Indonesians began to question the need for authoritarian government and rule by presidential decrees. Furthermore, economic growth was accompanied by increasing inequalities in the distribution of wealth and income, the rise of corruption by high-ranking military personnel and wealthy businessmen of Chinese descent, and the establishment of business monopolies by Suharto's children and relatives. The press had reported on such charges and on interministerial rivalries, prompting the closures of the three news magazines.

Consequences

Tempo's editors, Gunawan Mohamad and Fikri Jufri, challenged the government's closure in court. On May 3, 1995, the chief judge of state administrative court ordered their licenses to be restored. In the same month, the Supreme Court exonerated six men who were jailed for the 1993 killing of labor activist Marsinah. These outcomes enraged Suharto, who warned against such liberalism, which undermined *Pancasila* and thus the stability and security of Indonesia. Government censorship remained strong through Suharto's last years in office and continued into the twenty-first century.

Peng-Khuan Chong

Hong Kong Legislature Votes to Expand Democracy

Prompted by the British governor, Hong Kong's legislative council voted to extend the democratic basis of representation before the Chinese resumed ownership of the colony in 1997.

What: Politics
When: June 30, 1994
Where: Victoria, Hong Kong
Who:
Christopher Patten (1944-   ), British governor of the crown colony of Hong Kong
Li Peng (1928-   ), premier of China, an opponent of expanding Hong Kong's democratic institutions
Lu Ping (1927-   ), China's chief official dealing with Hong Kong
Martin Chu-Ming Lee (1938-   ), lawyer who led Hong Kong's reformist United Democrats
Allen Peng-Fei Lee (1940-   ), leader of Hong Kong's antidemocratic Liberal Party

The Legislators Decide

The issue before Hong Kong's legislative council in June, 1994, was how many more of its members should be elected by direct popular vote. At 5 A.M. on Thursday, June 30, 1994, after twenty hours of debate, the council finally approved British-appointed governor Christopher Patten's proposals to extend voting rights and representation in the colony. The governor's victory was carried by a margin of eight votes (thirty-two council members supported him, twenty-four were opposed, and two abstained).

Prior to the council's decision, eighteen of its sixty members were chosen by direct election. Another forty members were elected indirectly through functional constituencies such as lawyers, bankers, social workers, and teachers. Two other members were selected by a committee.

Overall, until the June vote, about two-thirds of the council's members were elected by only 190,000 of the colony's 3.7 million potential voters. By accepting Governor Patten's plan, the council increased the number of its directly elected representatives to twenty (the previous eighteen plus the two formerly chosen by an election committee), a number the Chinese were willing to accept by 1995. More significant was that by adopting Patten's initiatives, the council also extended the scope of the functional constituencies so that nearly all Hong Kong workers, about 2.7 million of them, received the vote.

The council's ratification of Patten's plan reflected an uneasy compromise among Hong Kong's political interests. The colony's advocates of full democracy, such as lawyer Martin Lee (leader of Hong Kong's United Democrats) and schoolteacher Emily Lau, voiced their disappointment. Both of them favored universalizing the franchise as well as having direct elections. Martin Lee, as a legislative council member, voted against Patten's proposals because they fell short of Lee's expansive democratic objectives. On the other hand, Allen Lee, who headed the colony's Liberal Party (a confusing party label because the Liberals were staunchly pro-Chinese and antidemocratic), opposed the decision because it carried democracy and direct elections too far.

Although the official response of the People's Republic of China, which was destined to acquire sovereignty over Hong Kong on June 30, 1997, was somewhat nuanced, it was overwhelmingly hostile to the extensions of democracy. Premier Li Peng and the primary officials entrusted with China's Hong Kong affairs, Lu Ping and Zhou Nan, castigated Patten, denounced the council's liberals and subversives, and charged Great Britain with violations of the Joint Declaration and of the Basic Law that both nations had agreed upon to guide the colony's reversion to China.

The Lease Expires

As a major world power, Great Britain formally acquired Hong Kong Island from China by the Treaty of Nanjing (Nanking) in 1842. Adjacent mainland Kowloon Peninsula was ceded to the British in 1860, and the so-called New Territories, also on the mainland, in 1898 were leased by Great Britain for ninety-nine years. Following World War II, Hong Kong rapidly became one of the world's most dynamic economies, featuring world-renowned textile industries and serving as a major import-export and financial center. It also became a home for hundreds of thousands of refugees fleeing communist China.

China reemerged as a world power after 1949, paralleling Great Britain's decline as a dominant force in international affairs. China's new status made retrocession of all of Hong Kong's 398 square miles to the People's Republic inevitable as leases on vital portions of the crown colony expired. In the meantime, the colony was governed by British-appointed governors and their executive councils. These tended to dominate the token, semiparliamentary legislative council. However effectively governed it may have been, Hong Kong was not a democracy during most of Great Britain's rule.

Mutual agreements for the transfer of Hong Kong to China were embodied in the Joint Declaration of 1984 and in the Basic Law that had been drafted by the Chinese as a constitution for post-1997 Hong Kong. Beijing meanwhile assured the citizens of Hong Kong that they could continue their way of life for fifty years after 1997 and that China would treat the former colony as an exception within the framework of "one country, two systems." That is, Hong Kong's customary free enterprise economy and, presumably, certain political liberties would be protected by the Communist regime.

In the light of these agreements and assurances, Governor Patten's predecessor did nothing, despite the urgings of democratic and anticommunist Hong Kong Chinese, to put more democratic institutions in place before Communist China resumed ownership in 1997. Even before Patten formally was installed as governor on July 9, 1992, however, Beijing had begun pressing him vigorously for more concessions. These pressures led to scores of negotiating sessions. Although China primarily sought to curtail the growth of democratic sentiments and institutions in Hong Kong, the bargaining chips it presented to Patten were economic. China's leaders wanted assurances from him that expenditures on Hong Kong's major new airport—a jointly financed and increasingly expensive Sino-British project—would be kept in check and that the colony's fiscal reserves that were to be transferred to China would be kept in excess of HK$25 billion. It was in this diplomatic context that Governor Patten proposed the democratic initiatives sanctioned by the legislative council.

Consequences

As China's encouragement of capitalist practices and the burgeoning of its special economic zones (including Guangdong Province, adjacent to Hong Kong) produced amazing economic growth in the People's Republic, support in Hong Kong for Patten and for more democracy waned. Patten's critics hoped to avoid antagonizing an ostensibly reforming China by further democratizing the colony, particularly because China seemed likely to leave Hong Kong's economy intact. Meanwhile, through 1995 Governor Patten and Chinese officials continued negotiating over the airport project and wrangling over creation of a new court, even as the Chinese created a shadow government for Hong Kong and elements of the Chinese army were allowed to occupy areas inside the colony. As 1997 drew closer, Hong Kong's economy appeared safe from communist abuses. Citizens of Hong Kong were politically polarized, however, and there were few indications that, once in charge, Beijing would tolerate Hong Kong's democracy.

Clifton K. Yearley

Rwandan Refugees Flood Zaire

In one of the worst crises in modern history, millions of Rwandans fled ethnic massacres in their homeland following the assassination of their president.

What: Civil war; International relations
When: July, 1994
Where: Rwanda and Zaire (now Democratic Republic of Congo)
Who:
Juvénal Habyarimana (1936-1994), Hutu president of Rwanda from 1973 to 1994
Cyprien Ntaryamira (1955-1994), Hutu president of Burundi in 1994

The Flood of Refugees

About two million desperate refugees poured across the border from Rwanda into neighboring Zaire after a civil war ended in victory for the Rwandan Patriotic Front. Fighting had renewed after the assassination on April 6, 1994, of President Juvénal Habyarimana of Rwanda and President Cyprien Ntaryamira of Burundi, Rwanda's neighbor to the south. The two leaders were killed when their airplane was shot down near Kigali, the capital of Rwanda. Both men were Hutu, the majority group in each country. The Hutu had fought the Tutsi for political and economic superiority for generations. The two presidents were killed returning from a conference seeking solutions to the conflicts tearing apart both East African nations. Evidence suggested that the airplane crash was an assassination by Hutu extremists opposed to any discussions with the Tutsi.

News of the killings set off a rampage by Hutu soldiers, who hacked to death more than 500,000 Tutsi in Kigali and surrounding areas. More than 250,000 Tutsi fled across the border to Tanzania, another country bordering the tiny Rwandan Republic. Violence between the Hutu majority and the Tutsi minority had led to the deaths of hundreds of thousands in Burundi during the 1970's and 1980's. This time, Burundi remained calm.

The massacre in Kigali renewed a civil war in Rwanda. The Tutsi-controlled Rwandan Patriotic Front defeated the Hutu regular army and quickly gained control of two-thirds of the country. By early July, the Tutsi had attacked and gained control of Kigali. By July 13, the Hutu government fled the city. Tutsi militia began burning the city and killing Hutu civilians. During early July, about two million refugees, mostly Hutu, flooded across the border into Zaire. The Hutu government withdrew to a safe zone in the southwestern part of the country, established by French forces in June. The French, unable to maintain order, withdrew by the end of August. The huge number of refugees posed major problems for relief workers from the International Red Cross and United Nations. The British aid group Oxfam called the crisis a disaster on a scale not witnessed in modern times. Medical workers in the camps had to deal with food shortages, a cholera epidemic that killed thousands every day, and continuing violence as Tutsi rebel troops fired mortar shells from the Rwandan side of the border into the camps, killing hundreds of people.

Causes of the Ethnic Violence

In 1993, the Rwandan population was estimated at nearly eight million, making Rwanda the most densely populated nation in Africa. About 90 percent of the population was Hutu, 9 percent was Tutsi, and 1 percent was Twa, a Pygmy people. The Twa had the lowest status of the three. All Rwandans spoke Kinyarwanda, a branch of the Bantu language. Almost two-thirds of the people were Roman Catholics, with the others following traditional tribal religions. The average Rwandan earned less than $300 a year.

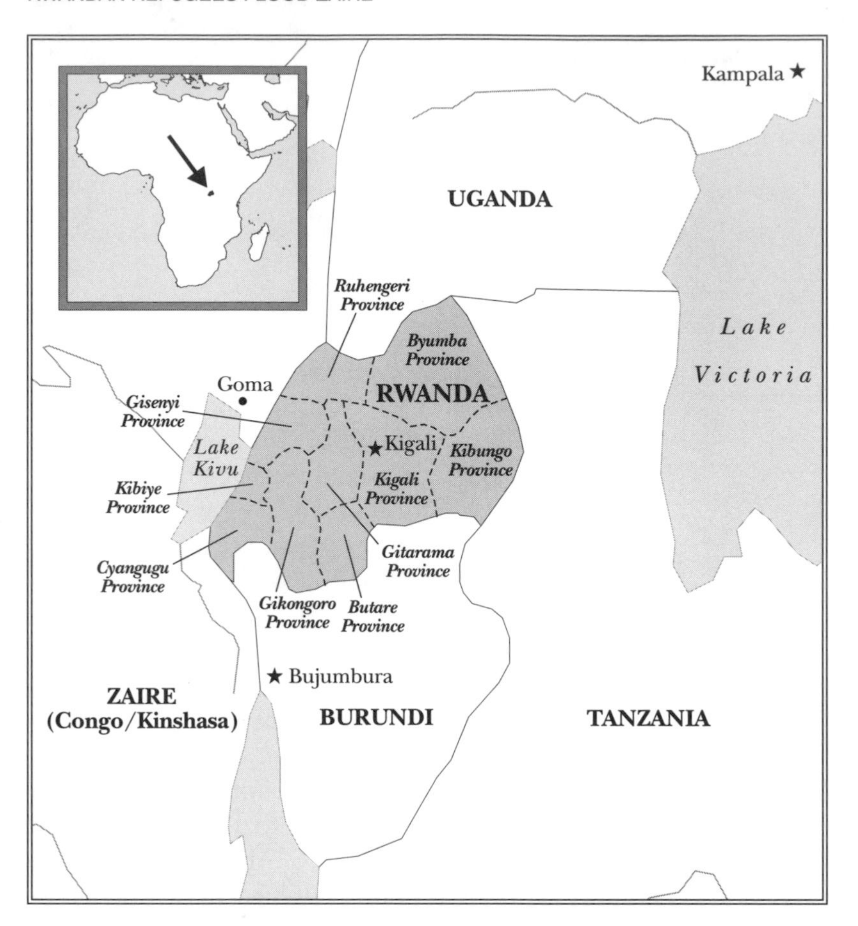

More than 90 percent of Rwandans worked in agriculture, primarily as subsistence farmers. Only a few grew enough sweet potatoes or corn to feed their own families. The war reduced even this low standard of living.

The roots of Rwanda's bloody conflicts date to the first contacts between Hutu farmers and Tutsi (formerly Watutsi) cattle herders and warriors. The Hutu were forced into inferior positions in a society dominated by a Tutsi warrior class. From 1899 to 1916, Ruanda-Urundi (as it was then called) formed part of German East Africa. During World War I, the British took control of the region, and in 1919, Belgium gained possession of it as a League of Nations mandated trust. The Belgians kept the two kingdoms separate and maintained their traditional social systems and forms of government. Both states had Tutsi kings. The Belgians did little to improve economic conditions.

In July, 1959, the Tutsi king of Rwanda died. The Hutu rose up in rebellion in Kigali, slaughtering thousands of Tutsi and throwing their bodies into the river. More than 20,000 deaths were recorded, and 200,000 Tutsi refugees fled the chaos to Burundi, which remained in Tutsi hands. For the first time in modern history, a Hutu government took control of Rwanda. When the Belgians pulled out in 1962, the Hutu retained power. Ethnic tensions remained high in both countries, reaching a peak of blood destruction in 1972, when more than 100,000 Hutu in Burundi were hacked and beaten to death by Tutsi mobs. Massacres took place again in August, 1988, when thousands of Burundians from each group were killed. Each time violence broke out in Burundi, tensions increased in Rwanda.

Land shortages were one cause of the trouble. The little land available remained in the hands of a small elite of Tutsi. Although in many places Hutu and Tutsi had lived together peacefully, even intermarrying on occasion, memories of past Tutsi oppression fueled the flames of resentment and were used to excuse Hutu mistreatment of Tutsi. Rwandan Tutsi feared that the government was planning to kill them all. They formed a rebel army, the Rwandan Patriotic Front, and began battling their way to power in an offensive that began in October, 1990. Habyarimana signed a peace accord with Rwanda Patriotic Front leader Alex Kanyarengwe on August 4, 1993. Talks to resolve the conflict, arranged by neighboring African states, appeared close to success before the murder of both countries' Hutu presidents set off a reign of terror that sent about one-third of the Rwandan population into exile.

Consequences

Refugees remained in their camps until the new Tutsi-dominated Rwandan government promised to restrain Patriotic Front soldiers. The new government tried to initiate war crimes trials for Hutu soldiers accused of killing Tutsi. Most

Hutu militiamen remained across the border in refugee camps, fearing that they would be killed if they returned home. Both Tutsi and Hutu believed that the other side was planning complete genocide. Some refugees returned to Rwanda and moved into camps there. The Tutsi government, claiming that these camps were havens for Hutu guerrillas, attempted to close them in early 1995 and to send inhabitants back to their villages.

On April 22, 1995, soldiers opened fire at the Kibeho camp when some inhabitants attempted to evade a screening process. As many as 2,000 of the 100,000 people in the camp were killed in the shooting. Whether the Rwandan government, under Tutsi leadership, could remain in control of a population that was predominantly Hutu appeared doubtful, and further conflict between the groups seemed likely.

Leslie V. Tischauser

U.S. Congress Advances Crime Bill

In one of the most important achievements of the Bill Clinton administration, the Democratic Congress advanced a crime bill that allocated $30.2 billion and banned assault weapons.

What: Law; Crime
When: July 28, 1994
Where: Washington, D.C.
Who:
Bill Clinton (1946-), president of the United States from 1993 to 2001
Bob Dole (1923-), Republican leader in the United States senate
Joseph Biden (1942-), chairman of the Senate Judiciary Committee

Conferees Agree on a Crime Bill

On July 28, 1994, conferees from the U.S. Senate and House of Representatives reached agreement on a crime bill that appropriated $30.2 billion over the ensuing six years. It allocated money to hire 100,000 new police officers and expand prison construction. Provisions of the bill also included expansion of the death penalty for federal crimes and a "three strikes" clause that would compel a life sentence for anyone convicted of a third serious felony under federal law. The most controversial aspect of the measure was the ban on nineteen different kinds of semiautomatic assault weapons.

President Bill Clinton called the law "a tough and serious legislative remedy to reduce violence and crime." Republican opponents attacked the prevention provisions of the bill, including the program to provide "midnight basketball" for inner-city youths. The assault weapons ban prompted heavy lobbying against the proposed law by the National Rifle Association (NRA) and other gun-owner organizations.

Agreement among members of the conference committee did not end the bill's legislative difficulties in the election-year Congress. The crime bill became a focal point of partisan differences during an election campaign season in which Republicans expected to make large gains and Democrats fought desperately to hold on to their majorities on Capitol Hill. President Clinton signed a modified version of the bill into law on September 13, 1994.

The Clinton Administration and Crime

The crime bill that was so controversial during the summer of 1994 had been pending in one form or another in Congress for almost a decade. By the beginning of 1994, crime had surpassed economic matters as a matter of major concern to Americans. Republicans and Democrats had been unable to agree on legislation to address mounting public concern about crime. Although government statistics indicated that violent crime had receded, the popular perception was that Americans lived in a dangerous and crime-ridden society.

To deal with the problem, President Clinton proposed in his State of the Union message of January 25, 1994, that Congress enact a "tough but smart" crime bill that mixed prevention and punishment. This identification with crime legislation reflected Clinton's posture as a "New Democrat," one not wedded to the older policies of his party, which had been seen as too soft on the crime issue. The president embraced the popular opinion that career criminals should receive certain and severe punishment. The crime bill set three felony offenses as the benchmark for defining a career criminal. The provision mandating life sentences for those convicted of a third felony offense became known as the "three strikes" provision. The crime bill also included a proposal to add 100,000 police officers and to provide prevention programs including so-called "midnight basketball" for inner-city youths. The latter initiative was one that President George Bush had endorsed.

During the spring of 1994, the crime bill

moved through Congress along the lines that the administration had hoped. In May, the House voted to ban nineteen different kinds of semiautomatic assault weapons. Under the leadership of Judiciary Committee chairman Joseph Biden, a Delaware Democrat, the Senate created a trust fund to pay for the bill by allocating money gained from reductions in the number of federal government employees. The measure also expanded the number of federal crimes covered by the death penalty.

After passing both houses of Congress in different versions, the bill went to a House-Senate conference committee. When the Senate and House conferees reached agreement in July, it seemed as if the Clinton administration was on the verge of a striking legislative victory.

The optimism was premature. On August 11, the House stunned the administration when it voted not to bring the conference report up for final debate. The NRA convinced conservative Democrats to vote against the bill because of the ban on assault weapons. Republican congressional leaders seized the opportunity to embarrass the president three months before the election. Members of the Congressional Black Caucus, unhappy with the harsh provisions of the bill concerning the death penalty, also voted not to bring the bill to the House floor. By a vote of 225 to 210, the bill was blocked. President Clin-

AP/Wide World Photos

President Bill Clinton signs the crime bill on September 13, 1994, under the watchful eyes of Marc Klaas (far right), whose daughter was kidnapped and killed, and Stephen Sposato (second from right), whose wife was killed by a gunman in the law firm in which she worked.

ton called it an occasion when "law and order lost, 210 to 225."

Over the next several weeks, the White House negotiated with the Republicans to find a way out of the impasse. Clinton agreed to remove some of the funds for crime prevention. Republican leaders had labeled these as wasteful government spending, or "pork." Enough concessions occurred on both sides to produce a victory for the White House when the House voted on August 21. The Republicans, led by Senator Bob Dole of Kansas, tried to block passage in the Senate, but that effort was defeated on August 25.

Consequences

The 1994 crime bill represented a typical legislative compromise on an important issue. Given the modest role of the federal government in fighting crime, neither the law's prevention provisions nor the harsh penalties it mandated were expected to have a significant impact on the incidence of crime in the United States. The law's larger importance lay in the political realm. Its passage allowed Clinton and the Democrats to rebut the Republican claim that their party was soft on crime. In the 1994 elections, however, the Republicans regained control of both houses of Congress.

Enactment of the assault weapons ban annoyed gun owners in general and particularly aggravated members of private militias, which received increased public attention in the early 1990's. Senator Dole promised in 1995 to repeal the assault weapons measure, but the bombing of the federal building in Oklahoma City in April, 1995, caused postponement of that strategy.

Lewis L. Gould

German Officials Seize Plutonium

In a sting operation, German police confiscated 363 grams of extremely toxic plutonium-239 that had been smuggled from Moscow to Munich.

What: International relations
When: August 10, 1994
Where: Munich, Germany
Who:
Justiniano Torres Benítez (1956-   ), Colombian who obtained the nuclear material in Moscow and transported it to Munich
Julio Oroz Eguia (1945-   ), Spaniard who helped Torres arrange the nuclear deal in Munich
Javier Bengoechea Arratibel (1935-   ), Spaniard who negotiated the plutonium transfer with German agents in Madrid and Munich
Rafael Ferreras Fernández (1954-   ), paid informer of the German police who initiated the sting operation in Madrid

The Munich Sting Operation

On August 10, 1994, the Bavarian State Criminal Police detected the presence of radioactive material in a suitcase owned by Justiniano Torres Benítez, a Colombian who had arrived in Munich on Lufthansa flight 3369 from Moscow. The suitcase contained 363 grams of a reactor fuel mixture of enriched uranium-238 and plutonium-239 as well as 201 grams of lithium. This represented the largest seizure of illegal and highly toxic plutonium in Germany.

The police in Munich arrested Torres and his two Spanish accomplices, Julio Oroz Eguia and Javier Bengoechea Arratibel. All three were charged with smuggling nuclear contraband into Germany. Torres, who married a Russian woman in 1991 and held Russian citizenship, was the key figure in obtaining the plutonium in Moscow. As a former representative of a helicopter firm, he had established connections with Russian military and civilian officials. Torres originally became acquainted with Oroz, his accomplice in Moscow and Munich, in the Nicaraguan embassy in Moscow. Bengoechea, a native of the Basque region of Spain, had helped negotiate the plutonium deal with German undercover agents in Madrid and Munich.

The arrest shocked Germany and the West because plutonium and lithium can be used for nuclear weapons. Although the amount of plutonium mix taken to Munich was not enough for a nuclear bomb, it would have been enough to poison Munich's water supplies. The Munich affair raised serious questions about the security of nuclear material located in the former Soviet Union. What frightened Western authorities in particular was the thought that a terrorist organization or state might obtain enough nuclear contraband to create a nuclear bomb.

Nuclear Contraband

The illegal sale of nuclear materials was made easier by the disintegration of the Soviet Union after December, 1991. The nuclear military sites in Russia, Belarus, Kazakhstan, and Ukraine seemed to be secure. Most of the nuclear materials smuggled out of the former Soviet Union came from civilian labs and atomic reactors. Many of the hundred thousand Russian workers in nuclear facilities were paid as little as $100 a month. Their low pay offered an incentive to engage in criminal activity. In 1993, Russian officials acknowledged eleven attempts to steal uranium. There were more than 950 radioactive storage areas in Russia alone. Lack of adequate security, economic difficulties, and the rise of organized crime in Russia made it possible for criminals to obtain nuclear material.

Although nuclear contraband has appeared in various parts of Europe, Germany has been affected more than any other single country outside the former Soviet bloc. Between 1991 and 1995, German police uncovered more than four hundred cases involving nuclear deals. In response, the German federal government and most German states have established special departments to investigate trade in atomic materials.

"Operation Hades," the sting operation that led to the Munich arrests, was directed by the Federal Intelligence Service. This agency, responsible for foreign intelligence, had established a special department to monitor money laundering and drug deals originating in the former Communist countries of Eastern Europe. In late 1993, Rafael Ferreras Fernández, a paid informant of this agency in Madrid, Spain, discovered that Russian plutonium was being offered on the black market.

In early July, 1994, after being informed that samples of nuclear material were in Munich, the Federal Intelligence Service informed the Bavarian State Criminal Police, which subsequently ran the local sting operation. By August 4, Torres, Oroz, and Bengoechea had offered Bavarian undercover agents four kilograms of plutonium for a price of $265 million. Torres traveled to Moscow to obtain the plutonium. On August 10, 1994, he returned to Munich on a flight from Moscow. The Bavarian police knew of his arrival because they had tapped the telephone line of Torres's accomplices in Munich. After landing in Munich, Torres was arrested, along with his two accomplices. In July, 1995, all three were sentenced. None of the sentences exceeded five years.

Consequences

The German sting operation irritated Russian authorities. Yevgeni Mikenin, the Russian deputy atomic minister, blamed the Munich operation for damaging Russia's international image. Mikenin and several other Russian spokespeople declared that not one gram of plutonium was missing from Russian facilities. They also suggested that the German sting operation was an attempt by the West to gain international control over Russia's plutonium inventory by demonstrating that Russia could not protect the vital material itself.

The Munich affair also caused a political storm in Germany. The journal *Der Spiegel* published accounts in April and May, 1995, that suggested that German police agencies knew before August 10, 1994, that plutonium was going to be transported to Munich on a public carrier. The journal accused the German police of willingly risking the health of the population of Munich. Furthermore, the German political opposition and press asked whether police sting operations have actually created a market for illegal plutonium. In May, 1995, the German government created a plutonium investigation committee, scheduled to present a report later that year.

Although German apprehensions about police sting operations are understandable, nuclear contraband represents a real danger. On the initiative of the general secretary of the United Nations, Jacques Attali investigated this problem and published his findings in a book titled *Economie de l'Apocalypse: Trafic et prolifération nucléaires* (Paris, 1995). Attali argued that at least thirty kilograms of plutonium had been stolen, much of it from facilities in the former Soviet Union. This is enough material to construct two or three primitive nuclear devices. Attali concluded that only with financial aid from the West can the countries of the former Soviet Union adequately secure their nuclear materials. Without financial help, it may take ten to fifteen years to eliminate the supply of contraband nuclear material.

Johnpeter Horst Grill

Lesotho's King Moshoeshoe II Dissolves Government

King Letsie III declared that he was dissolving the nation's first democratically elected government and removing from office Prime Minister Ntsu Mokhehle.

What: National politics
When: August 17, 1994
Where: Maseru, Lesotho
Who:
NTSU MOKHEHLE (1918-1999), prime minister of Lesotho from 1993 to 1994
BERENG SEEISO (1963-), king Letsie III of Lesotho from 1990 to 1994
MOSHOESHOE II (1938-), king of Lesotho from 1966 to 1990 and again from 1994 to 1996

A Democracy Falls

On August 17, 1994, King Letsie III of Lesotho announced on state-controlled radio stations that he was removing the country's prime minister, Ntsu Mokhehle, and dissolving his parliament. Ever since Lesotho had become an independent nation in 1966, the king had been only a ceremonial officer, with no real power. Mokhehle had won his position in a free democratic election, and his Basotholand Congress Party had won all the 243 seats in Lesotho's legislature. The king chose to take all that away.

King Letsie III declared that he was removing the prime minister because so many citizens were unhappy with the way the country was being run. Members of trade unions wanted higher pay and better benefits, and members of the military wanted more authority and more money. In fact, many people in Lesotho supported Mokhehle, and thousands of them had staged a rally in the capital city, Maseru, on August 15. King Letsie III addressed the demonstrators, urging them to disband. Two days later, he dissolved the government. Many people, inside and outside the country, believed that his real goal was to return his father, the former King Moshoeshoe II, to power.

Mokhehle was surprised to hear the radio announcements. He reminded King Letsie III that the country's constitution did not give him the power to rule, and he urged people to remain calm until the matter could be resolved. He promised to keep his elected position and expected the military to support him and the constitution. The same day, August 17, demonstrators marched again at the royal palace, protesting the king's actions. Some of them threw stones at the palace. This time members of the military who

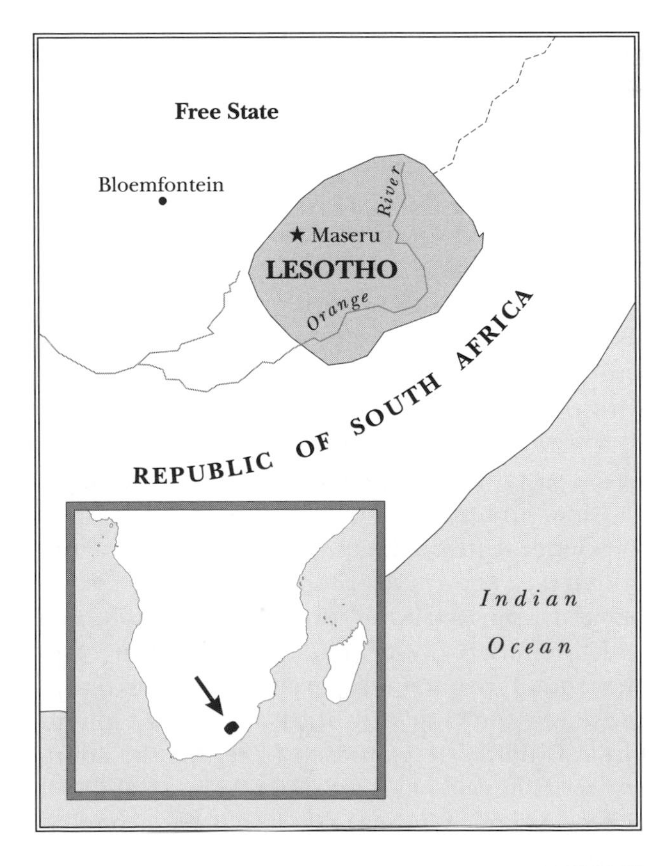

were loyal to the king fired into the crowd, killing four people and injuring ten more.

A New Democracy Emerges

The story of Lesotho, one of the poorest countries on Earth, has always been one of conflict, most of it among factions within the country itself. It became a British protectorate in 1868 and was then called Basutoland. Nearly one hundred years later, on October 4, 1966, the colony became an independent country with its own constitution. Moshoeshoe II, who had been paramount chief, became Lesotho's first king. The king had no real power to run the country. A national assembly and a senate made legislative decisions. The cabinet, run by Prime Minister Chief Jonathan, had executive power.

In 1966, King Moshoeshoe II tried to take more power, but in 1967 his attempt was stopped by Chief Jonathan. The king signed an oath that he would obey the constitution. Chief Jonathan was more interested in his own power than in the constitution. In 1970, when an election was held and he lost his position to Ntsu Mokhehle, Chief Jonathan suspended the constitution and used the military to hold on to his power.

AP/Wide World Photos

Moshoeshoe II.

For the next fifteen years, Chief Jonathan and the military ruled the country. Mokhehle left Lesotho to live in exile, but he and his followers staged several terrorist attacks in Lesotho. Chief Jonathan announced several times that free elections would be held, but he found various ways to block them. Lesotho's relations with other countries grew more tense. Nations that wanted to provide aid were hesitant because of the instability of the country's leadership.

Chief Jonathan was overthrown in 1986 by Major General Justin Lekhanya, the leader of the military. Lekhanya dissolved the national assembly and returned King Moshoeshoe II to office with increased power. Lekhanya and King Moshoeshoe II began to disagree in 1990. Lekhanya took away the king's authority and exiled him to Great Britain. He promised to return the country to civilian rule by 1992, after he had stabilized the economy. While these few men struggled over who would rule, and while they increased their personal fortunes, the people of Lesotho became poorer.

When King Moshoeshoe II was sent away, his son Bereng Seeiso was named King Letsie III. He signed an oath that he would not get involved in political matters but would hold the position of king only ceremonially. For a time he kept his promise.

After three more years of struggle, and after Lekhanya himself was overthrown, Lesotho finally held free elections. On March 27, 1993, Mokhehle was elected prime minister. He was sworn in on April 1, ending twenty-three years of military rule. A new constitution was drawn up, and King Letsie III promised to obey it. Under this constitution, the king would be head of state, with no power to govern.

Members of the military were unhappy about losing their power, and many of them were not loyal to the new government. Prime Minister Mokhehle struggled to hold on to his position and govern his nation.

Consequences

Reaction to the overthrow of Prime Minister Mokhehle was swift. Donor nations, including the United States, immediately suspended aid to Lesotho. The United States refused to recognize King Letsie III's new government. Human rights organizations worldwide condemned Letsie's actions.

On September 14, 1994, King Letsie III and Mokhehle signed a new agreement, arranged by diplomats from the nation of South Africa, which completely surrounds Lesotho. Under the new agreement, Mokhehle and the Basotholand Congress Party regained their power. King Letsie III abdicated his throne to his father, the former King Moshoeshoe II. The accord was signed in Pretoria, South Africa, and was considered to be South African president Nelson Mandela's first important foreign policy success.

Cynthia A. Bily

United States Changes Policy on Cuban Refugees

In a decision reflecting a broad shift in attitudes toward immigration, President Bill Clinton ended the long-standing practice of automatically granting asylum to Cubans fleeing their country.

What: International relations; Human rights
When: August 19, 1994
Where: Washington, D.C.
Who:
Bill Clinton (1946-), president of the United States from 1993 to 2001
Fidel Castro (1926 or 1927-), leader of Cuba from 1959
Lawton M. Chiles, Jr. (1930-1998), governor of Florida from 1991 to 1998
Janet Reno (1938-), attorney general from 1993 to 2001

Reversing Refugee Policy

During the first half of August, 1994, thousands of Cubans, determined to flee their island nation's poverty and repression, set out from the port of Cojimar in makeshift rafts, braving shark-filled waters in an effort to reach southern Florida, ninety miles away. There, they assumed, freedom and a better life awaited them.

For almost three decades, the United States government had granted asylum automatically to all Cubans who reached United States soil. As more and more Cuban refugees landed at Key West, Governor Lawton Chiles, Jr., of Florida pleaded with President Bill Clinton to do something about the influx. Chiles, running for reelection, feared that Florida taxpayers would rebel against the costs that a fresh wave of refugees might impose on the state.

On the evening of August 18, 1994, Clinton's attorney general, Janet Reno, announced at a hastily assembled news conference that Cuban refugees would no longer be entitled to automatic asylum. On the following afternoon, Clinton, in a televised press conference, repeated Reno's statement. He warned that Cubans intercepted at sea by the U.S. Coast Guard would henceforth no longer be taken to Florida; instead, they would be transferred to the U.S. naval base at Guantánamo Bay, at the eastern tip of Cuba, for detention for an indefinite period. In the future, Clinton proclaimed, Cubans who claimed refugee status would have to prove that they were threatened by persecution. Each individual case would be examined by the American interests section of the Swiss embassy in Havana. On September 9, 1994, after negotiations between the Cuban and United States governments, Fidel Castro, Cuba's leader, promised to stop further attempts by rafters to flee Cuba. In return, the United States agreed to accept about twenty thousand immigrants from Cuba per year.

Anticommunism Versus Xenophobia

In making his decision, Clinton responded to two strands of American public opinion: animosity toward immigrants and refugees in general, and a deep-rooted hostility to Cuba's leader, Castro. After taking power in January, 1959, Castro expropriated American-owned property without compensation and converted Cuba into a Communist state allied with the Soviet Union, America's archrival in the Cold War. In 1961, the United States broke diplomatic relations with Cuba, and in 1963, it imposed a trade embargo on the island nation. Castro's espousal of Communism was only one reason that so many Americans disliked him; another was the humiliation of having to tolerate an outspoken foe of the United States on an island that had been within the United States sphere of influence before 1959. The Cuban exile community that developed in Miami, Florida, provided a focus of anti-Castro sentiment on American soil.

From 1959 to 1962 (when commercial airplane flights from the island were halted), about 200,000 Cubans fled their homeland. In late 1965, special "Freedom Flights" were organized with Castro's cooperation. Although registration for these flights was closed in 1966, the flights themselves continued until 1973. The last mass influx of Cubans before 1994 was the Mariel boat lift of April to September, 1980, which took about 125,000 Cubans to American soil. Between 1980 and 1994, Castro forbade emigration.

During the economic boom years of the 1960's and early 1970's, neither the extreme liberality of the United States government's asylum policy toward Cubans nor the extensive financial assistance Washington offered Cuban refugees between 1962 and 1976 aroused much resentment among Americans. The Cuban refugees' flight to the United States was widely hailed as a propaganda victory for democracy in its worldwide struggle with communism, and their climb up the socioeconomic ladder in the United States was widely applauded as a triumph of hard work and determination.

As the American economy came to be beset by frequent short-term downturns and various long-term problems, many people in the United States came to view immigrants and refugees from Latin America as unwelcome competitors for jobs and public services. As president, Clinton heard widespread calls for cutbacks in legal immigration and witnessed widespread alarm over illegal immigration.

In 1994, Clinton still had bitter memories of the 1980 Mariel boat lift. President Jimmy Carter had ordered those refugees who lacked close relatives or other sponsors in the United States to be interned in camps across the United States until their fate could be determined bureaucratically. In June, 1980, a riot took place in a camp in Arkansas, where Clinton was then governor. Clinton later came to believe that popular indignation over the Mariel boat lift had cost him his chance of being reelected to the governorship in

AP/Wide World Photos

The internment camp at Guantánamo Bay in September, 1994.

November, 1980, and had also contributed to the defeat of Carter's bid for reelection to the presidency in the same year.

In 1992, the U.S. Congress passed the Cuban Democracy Act, which further tightened a thirty-year-old embargo. Serious shortages of oil and of medicine began to be felt in Cuba. Signs of malnutrition began to appear among its people and, on August 5, 1994, a riot erupted in Havana. The riot persuaded Castro to permit limited emigration by those who wanted to leave. The need to keep already bad relations with the United States from becoming worse moved him to crack down on emigration once again in September, 1994.

Consequences

In a second agreement with Castro, signed on May 2, 1995, Clinton promised that the Coast Guard would intercept at sea all Cubans fleeing their country and return them to Cuba. Those Cubans already interned in Guantánamo would be allowed to enter the United States. By early August, 1995, about half of the Guantánamo internees had been admitted to the United States, and the internment camp was slated to be closed by March, 1996.

The 1994 Cuban influx and Clinton's effort to stop it probably helped shift the debate in the U.S. Congress against advocates of free immigration and in favor of restrictionists. Proposals for reducing the total number of immigrants and refugees admitted to the United States, cracking down more severely on illegal immigration, and excluding legal aliens from certain welfare state benefits seemed to have better prospects of being enacted into law.

Paul D. Mageli

Provisional Wing of IRA Declares Cease-Fire

After twenty-five years of violence, the IRA's Provisional faction declared a cease-fire, raising hopes for a peaceful solution to political disputes in Northern Ireland.

What: Civil strife
When: August 31, 1994
Where: Dublin, Ireland
Who:
Gerry Adams (1948-), head of Sinn Féin, the IRA's political wing
John Major (1943-), prime minister of Great Britain from 1990 to 1997
Albert Reynolds (1933-), prime minister of Ireland from 1992 to 1994
Bill Clinton (1946-), president of the United States from 1993 to 2001

The IRA's Cease-Fire

On Wednesday, August 31, 1994, the Provisional Wing of the Irish Republican Army, the "Provos," announced that there would be an immediate cessation of all military operations in its armed struggle to separate the six counties of Northern Ireland from the United Kingdom and to unify all of Ireland within the Irish Republic. The IRA stated that it was willing to suspend its reliance upon armed force, which critics referred to as terrorism, and join the democratic process of establishing peace in Northern Ireland.

Many observers expressed optimism that the twenty-five years of violence in the north was over. The optimism was particularly evident among the Catholic community in Northern Ireland, the nationalists, who desired a single Ireland under the Irish Republic. Public officials also expressed optimism, particularly those in the Republic of Ireland, including the Republic's prime minister or taoiseach, Albert Reynolds, as well as President Bill Clinton of the United States. Others, including the Protestant majority in the north and some British politicians, were more cautious or even pessimistic. More than a few feared a conspiracy that could lead to Britain's abandonment of Northern Ireland's pro-British unionists or loyalists to the nationalists, the Catholics, and the Irish Republic.

The IRA's announcement gave hope that the endemic violence might be over, but there was reason for caution. Many doubted that the political spokespeople for the IRA, particularly Gerry Adams, could commit and control the IRA membership, whose tradition was that only force could evict the British from the north. Others asked whether the Protestant unionist-loyalist paramilitary groups, the mirror image of the IRA, would also agree to a cease-fire. Finally, there was a question of whether a political solution could be found to satisfy both the Protestant majority's desire to remain within the United Kingdom and the Catholic nationalist minority's—and the IRA's—wish to join the Republic of Ireland. Hope was balanced by fear and optimism by pessimism. Most agreed that the IRA's statement was only a beginning to any permanent solution.

Irish, British, or Both?

Conflict between the Irish and the English has existed for centuries. In the early seventeenth century, English rulers established "plantations" in northern Ireland, known as Ulster, and encouraged settlers from Protestant England and Scotland to colonize the region. Many did, and soon a majority of the population was Protestant, with the Catholic minority largely becoming a displaced underclass.

In 1916, a faction of Irish nationalists attempted an insurrection against British rule, which had been established officially over all of Ireland in 1800. The insurrection failed, but by 1919 civil war had broken out between the nationalists—known as Sinn Féin, with the Irish Republican Army as its military wing—and the British government. A peace treaty signed in 1921 established an independent Irish Free State within the British Empire.

The unresolved problem was Ulster. The new Irish state was overwhelmingly Catholic, but in the north the majority was Protestant. Religion is not the only key to understanding the Irish conundrum. Historical perceptions and culture and class differences also play roles. In 1922, instead of joining the Irish Free State, six of the nine counties of Ulster became Northern Ireland, an integral part of the United Kingdom of Great Britain, with the Protestants in control.

The IRA had split over the treaty establishing the Free State because it was not yet a republic and it was part of the British Empire. The Free State crushed the IRA's new rebellion, and the remnants of the IRA went underground. Sporadic violence continued in the Free State and Northern Ireland, but the IRA had become a nuisance on both sides of the Irish border.

The situation changed in the 1960's. The Catholic minority in the north, partially inspired by the Civil Rights movement in the United States, demanded an end to the discrimination that it had long suffered. Peaceful demonstrations were overtaken by violence. The British government sent troops, and in reaction the moribund IRA was reborn. Soon even the IRA divided over tactics and goals, with the Provisionals becoming the largest faction. Violence led to more violence, and by 1994 more than three thousand persons had died.

There were attempts to end the violence. Talks were held, sometimes in secret, even between the IRA-Sinn Féin and the British government. Little came of the talks, but contacts continued as the death count increased, with no military solution in sight. The IRA's August 31, 1994, cease-fire announcement was a dramatic departure from its previous armed stance. The decision to end armed resistance was the result not only of the failure of the IRA's armed campaign but also of considerable public pressure and many previous private discussions.

Consequences

Permanent peace was still only a hope. The unionist or Protestant paramilitary groups feared a sell-out by Britain's prime minister, John Major, but in October they also joined the cease-fire. In addition, the border was opened between north and south, army patrols ended in Belfast, and in May, 1995, for the first time in twenty-three years, British and Sinn Féin representatives met officially. The special relationship between Great Britain and the United States came under strain when, over objections by the British government, the Clinton administration welcomed Sinn Féin's Gerry Adams to the United States.

There was relief that the killing had ended, but questions remained. Would the Protestant-unionist majority in Northern Ireland ever consent to join the Irish Republic, and would the IRA and other radical nationalists accept anything less? Prospects for long-term peace were dimmed when the IRA claimed responsibility for a bombing in February, 1996, breaking the cease-fire.

Eugene S. Larson

World Delegates Discuss Population Control

An international conference sponsored by the United Nations emphasized links among population control, economic development, and the advancement of women.

What: Social reform; Economics; Human rights; Environment
When: September 5-13, 1994
Where: Cairo, Egypt
Who:
John Paul II (Karol Wojtyła, 1920-), pope of the Roman Catholic Church from 1978
Thomas Robert Malthus (1766-1834), British economist and demographer
Timothy Wirth (1939-), U.S. undersecretary of state and a leader of the U.S. delegation to Cairo

The Cairo Conference

The United Nations sponsored a conference in Cairo, Egypt, in September, 1994, with the intent of creating a twenty-year master plan for population control. Although conferees agreed to roughly 90 percent of the draft document in advance of the conference, debate and controversy reflected fundamental differences on a variety of difficult issues. These issues included reproductive rights, abortion, women's rights, and family structure. Press reports focused on dissent from nations with traditional views about the role of women, family life, and reproductive freedom. In particular, the delegation from the Vatican (represented as a sovereign state) was joined by leaders from Islamic states in opposing draft language believed to be destructive to the traditional family unit. Delegates particularly opposed liberal access to abortions.

The issue of abortion, described in paragraph 8.25 of the plan, raised the ire of the Islamic leaders, Vatican delegates, and Pope John Paul II himself. The plan conceptualizes abortion as a public health issue, focusing particularly on the potential risk of unsafe abortions to women's health. Taking into account the concerns of the Vatican and allies, the adopted language holds that "In no case should abortion be promoted as a method of family planning." It added that "women who have unwanted pregnancies should have ready access to reliable information and compassionate counseling."

With compromise language on abortion in place, the plan passed unanimously. Although emphasizing generalized development over stringent numerical goals, the plan advances the hope of stabilizing the rate of population increases. Earth's population was 5.7 billion in 1995. The U.N. Population Fund estimated that the world's population would grow to between 7.8 billion and 12.5 billion by the year 2050. The plan adopted in Cairo set the target population at the lower end of that range.

One of the themes of the conference was the link between population control and development issues. Although conferees came with different agendas and philosophical approaches to population control, they agreed on a plan that supports the notion that population growth is significantly tied to poverty, underemployment, inequality in treatment of women and men, and overconsumption of dwindling resources.

Interconnected Issues

The plan's theoretical approach is significant in that it departs from the influential thought of Thomas Malthus, an economist of the late eighteenth and early nineteenth century who postulated that population growth would outstrip natural resources and food supply. Many Malthusians promote large-scale contraceptive delivery systems as means of decreasing population growth rates. They are criticized for advocating "top-down" population control policies that conceptualize population growth as the cause of various problems such as food scarcity, illiteracy, inequality of women and men, and environmental degradation.

In their zeal to control population, some top-down population control plans influenced by Malthusian theory aim to persuade or even coerce people who live in poverty to have fewer children. Development theorists and feminists who attended the Cairo conference generally agreed that the causal direction is reversed: High rates of childbearing and population growth may be caused by poverty rather than being a cause of poverty. This view depicts population growth as symptomatic of underdevelopment. Thus, poverty, illiteracy, underemployment, high infant mortality rates, and inequality of women and men may increase incentives to have more children.

In developing nations, where infant mortality is much higher than in industrialized nations, ninety deaths occur for every thousand live births. Industrialized nations experience twenty deaths for every thousand live births. High infant mortality often is associated with large families. Families experiencing the death of a child may have a greater desire for more children, possibly to make certain that some survive.

In addition, in developing nations, parents who have many children may advance their own survival. Children provide necessary labor in agricultural areas and supplement incomes in urban areas by working outside the home. They may also enable parents to work by caring for siblings while parents work. Similarly, in countries without formal systems of old-age benefits, children may provide care to aging parents. In industrialized nations where income and educational levels are relatively high, various incentives discourage formation of large families. Children attend school rather than providing labor, and the costs of rearing a child are higher.

Top-down, coercive population control policies have been criticized as emphasizing contraceptive methods that are inconsistent with cultural norms and resources. Hormonal methods and intrauterine devices (IUDs) pose risks to healthy women in industrialized nations with access to good medical care. Health risks are even higher among women in developing nations, who may be nutritionally deprived and have erratic access to health care. These contraceptive methods require ongoing health assessments to ensure effectiveness and safety. Less coercive methods, such as barrier devices (condoms or diaphragms), have advantages of being safer and interfering less with culturally traditional birth control methods, such as breast feeding, which inhibits conception. They are deemphasized by some population planners because they are not as effective.

The Cairo plan emphasizes empowerment of women as a means of promoting economic development and controlling population. In the plan, women's empowerment is promoted through education, economic opportunity, and reproductive choice. These life improvements, in turn, encourage women to have fewer children. The blueprint outlines several major objectives. Education for girls is stressed. The plan points out that well-educated women have fewer children and that education increases earning power. Second, the plan extends family planning services to more women. An estimated 100 million women in 1994 lacked access to family planning. Third, the plan emphasizes health care services and reproductive care for women. The plan aims to reduce infant mortality and health risks of women in childbirth and abortion.

Consequences

The Cairo plan recognized the complexity of the population problem and the ways in which social, economic, and political phenomena affect population growth rates. The plan embraced the need for economic development as a precursor to reduced population growth but also promotes women's empowerment as a means of achieving this objective.

Mary A. Hendrickson

Haiti's Military Junta Relinquishes Power

Haiti's military junta, headed by Raoul Cédras, agreed to relinquish power, thus permitting the restoration of democracy in Haiti.

What: National politics
When: September 18, 1994
Where: Port-au-Prince, Haiti
Who:

RAOUL CÉDRAS (1949-), military leader of Haiti in 1991

JEAN-BERTRAND ARISTIDE (1953-), president of Haiti in 1991 and 1993 to 1996

BILL CLINTON (1946-), president of the United States from 1993 to 2001

JIMMY CARTER (1924-), president of the United States from 1977 to 1981, and personal emissary of President Clinton to Haiti in 1994

COLIN POWELL (1937-), chairman of the Joint Chiefs of Staff from 1989 to 1993, and personal emissary of President Clinton to Haiti in 1994

SAM NUNN (1938-), U.S. senator from Georgia and personal emissary of President Clinton to Haiti in 1994

Restoration of Democracy

During the summer of 1994, U.S. president Bill Clinton became increasingly irritated by the refusal of the Haitian military junta composed of Raoul Cédras, Philippe Biamby, and Michel François to honor the Governor's Island Agreement of 1993. In that agreement, the military leaders had promised to give up the political power they had seized in a September, 1991, coup and to permit the return to power of Jean-Bertrand Aristide, who had been democratically elected president of Haiti in December, 1990.

It became obvious to Clinton and Aristide that only military force would make the junta relinquish its usurped power. In a last effort to avoid the bloodshed that would inevitably result from an American invasion of Haiti, Clinton sent to the Haitian capital of Port-au-Prince a distinguished delegation composed of former U.S. president Jimmy Carter, General Colin Powell, and Senator Sam Nunn. The delegation arrived in Port-au-Prince on Saturday, September 17, 1994. President Clinton had the members of his delegation inform the leaders of the junta that unless an agreement were reached by noon on September 18, a U.S. invasion would begin on that day.

General Powell described bluntly the extent of American military force that would be directed against Haiti, but the junta leaders continued to stall. Once Biamby learned that American airborne troops had definitely left the U.S. mainland for Haiti, however, the junta leaders quickly accepted the conditions set by President Clinton. This agreement, which granted a general amnesty to the junta leaders and their assistants, required Cédras, Biamby, and François to resign by October 15, 1994, the day on which Aristide would return to Port-au-Prince to resume his position as the president of Haiti. Multinational forces arrived in Haiti to reestablish peace and to prevent reprisals against Aristide's supporters by the violent thugs who had beaten and killed opponents of the junta since the military coup of 1991.

Dictatorship in Haiti

Although Haiti obtained its independence in 1804 and is thus the second-oldest republic in the Americas, Haitians have endured generations of political and social exploitation by various despots and military tyrants. Strong popular opposition to his rule persuaded Haitian president Philippe Sudre Dartiguenave to ask the American government of Woodrow Wilson to restore order in Haiti, resulting in an occupation of Haiti by American Marines from 1916 until 1934.

AP/Wide World Photos

Raoul Cédras.

Between 1957 and 1986, Haiti was ruled by two brutal dictators named Duvalier. François Duvalier (called Papa Doc) governed from 1957 until his death in 1971. He was succeeded by his son, Jean-Claude Duvalier (called Baby Doc). Thousands of opponents of the Duvaliers were killed by the Tontons Macoute, organized criminals who worked for the Duvaliers. Hundreds of thousands of Haitians went into exile in order to avoid torture or death. When Jean-Claude Duvalier left for France in 1986, the Haitian economy was in ruins. The Duvaliers, their henchmen, and a small group of corrupt businessmen and military officers had become extremely rich while more than 90 percent of Haitians lived in abject poverty.

Even after Jean-Claude Duvalier's resignation, democracy did not flourish in Haiti. A series of military leaders tried unsuccessfully to prevent civilian control of the Haitian government and military, but after four more years of dictatorship, internationally supervised presidential elections were finally held in December, 1990. Aristide, a Roman Catholic priest whose outspoken opposition to the dictatorship of Jean-Claude Duvalier had resulted in his temporary exile to Quebec, won more than 70 percent of the vote. He was inaugurated as president in February, 1991. He dismissed corrupt officers from the army, put an end to rampant bribe taking, and began to implement civilian control over the Haitian military. Believing that they could not become rich in a democratic Haiti, soldiers under the command of Cédras, Biamby, and François overthrew the government, forced Aristide into exile in the United States, and installed themselves in power.

Foreign governments continued to recognize Aristide as the legal head of the Haitian government. The Organization of American States and the United Nations imposed sanctions and a strict embargo on Haiti. International trade with Haiti almost ceased. The Haitian economy deteriorated, and life in Haiti became almost intolerable. At least three thousand Haitians were killed during the reign of Cédras.

During the summer of 1994, thousands of Haitians fled Haiti in boats, hoping to make it to Florida. President Clinton concluded that the United States could not accommodate so many Haitian refugees. He decided that the restoration of democracy in Haiti would persuade Haitians to stay in their homeland. His efforts, which were highly controversial in the United States, proved successful. Aristide reassumed his presidential power in Haiti on October 15, 1994.

Consequences

After his return to power in October, 1994, Aristide had to solve serious economic problems while reestablishing the rule of law and persuading foreign companies that it was now safe to invest in Haiti. Massive international aid helped to rebuild the infrastructure in Haiti. American and other foreign soldiers remained in Haiti for several months after the restoration of democracy in order to train Haitian soldiers and police officers to respect constitutionally protected rights.

Edmund J. Campion

Congress Rejects President Clinton's Health Plan

President Bill Clinton's national health care bill, offered as the first significant effort in the field since Harry S. Truman's Fair Deal in the 1940's, suffered a congressional defeat in 1994.

What: Medicine; Social reform
When: September 26, 1994
Where: Washington, D.C.
Who:
Bill Clinton (1946-   ), president of the United States from 1993 to 2001
Hillary Rodham Clinton (1947-   ), first lady of the United States from 1993 to 2001
George J. Mitchell (1933-   ), Senate majority leader from 1988 to 1994
Bob Dole (1923-   ), Senate minority leader from 1986 to 1994, and Senate majority leader in 1995

Health Care Plan Rejected

On September 26, 1994, Senate Majority Leader George J. Mitchell (D-Maine) announced that Congress would not vote on President Bill Clinton's health care plan during the current session. Mitchell asserted that the "insurance industry on the outside and a majority of Republicans on the inside proved to be too much to overcome."

Under the proposed Health Security Act of 1993, employers faced mandatory insurance payments for all employees. Guaranteed medical coverage for the unemployed constituted the plan's second basic provision. When opposition from conservatives and the private sector mounted against the program, the president attempted to save it by allowing discounts for small businesses and setting spending limits on health care within the national budget. These efforts failed in 1994.

Bill Clinton, who introduced the program a year earlier, called it a temporary defeat and argued that he would "keep up the fight" and prevail "for the sake of those who touched us during this great journey." President Clinton and First Lady Hillary Rodham Clinton, who headed the national health care task force, pledged to bring the plan back before Congress to secure its passage in 1995.

Republicans, sensing victory in the November, 1994, elections, disagreed with the explanations offered by Democrats in Congress and the White House. Senate Minority Leader Bob Dole (R-Kansas) noted that the electorate "feared an overdose of government control. We saw democracy in action." The Republicans informed Clinton that they would block his international trade agreement unless he withdrew the bill. With Democrats telling him that they could not force cloture on a Republican filibuster, the president complied with the request to end the current drive for national health care.

Other Democrats such as Mrs. Clinton blamed special corporate interests in the insurance industry for the bill's downfall. Senator Mitchell joined Mrs. Clinton in bemoaning the bill's defeat and identified enemies who were "very adept at conducting campaigns of misinformation."

Insurance company executives counterattacked. Charles N. Kahn III, executive vice president of the National Health Association of America, the organization that produced the "Harry and Louise" television commercials aimed at shifting public opinion against the bill, defended his industry's position. Insurance companies, he stated, only wanted to "raise questions. Our purpose was to change the product, not harm it."

Health Coverage for All?

The drive for national health insurance can be traced to President Harry S. Truman's Fair

Deal. President Truman proposed a comprehensive medical insurance program to Congress on November 19, 1945. A conservative congressional coalition of southern Democrats and Republicans, who gained control in the November, 1946, elections, joined the American Medical Association and the insurance industry in blocking the bill.

President Dwight D. Eisenhower, who succeeded Truman, signed a bill creating the Department of Health, Education, and Welfare on April 11, 1953. Generally, however, Eisenhower opposed expanding the health role of the federal government.

Eisenhower's Democratic successors, John F. Kennedy and Lyndon B. Johnson, expanded the federal government's role in extending health coverage. In 1965, President Johnson signed the law providing coverage to the aged (Medicare) and the destitute (Medicaid).

The presidents who succeeded Johnson, however, followed more conservative approaches. One major factor considered by these presidents was the steadily increasing national debt, which grew from $979 billion in 1981 to nearly $3 trillion in 1989. Annual deficits exceeding $200 billion became common during the presidency of George Bush (1989-1993).

During the 1992 presidential campaign, Governor Bill Clinton (D-Arkansas) called for a national health insurance plan for all employees. Under Clinton's plan, employers would finance the insurance. President George Bush countered with a proposal that included deductions and tax credits. Congress, dominated by Democrats, rejected Bush's plan. Bush lost to Clinton in the 1992 presidential race by an electoral college vote of 168-370.

Clinton entered the presidency as a new Democrat who offered a "third way" between traditionally conservative and liberal health programs. On September 22, 1993, Clinton proposed a bill that provided coverage for most Americans through purchasing organizations known as health alliances. Eighty percent of the funding originated with employers. The Clinton plan offered coverage to all Americans by 1999.

Opposition to Clinton's Health Security Act originated from numerous sources. Fellow Democrat Jim Cooper of Tennessee proposed a program funded by "market forces" on October 6, 1993. The National Governors Conference rejected the president's program in January, 1994. The following month, the Business Roundtable, including two hundred of the nation's largest corporations, announced its disapproval. Nearly all the plan's opponents listed employer payments as the major obstacle.

Eventually, Congress bowed to the pressure. In July, 1994, Democrats began retreating from the Clinton bill in the House of Representatives, countering with a plan that reduced employers' costs. One month later, however, Senate Democrats broke with the president, proposing an al-

AP/Wide World Photos

First Lady Hillary Rodham Clinton meets with Democratic senators to discuss health care reform.

ternative program based on voluntary participation.

As Congress prepared to end its 1994 session, leaders from both parties conceded that a compromise bill was all but impossible prior to the November elections. When one last bipartisan effort led by moderates failed to produce even minor support, Senator Mitchell conceded defeat.

Consequences

The defeat of the Health Security Act, passage of which seemed ensured under a presidency and Congress dominated by Democrats, served as a harbinger of the future. On one hand, it was a test of President Clinton's leadership. On the other, it signaled a shift in public opinion away from an intrusive federal bureaucracy, symbolized by the Republicans' "Contract with America" that voters endorsed in the November, 1994, elections.

Although health care remained a pressing concern for most Americans, it did not override the equally important need to balance the federal budget. As the national debt approached $5 trillion, an anxious electorate decided to try a more conservative approach.

J. Christopher Schnell

The Bell Curve Raises Charges of Renewed Racism

The Bell Curve *argued that social programs must recognize the significance of inherited intelligence; opponents of the book suggested that its arguments were based on racist politics.*

What: Education
When: October 19, 1994
Where: United States
Who:
Richard J. Herrnstein (1930-1994), Harvard psychologist and coauthor of *The Bell Curve*
Charles Murray (1943-), political scientist and coauthor of *The Bell Curve*
Stephen Jay Gould (1941-), Harvard paleontologist and a critic of *The Bell Curve*

Publication and Reaction

The Bell Curve: Intelligence and Class Structure in American Life raised a furor when it appeared in bookstores on October 19, 1994. Richard J. Herrnstein and Charles Murray's book claimed that the evidence for differences in intellectual ability, as measured by intelligence (IQ) tests, is overwhelming, and that these differences occur not only among individuals but among social and racial groups as well. Further, it suggested that efficient social policy must recognize and be crafted in the light of these differences. The authors were particularly critical of affirmative action policies that ignored racial differences in ability. Barring development of appropriate social policies, they predicted the development of a stratified society based on intellectual ability and consisting of a well-to-do upper class and a very large and poor lower class. The book did not clearly define how to achieve the social policies that could avert such stratification but did suggest a return to a neighborhood-based society in which each person's contribution is valued by other members, no matter what intellectual depth it requires.

The book elicited a firestorm of criticism and charges of racism. The most caustic critics accused the authors of intentionally promoting right-wing, racist politics. Others suggested that their conclusions were merely incorrect and not necessarily evil in intent. A third group thought that the differences in test scores were real but that it was not helpful to say so.

The book does not read like a racist tome. It offers a carefully reasoned presentation, highly sympathetic to all the individuals and groups considered. Some of the commentary on the book agreed with its arguments and conclusions, viewing the book's critics as unreasoning and unreasonable.

The most consistent criticisms of the work, however, seem to be reasonable. One of these suggested that the meaning of IQ test results is unclear for at least two reasons. First, although the tests are reasonably good predictors of academic success, they are poor predictors of occupational, social, or political success. Second, although differences are consistently measured among races and among cultural groups, it has not been established that they result from differences in innate academic ability rather than cultural biases in the test or differences in subjects' learning environments. One common criticism questioned the immutability of IQ. Many psychologists believe that training can overcome many of the deficiencies indicated by the tests. In *The Bell Curve,* Herrnstein and Murray took the alternative position on each of these points.

Does IQ Determine Destiny?

The history of arguments over IQ testing is outlined in *The Bell Curve* and, from the argument's other side, in *The Mismeasure of Man* (1981) by Stephen Jay Gould. These books, along with *The Bell Curve Wars* (1995), edited by Steven Fraser, and *The Bell Curve Debate* (1995), edited by Russell Jacoby and Naomi Glauberman, give a fairly complete picture of that history and of the debate renewed by *The Bell Curve*.

Alfred Binet, a French psychologist, developed the first important psychological tests of reasoning ability in the early 1900's. Binet's purpose was to identify children with problems in reasoning ability so they could be helped. A number of investigators developed extensions of Binet's test, notably Lewis Terman, who produced the Stanford-Binet test. Terman, in contrast to Binet, believed that the results of the tests indicated an individual's potential. A low IQ meant limited future options. There was no way of improving a person's ability: It was determined genetically.

With this attitude in place early in the twentieth century, extensive testing generated a host of clearly (in hindsight) prejudiced conclusions about the innate intelligence and resultant social position of different groups. The poor, for example, were thought to be poor because of inherent inability, and there was nothing anyone could do. According to such arguments, black Americans, Hispanic Americans, American Indians, and many immigrants were more heavily represented among the poor because they were, on average, of low innate ability.

As the century progressed, other explanations were proposed for the differences in IQ scores. These included cultural bias in the tests and differences in learning environments among individuals and groups. In addition, many educators and psychologists came to believe that good educational programs can overcome many of the deficiencies indicated by the tests. This interpretation became more popular as the political and cultural climate changed and was in part responsible for the advance of civil rights and equal opportunity legislation later in the century.

The Bell Curve was published during a major political swing toward conservatism, generally as-

AP/Wide World Photos

Charles Murray.

sociated with the earlier interpretation of IQ tests. The book revived the ideas that a low IQ restricted a person's potential and that little could be done to change a person's IQ. Although it argued for a place in society for everyone, it also suggested that because of inherited intellectual ability, many individuals would be restricted to menial positions and that nonwhite people would be overrepresented in that group.

Consequences

In the eyes of its detractors, the book turned back the clock to a time of socially justified racial and economic inequality. In the eyes of its defenders, it introduced sanity to the arguments concerning affirmative action and other social programs. Its greatest shortcoming may be that it denied the potential of individuals to improve their intellectual skills if given the opportunity. The book served to refocus American society on affording that opportunity to all. It will not have the impact its detractors fear because too few people in American society are willing to assume that a single test, the meaning of which is unclear, is capable of defining the intellectual or overall worth or life chances of a person.

Carl W. Hoagstrom

North Korea Agrees to Dismantle Nuclear Weapons Program

On October 21, 1994, the United States and North Korea agreed to follow a step-by-step plan to end North Korea's nuclear weapons program and divert it to peaceful purposes.

What: Military; International relations
When: October 21, 1994
Where: Geneva, Switzerland
Who:
Bill Clinton (1946-　　), president of the United States from 1993 to 2001
Kim Il Sung (1912-1994), president, Democratic People's Republic of Korea (North Korea) from 1972 to 1994
Kim Jong Il (1942-　　), Kim Il Sung's son and successor

The Framework Agreement

The "framework agreement" made between the United States and North Korea on October 21, 1994, was a turning point in the negotiations over North Korea's nuclear program. It outlines a series of steps that divert North Korea from nuclear weapons production to safer, domestically useful electrical energy production.

The agreement addresses North Korea's nuclear program of the past, present, and future. For the past, although it leaves some key issues for the end of the process, it freezes what has been done and shuts down North Korea's existing reactors under the supervision of the International Atomic Energy Agency (IAEA). For the present, it supplies North Korea's energy needs by providing 500,000 tons of fuel oil per year for furnaces and light industry. The oil is not to be used for military purposes. It also provides for "liaison offices" in Washington and P'yongyang where the two governments can keep up a running dialogue. Basic commercial contacts are hoped to provide better trade relations. The contacts will be in such areas as telecommunications and banking.

For the future, the agreement provides for a consortium, the Korean Peninsula Energy Development Organization, to construct two new thousand-megawatt light water reactors (LWRs). These will replace Chernobyl-type gas-graphite reactors from which the North Koreans can extract plutonium for weapons manufacture and export, in violation of the Nuclear Non-Proliferation Treaty (NPT).

The agreement provides for the storage of existing nuclear waste from a reactor, operating since 1986, that could be reprocessed into plutonium as well as for shipment of the waste out of North Korea to another country prior to reprocessing. This means that North Korea has surrendered the ability to make bombs in the future, at least without outside assistance.

The agreement also provides for North Korea to allow IAEA inspection of two suspicious nuclear waste dumps. Those investigations may reveal whether the country had a secret and illegal weapons program prior to 1993. When these steps have been accomplished, the United States will deliver the core components and fuel for the two new LWRs so they can begin operation.

The Problem and the Negotiations

North Korea's nuclear program has been a matter of world concern since the detection in 1989 of a new building under construction that was almost certainly a plant designed to reprocess spent reactor fuel into plutonium, the basic component of nuclear weapons. North Korea claimed that its nuclear program was solely for peaceful purposes such as making electricity. Advanced weapons systems such as missiles, however, are among its most profitable exports, and the production of atomic weapons, whether for national security or for export, was regarded as a prime threat to world peace and stability. It was particularly feared by South Korea, the north's archrival, and Japan, its former colonial master, both of which are allies of the United States.

Nuclear threats are nothing new in Korea. The United States threatened to use atomic bombs during the Korean War and made nuclear weapons an integral part of its defense arrangement with South Korea. The Soviet Union, on the other side, maintained the nuclear balance by providing a nuclear shield for North Korea during the Cold War.

North Korea's domestic nuclear program began in 1965 with a small Soviet research reactor. A second, larger reactor went into operation in 1986 at the town of Yongbyon. In 1989, it was shut down for maintenance, and part of its fuel was unloaded and possibly reprocessed to make a small amount of plutonium. The world community is unsure of this because it is unclear whether North Korea had reprocessing capability at the time. North Korea denies that it made plutonium or that it ever had a nuclear weapons program. In 1993, however, satellite intelligence discovered two "hot" radioactive sites near Yongbyon that suggest that the unloaded fuel was reprocessed in 1989. The U.S. Central Intelligence Agency announced that the world should assume that North Korea had one or two atomic bombs.

The discovery of the two "hot" sites started a furor. The IAEA demanded access to inspect the sites, and the North Koreans refused. When the IAEA threatened to take the issue to the United Nations for possible sanctions against North Korea, the government of President Kim Il Sung announced renunciation of the NPT and withdrawal from the IAEA. Delicate negotiations followed, with a goal of convincing the North Koreans to change their minds. Ultimately a dialogue began with the United States, leading to the framework agreement signed by representatives of the Bill Clinton administration and the North Korean government of Kim Jong Il on October 21, 1994.

Consequences

After the agreement, North Korea froze its nuclear program and subjected all of its known nuclear facilities to continual IAEA surveillance. The United States shipped fifty thousand tons of fuel oil, though it later discovered that the North Koreans had violated the agreement by diverting some of it to steel production. Diplomats visited each other's capitals in search of suitable property for the liaison offices, without early success. The North Koreans accepted direct-dial telephone service and credit cards.

The most difficult problem, however, was the LWRs. The Americans envisioned the Korean Peninsula Energy Development Organization as a conduit for South Korean funds. Because it had so much to gain from the agreement, South Korea's contribution was anticipated to be $3 billion, nearly 75 percent of the $4.2 billion total cost of the agreement. The South Koreans insisted that they design and build the LWRs, but the North Koreans, sensitive about the technological gap between north and south and anxious not to lose face, demanded that the reactors come from the United States or Germany. Neither of those countries was interested in spending $3 billion on the project. At a series of meetings in Berlin, Germany; Geneva, Switzerland; and Kuala Lumpur, Malaysia, the Americans insisted that there was no alternative to South Korean financing and construction for the LWRs. North Korea finally accepted this position on June 13, 1995.

Donald N. Clark

Israel and Jordan Sign Peace Treaty

The Middle Eastern states of Israel and Jordan formally ended forty-six years of declared war with a peace treaty anticipating an era of cooperation in a variety of enterprises.

What: International relations
When: October 26, 1994
Where: Israel-Jordan border
Who:
Yitzhak Rabin (1922-1995), prime minister of Israel from 1974 to 1977 and 1992 to 1995
Hussein ibn Talal (1935-1999), king of Jordan from 1952 to 1999
Bill Clinton (1946-), president of the United States from 1993 to 2001

The Peace Agreement

Climaxing months of negotiations, Prime Minister Yitzhak Rabin in Israel and King Hussein of Jordan signed a peace treaty between the two Middle Eastern nations on October 26, 1994. President Bill Clinton of the United States, who had played an important role in the peace process, was the featured guest at the signing ceremony.

Eight Islamic countries sent representatives to the Arava border crossing just north of the Gulf of Aqaba for the formal ending of a forty-six-year state of war between Israel and Jordan. The agreement came about despite King Hussein's earlier insistence that Syria, Lebanon, and the Palestine Liberation Organization (PLO), unrepresented at the event, had to be part of any peace agreement.

The treaty provided for a variety of interactions between the two nations that had been impossible since 1948. It returned to Jordanian control certain territories seized by Israel in the 1967 Six-Day War. It opened the borders to commerce and travel by private citizens and normalized diplomatic relations between the two nations. It settled longstanding disputes over water rights in this arid region, and it ensured that a number of other matters—environmental protection, energy, tourism, and avenues of transportation and communication among them—would thereafter be approached cooperatively.

Economic motives were paramount in moving the peace process to a successful conclusion. Even though direct trade between Israel and other Arab states was still prohibited, Israel would be in a position to increase its exports of manufactures and agricultural products to Jordan. Meanwhile, Jordanians looked forward to the use of routes to Israeli Mediterranean ports such as Haifa instead of being forced to ship goods via the longer sea route through the Gulf of Aqaba.

Retreat from Confrontation

The previous hostile relationship between the modern states of Israel and Jordan can be thoroughly understood only in the larger context of centuries of religious, ethnic, and political conflict. The political situation from the end of World War I to the new peace accords can be summarized as follows.

The Conference of San Remo in 1920 allotted control of Palestine (the area covered by modern Israel and Jordan) to Great Britain. The following year, a new government of the eastern portion of Palestine was established under Abdullah ibn Hussein. The area, called Transjordan, was specifically excluded from any future establishment of a Jewish homeland. In 1946, Abdullah was designated king. In 1948, following the elimination of the already reduced Palestine by the creation of the independent state of Israel, the five angry adjacent Arab nations, including Transjordan (which shortly became known as Jordan), seized a portion of the newly designated Israel, including part of Jerusalem. Much of this land remained in Jordanian hands as a result of an armistice signed in 1949.

These political shifts caused difficult refugee problems and resentment on Israel's part. By virtue of the Six-Day War, a lightning military offensive in June, 1967, Israel reoccupied this territory as well as the West Bank of the Jordan River, thus threatening the future economic viability of Jordan. Jordan subsequently accepted a 1974 Arab summit conference's choice of the PLO as the representative of Arabs remaining in the Israeli-occupied West Bank, thus making Jordan mainly a spectator in the continuing Israeli-Palestinian conflict. Jordan continued to proclaim that a state of war with Israel had existed since 1948.

International efforts toward reconciliation of Israel and its Arab neighbors in subsequent years met with limited successes, but following accords between the PLO and Israel in 1993, talks between Jordanian and Israeli representatives began to bear fruit. King Hussein realized that maintaining hostility toward the more prosperous country to his west hindered his efforts to rejuvenate Jordan's feeble economy. It was more beneficial to have peaceful access to the land that Jordan had coveted than to gain political control over it. In addition, the PLO's control of the issue of Palestinian refugees within Israel's borders was undermining Jordan's importance in the larger peace process.

In the summer of 1994, Hussein took advantage of Israel's willingness to discuss for the first time the matter of the disputed border. Sufficient progress was made that by July 25, with President Clinton serving as host, Hussein and Rabin felt free to meet publicly for the first time and sign the Washington Declaration, which paved the way for the October peace treaty.

Although Jordan renounced its claim to West Bank hegemony, it regained 128 square miles of land between the Gulf of Aqaba and the Dead Sea and a small parcel south of the Dead Sea. Much of this land was unsettled but of symbolic importance to Jordan. The Washington Declara-

AP/Wide World Photos

As President Bill Clinton looks on, King Hussein of Jordan (left) and Israeli prime minister Yitzhak Rabin shake hands after signing a peace agreement.

tion also recognized Jordan's right to custodianship of Islamic shrines in Jerusalem. The declaration and the subsequent treaty angered the PLO and drew the disdain of Israel's other neighbors, but international political experts judged it an important step on the road to general peace and stability in an area that had known little of either for many decades.

Consequences

Implications of the treaty for general peace in the region were soon evident. On January 26, 1995, Jordan and the PLO agreed to cooperate on a wide variety of matters, including the custody of Islamic shrines, security concerns, financial policy, postal links, and educational and cultural affairs. Jordanian currency was to become legal tender in the Gaza Strip, in Jericho, and in any other areas to which Palestinian self-rule might be extended. In March of 1995, Israel and Syria agreed to resume direct talks toward a peace treaty.

On January 30, 1995, Israel formally turned over to Jordan the lands specified in the treaty. Other provisions of the Israeli-Jordanian accords continued to be implemented.

Robert P. Ellis

Californians Vote to Limit Aid to Undocumented Immigrants

Sixty percent of California's voters approved a ballot proposition to limit the public services available to undocumented immigrants.

What: Social reform; Economics; Politics; Political reform
When: November 8, 1994
Where: California
Who:
PETE WILSON (1933-), Republican governor of California from 1991 to 1999
CHANG-LIN TIEN (1935-), chancellor of the University of California, Berkeley from 1990 to 1997
JACK W. PELTASON (1923-), president of the University of California system from 1992 to 1995

A Vote to Restrict Services

On November 8, 1994, approximately 60 percent of the voters of California marked their ballots in favor of Proposition 187, the so-called Save Our State or SOS initiative. The proposition had been drafted by a conservative Orange County businessman. This legislation would end state-funded education and welfare benefits for illegal aliens. It also limits publicly funded medical assistance available to illegal aliens. They could be treated only in life-threatening emergencies requiring immediate attention. Under provisions of the proposition, teachers and physicians must report illegal aliens to the immigration authorities.

The state of California, hard pressed by an economic downturn and by a series of natural disasters including earthquakes, fires, and floods, was home to an estimated 1.6 million illegal aliens. Many of them held minimum-wage jobs that most Americans were reluctant to fill. The estimated annual cost to California for services related to illegal immigrants exceeded $3 billion. How much of this was offset by various taxes paid by these workers was a subject of debate.

Both supporters and opponents of Proposition 187 agreed that this legislation was unconstitutional, violating both the equal protection guarantees of the United States Constitution and hundreds of antidiscrimination laws. The primary motivation of those who supported the initiative was to get the matter of providing public services to illegal aliens into the courts. Their ultimate aim was to overturn some legislation and to negate some of the related decisions handed down by the United States Supreme Court.

Underlying the appearance of Proposition 187 on the ballot in 1994 was the campaign of Governor Pete Wilson, who rode a wave of conservatism into the state house in 1991. Wilson, a two-term United States senator with presidential ambitions for 1996, proposed legislation targeting illegal aliens. His programs were popular among voters who faced diminished employment opportunities, increased taxes, and decreased public services. Illegal aliens were identified as a cause of these problems.

Wilson campaigned for the passage of Proposition 187 realizing that the national publicity such a campaign would generate could benefit him substantially among conservative voters, whose ranks were growing nationally. Despite its obvious defects, the proposition that six of every ten California voters approved enjoyed considerable popularity nationwide.

Economic Effects

Agriculture and mining had been mainstays of California's economy since the sprawling territory achieved statehood in 1850, becoming the thirty-first of the United States. Immigrants, legal and otherwise, helped to build the state and became a fundamental part of its economy. California had long been one of several significantly multiethnic states. Many people count this among California's greatest assets.

California's major industries, including defense and aerospace, began to feel severe economic pressures with the cessation of the Cold War during the late 1980's and early 1990's. Voters faced with uncertain economic futures resented having to pay taxes to help support those whose presence in their state and country was illegal. These undocumented immigrants contributed substantially to the state's economy, however, by taking jobs that their legally documented counterparts were often unwilling to take and by paying taxes for unemployment insurance and supplemental security income, benefits to which they did not have access.

Passage of Proposition 187 did not result in the disappearance of illegal immigrants from California. The conditions under which they remained there, however, were increasingly difficult. The full enactment of this initiative was blocked by the courts as test cases worked their way through the judicial system. Less than a week after the election, various groups had begun to test the law, and on November 14, 1994, a federal judge temporarily blocked enactment of the measure.

It is argued that Proposition 187 not only violated the Constitution and various laws but also created unacceptable conflicts between state law and professional ethics and responsibilities. Physicians are professionally obligated to treat any patient who requires treatment. A law mandating that they turn in patients who seek their professional services is in violation of the standards by which the medical profession traditionally has been guided.

Consequences

The repercussions following the passage of Proposition 187 were enormous. The vote had implications far beyond the matter of what services should be available to illegal aliens and their dependents. It reflected a major shift in the thinking of many Americans about the kind of nation the United States is becoming and suggested an undercurrent of racism. The proposition expanded debates on immigration at the national level.

From the late 1960's to the early 1990's, Americans who had been unfairly disenfranchised became beneficiaries of legislation that accorded them the rights guaranteed by the Constitution. Those who had been discriminated against because of race, gender, religion, sexual preference, or physical disability were now protected. Such programs as affirmative action required preferential treatment for those falling into the above categories in terms of employment, educational, and other opportunities in organizations that receive government funding. These programs were intended to reverse the effects of past discrimination.

In late July, 1995, the Board of Regents of the University of California voted to abolish affirmative action. They decided to end racial preferences in hiring and contracting by January, 1996, and end preferences in admissions to the nine campuses of the university system by January, 1997. Jack W. Peltason, president of the system, agreed to comply with the mandate, adding that the system sought to reflect California's ethnic diversity in the populations of its nine campuses. Chancellor Chang-lin Tien of the University of California, Berkeley, had never strongly supported affirmative action. He accepted the regents' recommendation that the universities in the system do everything they could to achieve diversity but without using preferences for admission based on race, gender, or ethnicity.

R. Baird Shuman

Republicans Become Congressional Majority

In an election that marked a turning point in American politics, the Republican Party, with its "Contract with America," seized power and the legislative initiative.

What: National politics
When: November 8, 1994
Where: United States
Who:
Bill Clinton (1946-), president of the United States from 1993 to 2001
Newt Gingrich (1943-), Speaker of the House in 1995
Bob Dole (1923-), Senate majority leader in 1995
Rush Limbaugh (1951-), influential conservative syndicated radio personality

The Election

The party holding the presidency generally loses some representation in the off-year elections. Rarely, however, has the turnover been as extreme as in the 1994 elections. The Republicans seized control of both houses of Congress, picking up 52 seats in the House, making the breakdown 230 Republicans, 204 Democrats, and 1 independent. In the Senate, they gained 9 seats for a controlling 53-47 majority. Not one Republican incumbent was defeated. Although some of the reasons for the change were that a number of incumbent Democrats had retired, that there was deep-seated dissatisfaction with the president, and that the Republicans had run a number of charismatic candidates, the importance of the issues as embodied in the Republicans' "Contract with America" cannot be overlooked. Many Republican candidates allied themselves with the Contract with America, a legislative agenda championed by Congressman Newt Gingrich.

The switch to the Republicans in the national arena was also evident on the state level. Democrats lost governorships in every contested large state and saw their numbers decline in many state legislatures. Additionally, after the election, a number of Democrats switched party affiliation, strengthening the Republican control in both houses of Congress and several state legislatures.

The Contract with America provided a blueprint for the first days of the new Congress. Gingrich, a historian, borrowed ideas from the first hundred days of the Franklin D. Roosevelt presidency. As the new Speaker of the House, he made good on his promise to bring to a vote all the issues contained therein. Although he was not successful in obtaining passage of all the bills, and although the Senate, led by Bob Dole, was much more deliberate and slower, the Republicans changed the political rhetoric and the terms of debate. Furthermore, a number of congressional committees were eliminated or merged, and staffs were cut back. This resulted in more openness and flexibility in the legislative process. New bills were made available immediately on the Internet, and, under new rules, they could not be bottled up in committee as easily.

What makes the election so important is the magnitude of the switch. Nevertheless, it is important to remember that the 1994 Republican majorities in Congress were relatively quite slim. To regain control in 1996, the Democrats would need to capture fourteen Republican seats in the House and four in the Senate. Although this appeared possible, it would require a united Democratic Party.

Rising Conservatism

Although the post-World War II period has seen the election of more Republican presidents than Democrats, until the 1994 election, the coalition governing the nation was essentially unchanged from the one that had elected Democrat Franklin D. Roosevelt in 1932. Although the

AP/Wide World Photos

The new Speaker of the House, Newt Gingrich (left), and the new Senate majority leader, Bob Dole, celebrate the Republican dominance in the Congress.

Republicans had won a congressional election decisively in 1946, that was the last time they had held a true working majority. In 1952, they gained a slim majority in both houses, and in the 1980 Ronald Reagan landslide, they took control of the Senate. In each case, however, in the next contest they returned the power to the Democrats before enacting major legislative changes or procedural reforms.

Diverse local issues had been driving congressional elections. These issues often had little to do with the matters actually debated in Congress. On September 27, 1994, House Minority Leader Gingrich announced the Contract with America. Almost all the Republican candidates for the House agreed to support the document if elected. It set much of the agenda for the campaign.

Among the issues and themes highlighted in the contract were dissatisfaction with both the size and lack of efficiency of the government. The document expressed the feeling that government programs were not helping the persons for whom they were designed. The contract addressed other issues, including the need for budget cuts leading to a balanced budget; disillusionment with President Bill Clinton and his health care plan, which would have increased the size and influence of the bureaucracy while cutting into the average person's medical options; corruption, as symbolized by House Ways and Means Committee chairman Dan Rostenkowski, who was subsequently indicted for fraud; and the arrogance of the Democratic leadership of House Speaker Tom Foley, who refused to allow many pieces of legislation to reach the floor of the House for debate.

These issues were discussion topics for numerous radio talk show hosts who blanketed the airwaves of the country. Although the political proclivities of the new wave of hosts covered the political spectrum from left to right, the most popular ones were conservative. Rush Limbaugh became the most successful and influential. Exposing government excesses and tweaking the liberal establishment, he built a loyal group of listeners on the more than six hundred stations that carried his show. Most of his listeners were conservatives. This type of media support certainly aided the Republicans' efforts.

Consequences

Historians have been trying for a number of years to identify a modern election to describe as a "watershed." Richard Nixon and Reagan had won in landslides, and conservatives Reagan and George Bush had held the presidency for three terms, but in each case only the executive had been changed. Congress had remained solidly in the hands of the Democrats, with the exception of the 1981-1983 senatorial hiatus.

The election of 1994 changed that. The Democrats kept the White House in 1996, but the Republicans maintained their control of both the House and Senate through the 1996, 1998, and 2000 elections. These electoral successes seemed to ensure that 1994 would indeed be remembered as a landmark election. Pundits may argue that Reagan's 1980 election marked the beginning of an era, with 1994 reaffirming the change. This would be the first time that a congressional, rather than a presidential, election would have been credited with such importance.

Theodore P. Kovaleff

Iraq Recognizes Kuwait's Sovereignty

The Iraqi government, in an effort to persuade the United Nations Security Council to lift economic sanctions imposed after the invasion of Kuwait, officially recognized the sovereignty of Kuwait.

What: International relations
When: November 10, 1994
Where: Baghdad, Iraq
Who:
SADDAM HUSSEIN (1937-), president of Iraq and chairman of the Revolutionary Command Council from 1979; president for life beginning in 1990
TARIQ AZIZ (1936-), Iraqi deputy prime minister from 1991
JABIR AL-AHMAD AL-JABIR AS-SABAH (1926-), emir of Kuwait from 1977
ANDREI KOZYREV (1951-), Russian minister of foreign affairs from 1990

Recognizing Kuwait

Iraq's government claimed that Kuwait was a province or governorate of Iraq to justify its invasion of the oil-rich neighboring desert country on August 2, 1990. Iraq continued to refer to Kuwait by those terms publicly even after being driven out by a United Nations-sponsored military coalition in February, 1991, during the Persian Gulf War. It was not until November 10, 1994, that Iraq formally recognized Kuwait as an independent nation. Following active intermediation by Russia, especially that of Foreign Minister Andrei Kozyrev, President Saddam Hussein of Iraq signed a decree recognizing Kuwait's sovereignty, territorial integrity, and political independence within borders newly demarcated by a United Nations commission. Hussein signed the decree as his country's president and as chairman of the Revolutionary Command Council (RCC), the ruling Baath Party's chief decision-making body. Iraq's National Assembly (Parliament) ratified the decree later that day.

The move was calculated to encourage the United Nations Security Council to lift the economic sanctions that it had imposed immediately after Iraq's seizure of Kuwait. The sanctions included an embargo on all imports except medical, food, and other humanitarian goods and on Iraq's oil exports beyond those necessary to pay for the goods it was allowed to import.

The embargo was not lifted at the Security Council's meeting of November 14-15, 1994, primarily because of opposition from the United States and Great Britain. Neither were the sanctions eased at the Security Council's subsequent regular reviews of Iraq's compliance with various Security Council resolutions. Russia, France, and China, the three other permanent veto-wielding representatives of the fifteen-member Security Council, had strong financial interests in resuming trade with Iraq and were more favorably disposed to ending the embargo.

The United States was the most insistent that Iraq meet provisions beyond recognition of Kuwait. These included Iraq's return of all Kuwaiti detainees and allowing unrestricted United Nations inspections to ensure the neutralizing and monitoring of all Iraqi weapons of mass destruction. Because these provisions had not been fully satisfied, the United States argued that no concessions should be made to reward Iraq for its gesture of recognizing Kuwait. Additionally, there were reports that the Bill Clinton administration, like the George Bush administration before it, was hopeful that the continued embargo would topple Hussein's regime.

The Embargo

Iraq invaded and occupied Kuwait on August 2, 1990, claiming that the latter was its nineteenth province. This claim rested on the fact that in 1899 Britain and Kuwait, nominally still part of the Ottoman Empire, had concluded a treaty. According to that treaty, in return for British protection, the ruler of Kuwait granted Britain special privileges. The treaty was signed in the light of Germany's imperial ambitions and its desire to extend the Berlin-to-Baghdad railroad to Kuwait on the Persian Gulf. Because the Ottoman sultan had accepted Kuwait's protectorate and because Iraq itself had no independent status before 1932, Iraq's long-standing claim to its oil-rich desert neighbor had few foreign sympathizers.

Iraq refused to abide by Security Council decisions ordering its withdrawal from Kuwait in August, 1990. The U.N. Security Council passed resolutions imposing stringent economic sanctions on Iraq's imports and exports, in particular on its sale of oil, the major hard-currency earner. A United Nations-sanctioned multinational force under United States command first bombed Iraqi positions and installations in January, 1991, and then launched a ground attack. That attack, called Operation Desert Storm, ejected the Iraqi forces from Kuwait in February, 1991. The Kuwaiti emir and other members of the ruling as-Sabah family returned from their refuge in Saudi Arabia. Iraqi authorities claimed dietary, public health, and economic hardships on the population as a result of the United Nations' economic sanctions.

In the meantime, the United Nations Iraq-Kuwait Boundary Demarcation Commission made a geographic and historical survey of the existing Iraqi-Kuwaiti border and recommended that it be moved to Kuwait's advantage by some 3,000 feet into territory previously held to be part of Iraq. Some previously disputed oil wells in the border-straddling Rumaila field were now located in Kuwaiti land. Through its Resolution 833 of May 27, 1993, the Security Council endorsed this new 124-mile international boundary as the final demarcation between the two countries and demanded that they respect its inviolability.

Iraqi claims to Kuwaiti territory had begun immediately upon Britain relinquishing its protectorate in June, 1961. Iraq's leader at the time, Brigadier-General Abdel Karim Kasim, made the case for the return of its "lost province" on the grounds that it had been arbitrarily separated by the British. The immediate dispatch of a British force to Kuwait in July, 1961, and its replacement by Arab League detachments in October, 1961, muted Iraq's claims to sovereignty over Kuwait. In 1963, following establishment of a Baath Party regime in Baghdad, Iraq recognized Kuwait's independence, but this was not formalized. Despite occasional tensions between the two countries alternating with cooperation—especially during the Iraq-Iran war of 1980-1988—it was not until August, 1990, and Iraq's invasion of Kuwait that the territorial crisis came to a head.

Consequences

Iraq's formal recognition of Kuwait's sovereignty within its new United Nations-drawn borders led to the U.N. Security Council's eventual lifting of its economic embargo. The embargo was lifted only after the other prerequisites were deemed to have been met by Iraq. The country's international reputation continued to suffer as a result of allegations of Iraqi human rights violations, charges of involvement in an assassination plot against former U.S. president George Bush while he visited Kuwait in April of 1993, and other actions considered unacceptable by some in the international community. In addition, the United Nations Special Commission on Iraq, which monitored Iraq's nuclear, biological, and chemical weapons programs, was not satisfied with Iraq's status. The country's credibility had to increase before normal relations with it would be fully restored.

Peter B. Heller

Ukraine Agrees to Give Up Nuclear Weapons

Ukraine agreed to give up what had been a part of the Soviet strategic nuclear arsenal and became a signatory of the Nuclear Non-Proliferation Treaty of 1968.

What: International relations; Military capability
When: November 16, 1994
Where: Ukraine
Who:
Leonid Kravchuk (1934-   ), president of Ukraine from 1991 to 1994
Leonid Kuchma (1938-   ), president of Ukraine from 1994

Ukraine Gives Up Nuclear Weapons

Although the United States and Russia had ratified the first START (Strategic Arms Reduction Talks) treaty on October 1 and November 4, 1992, respectively, the treaty could not go into force until three nuclear successor states to the Soviet Union—Ukraine, Kazakhstan, and Belarus—agreed to give up their inherited nuclear arsenals and ratify the Nuclear Non-Proliferation Treaty (NPT) of 1968. On November 16, 1994, by a parliamentary vote of 301 to 8 with 20 abstentions, Ukraine ratified the NPT. With it, Ukraine pledged to eliminate the approximately thirteen hundred strategic nuclear warheads still on its territory.

The ratification of the NPT was no easy matter for Ukraine. Its independence was accompanied by distrust of Russia. Russian nationalists claimed stretches of eastern Ukraine as well as the Crimean peninsula, where ethnic Russians outnumbered Ukrainians. Leonid Kravchuk, president of Ukraine from 1991 to 1994, had been in charge of Communist ideology in the former Ukrainian Soviet Socialist Republic. He quickly became a steadfast nationalist who repeatedly was at odds with Russia. In May, 1992, for example, Kravchuk refused to attend a summit of the Commonwealth of Independent States in Tashkent. This was a calculated snub of Russian president Boris Yeltsin.

Ukrainian independence was complicated by the fact that it was now the world's third-largest strategic nuclear power, with approximately eighteen hundred warheads. Its parliament, the *rada*, agreed to hand over to the Russians its tactical, battlefield weapons. Strategic weapons were another matter. Ukrainian nationalists repeatedly compared their country with France, a nuclear power of similar size and population. For Ukraine to be taken seriously, they contended, it should remain a major nuclear power.

The Ukrainian government played unsuccessfully on the West's suspicions of Russia. Kravchuk's main problem, however, was Ukraine's economy. He saw the nuclear weapons issue as a means of obtaining economic assistance from the West. Initially, the United States pledged $175 million to make possible the task of denuclearization and another $155 million in economic assistance.

Nuclear Treaties

START I had been signed on July 31, 1991, by U.S. president George Bush and Soviet secretary general Mikhail Gorbachev. Following the demise of the Soviet Union, the bilateral treaty became a five-nation agreement by virtue of the Lisbon Protocol of May, 1992. The protocol demanded that the Soviet successor states give up their nuclear arsenals before START I could go into force. The republics of Belarus and Kazakhstan complied with little difficulty. Ukraine, however, held up START I until November, 1994.

At the end of 1991, Kravchuk had promised the U.S. and Russian governments the elimination of all strategic nuclear weapons. Ukraine did in fact shortly hand over to Russia 180 warheads In May, 1992, when Kravchuk signed the Lisbon Protocol, he pledged in a letter to Bush "the elimination of all nuclear weapons" within seven years of START I coming into force. On January 3, 1993, the day Bush and Yeltsin signed the START II treaty, Kravchuk restated his commitment to a nonnuclear Ukraine. Kravchuk was unable to keep his promise, however, when the *rada* blocked shipment of the missiles to Russia.

After the United States and Russia ratified START I, the *rada* reinterpreted the treaty as suggesting that Ukraine, at that time still in possession of 1,656 warheads, was required to destroy immediately only 42 percent of its arsenal. The rest would be destroyed only gradually, after thirteen conditions were met. One condition was additional economic assistance.

In March, 1994, Kravchuk met with President Bill Clinton. Again he spoke of his commitment to a nonnuclear Ukraine. He asked for additional money and came away from the meeting with a promise of more than double the original amount of assistance, $350 million each for disarmament and economic aid. The *rada* still refused to ratify the NPT. Moscow and Washington became increasingly exasperated at the stalling action. It was up to the next Ukrainian president, Leonid Kuchma, to deliver on Kravchuk's promises.

After Kuchma's election as president in July, 1994, he set out to lobby the *rada* for ratification. Unlike Kravchuk, a Communist and later nationalist ideologue, Kuchma was an aerospace engineer who for six years had headed Yuzhmash, the largest rocket factory in the Soviet Union. He argued that Ukraine could not afford nuclear weapons and would gain nothing by keeping them. At one point, he sought a show of hands from legislators for volunteers to build nuclear test sites in their districts should the nation remain a nuclear power. On the day of the vote, Kuchma reminded the *rada* that "Ukraine today has no choice. . . . The process of world disarmament depends on one decision today."

Consequences

START I was formalized at a signing ceremony in Budapest, Hungary, on December 5, 1994, during the Conference on Security and Cooperation in Europe. The road was clear for the ratification of START II.

The Ukrainian government did not receive the large sums of aid it had demanded. The NPT, however, did bring Ukraine financial benefits. The nuclear fuel in the warheads still belonged to Ukraine, and after the warheads were delivered to Russia, their highly enriched uranium would be extracted and processed into fuel rods for civilian nuclear reactors by the U.S. Enrichment Corporation, a government-owned company. Moreover, Russia wrote off Ukraine's debt for past energy deliveries.

The Ukrainian ratification of the NPT strengthened the global system of controlling nuclear weapons. The signatories of the NPT would have had difficulty appealing to other nations to forgo nuclear weapons if the nuclear club had suddenly increased by three. The *rada*'s ratification also meant that four nations (South Africa, in addition to Belarus, Kazakhstan, and Ukraine) had abandoned nuclear weapons, a hopeful sign in the process of achieving nonproliferation.

Harry Piotrowski

Angolan Government Shares Power with UNITA Rebels

The Angolan government and rebel UNITA leaders signed the third peace treaty since 1989; for the first time, however, UNITA was guaranteed a share of power in the government.

What: National politics; Civil war
When: November 20, 1994
Where: Angola
Who:
EUGENIO MANUVAKOLA, secretary general of the National Union for the Total Independence of Angola (UNITA)
VENÂNCIO DE MOURA, foreign minister of Angola
JOSÉ EDUARDO DOS SANTOS (1942-), president of Angola from 1979
JONAS SAVIMBI (1934-), founder and president of UNITA

A Lasting Peace Treaty?

On November 20, 1994, a peace treaty was signed in Lusaka, Zambia, by representatives of the Angolan government and UNITA (National Union for the Total Independence of Angola), ending almost twenty years of civil war. Angolan foreign minister Venâncio de Moura renounced violence on behalf of the government, and UNITA secretary general Eugenio Manuvakola welcomed "the spirit of national reconciliation" as representatives from nearly thirty African countries and observers from the United States, Portugal, and Russia applauded. Alioune Blondin Beye represented the United Nations (U.N.), which mediated the talks that led to the accord.

The comprehensive agreement, known as the Lusaka Protocol, consisted of ten sections covering all legal, political, and military issues that had been agreed to during the year-long talks. It was agreed that UNITA would be represented in the police and army and at all levels of the national government. The agreement called for Jonas Savimbi, the founder and president of UNITA, to become deputy president of the country and for a new round of presidential elections to be held. A U.N.-monitored cease-fire was established, with UNITA forces demobilized and integrated into the Angolan military. A commission including the United Nations, the United States, Russia, and Portugal was established to oversee the demobilization as well as other political aspects of the agreement.

Observers were hopeful but reserved because the agreement was not signed by President José Eduardo dos Santos, who was present at the ceremony, or by Jonas Savimbi, who pleaded that continued fighting made it unsafe for him to leave the country. The state-controlled newspaper *Jornal de Angola* wrote that the cease-fire would not be respected without the signatures of both leaders. Fighting continued around important regional centers such as the former UNITA capital of Huambo.

The Angolan army was the key to maintenance of the peace agreement. Two weeks after government capture of Huambo, troops went on a rampage, requiring the military police to discipline their own troops. The government's dilemma was whether to exercise military restraint in the hopes of enticing UNITA rebels into peaceful compromise, a strategy that had the risk of UNITA reorganization, or to strike hard for a decisive victory with more than $3 billion of newly acquired weaponry from Russia, Bulgaria, the Czech Republic, and North Korea.

The Portuguese Legacy

In 1483, Portuguese explorers established peaceful trading relations with the powerful Kongo kingdom, which governed much of what is now Angola. By the 1540's, cooperation had broken down as Portuguese merchants realized the profit potential of the slave trade. The authority of Kongo monarchs steadily eroded, and the kingdom all but disappeared by 1720.

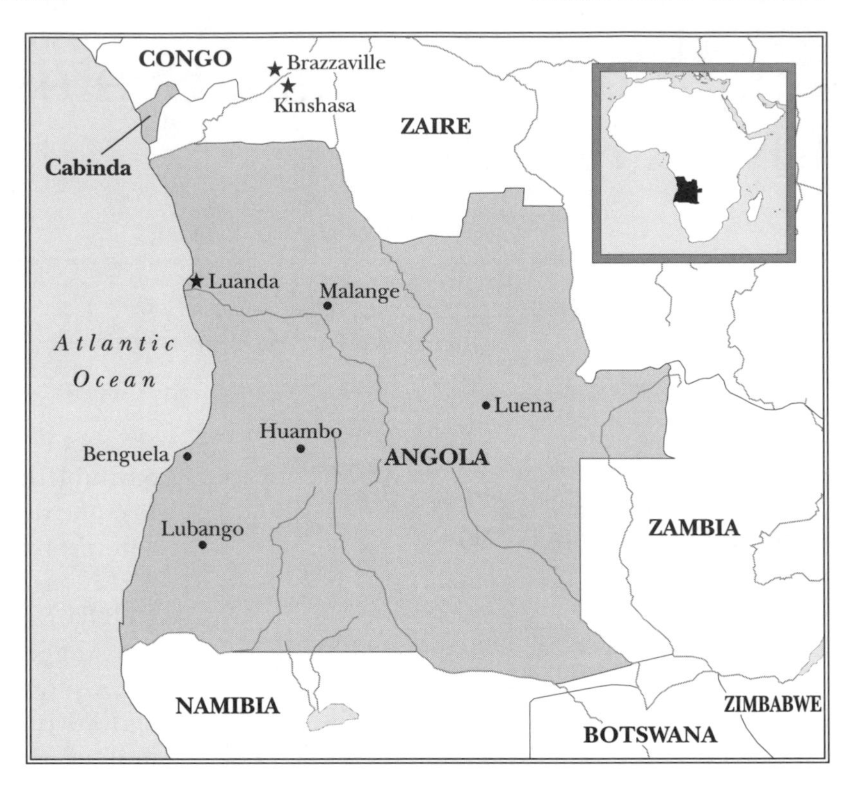

During the seventeenth and eighteenth centuries, Angola was turned into a wilderness as Portuguese traders used it as a supply base for a slave trade with Brazil. An estimated 1.3 million slaves were exported to the New World from Angola between 1526 and 1810.

The Portuguese did less than any European country to prepare their colonial territories for self-government. As France and Great Britain rapidly decolonized after 1957, expectations in Portuguese colonies were heightened. Between 1961 and independence in 1975, three Angolan independence movements fought a Portuguese army that numbered, at its height, some sixty thousand troops. The final stages of decolonization were accelerated by a military coup that occurred in Portugal on April 25, 1974. Independence was declared on November 11, 1975, prompting more than 300,000 whites to leave the country.

Even before the Portuguese departure, civil war had erupted among three rebel factions, each with an ethnoregional core. UNITA, headed by Savimbi and centered in the Ovimbundu-speaking south-central plateau, was backed by South Africa and the United States. The Popular Movement for the Liberation of Angola (MPLA), led by Agostinho Neto and based among Kimbundu speakers in Luanda, was supported by Russia. The National Front for the Liberation of Angola (FNLA), led by Holden Roberto and drawing support from the northern Kongo-speaking population, was supplied by China and Zaire.

By early 1976, the MPLA had been recognized by the Organization of African Unity as the legitimate government of Angola, and with the help of twenty thousand Cuban troops it controlled 90 percent of the country. Neto died in Moscow on September 10, 1979, and was succeeded by dos Santos as president. Angola's problems were complicated by the presence in southern Angola of South-West African People's Organization (SWAPO) troops, fighting for the liberation of Namibia from South Africa.

With the decline of Cold War tensions, by 1991 external support for the warring factions ended. On May 31, 1991, the MPLA regime and UNITA rebels signed the Bicesse Accords in Lisbon. The accords provided for a cease-fire and multiparty elections. In the absence of any democratic tradition or superstructure, the election of September 29-30, 1992, which endorsed control by MPLA, was contested by UNITA, leading to the bloodiest phase of the Angolan civil war.

Caught unprepared, Angolan government troops lost two-thirds of the country, including important economic centers, before finally recapturing most of the lost territory in 1993. Between September, 1992, and July, 1993, there were an estimated 100,000 war-related deaths. With both sides weary from the war, negotiations began in earnest in the fall of 1993, under the mediation of the United Nations.

Consequences

A year after the signing of the Lusaka agreement, the results were unclear. Although the cease-fire generally held, violence continued, including the shooting down of a U.N. helicopter by UNITA gunners. In April, 1995, the first contingent of a U.N. peacekeeping force of seven thousand arrived in Angola to prepare tent cities where UNITA troops were to turn in their weapons.

Most observers during 1994 and 1995 believed that neither the Angolan government nor UNITA was prepared to make the compromises necessary for a successful coalition government. At a UNITA congress in Bailundo in February, 1995, Savimbi demanded a veto over national policy as the price for UNITA participation in a national government. The government promptly rejected that demand. Savimbi's lack of support from traditional allies in South Africa and Congo (Zaire) suggested that he might be forced to cooperate.

The 1994 peace accord collapsed in 1998, and Dos Santos's government afterward refused to undertake new peace talks with UNITA. Dos Santos continued to hold the presidency through 2001 but announced in August of that year that he would not seek reelection in future elections.

John Powell

Western Hemisphere Nations Approve Free Trade Area of the Americas

The leaders of the Western Hemisphere met for a three-day summit in Miami to work toward establishing a free trade area for North, South, and Central America.

What: Economics; International relations
When: December 9-11, 1994
Where: Miami, Florida
Who:
Bill Clinton (1946-), president of the United States from 1993 to 2001
George Herbert Walker Bush (1924-), president of the United States from 1989 to 1993
Mickey Kantor (1939-), U.S. trade representative under President Clinton

Partnership for Prosperity

On December 9, 1994, the leaders of thirty-four nations of the Western Hemisphere met at the Villa Vizcaya on Biscayne Bay in Florida for the Summit of the Americas. The goal of the summit was the eventual establishment of a free trade area stretching from Alaska to the tip of South America. The three-day summit resulted in the issuance of a communiqué calling for the creation of a Partnership for Development and Prosperity in the Americas. The nations agreed to work to form a Free Trade Area of the Americas by the year 2005. Cuba was the only nation in the Western Hemisphere not invited to attend.

The nations agreed to strengthen and support the democratic process and to support constitutional governments throughout the hemisphere by working through the Organization of American States. They also agreed to simplify government and supported the establishment of independent court systems. Particular emphasis was placed on better meeting the needs of women, indigenous people, children, the disabled, and the elderly.

Major areas of negotiation were control of narcotics and protection of the environment. The summit communiqué included strong language against all aspects of the drug trade and condemned terrorism and corruption. Environmental concerns resulted in agreement to support economic growth that is compatible with the environment. The nations further agreed to ban the use of leaded gasoline throughout the hemisphere by the year 2000.

Delegates recognized the need for additional investment and creation of infrastructure. Financing from private and international sources was expected to be used to meet these needs. Concern was expressed about the high debt loads carried by some of the countries. Delegates set a goal of identifying barriers to investment and eventually developing a hemispheric investment code.

The central message of the summit communiqué was the call to develop the Free Trade Area of the Americas by the year 2005. It called for the elimination of barriers, subsidies, and unfair trade practices. The process to be used to accomplish this goal was regularly scheduled meetings to negotiate differences. Key areas to be considered at future discussions were investments, services, intellectual property, dumping, and export subsidies. Existing trading blocs within the hemisphere would be expanded to create the Free Trade Area of the Americas by the target date. As a first step in this process, Chile was invited to join the North American Free Trade Agreement (NAFTA).

The Movement for Worldwide Free Trade

The summit occurred immediately following resolution of the Uruguay Round of the General Agreement on Tariffs and Trade, which set world trading rules. Throughout the world, trading blocs were being formed that eliminated tariffs and other barriers to trade among members. The economic growth and political stability of Latin America led to substantially increased trade among Latin American countries and between them and the United States.

In 1990, U.S. president George Bush, in the Enterprise of the Americas Initiative, proposed the establishment of a hemispheric free trade area. NAFTA, a first step toward that goal, went into effect on January 1, 1994. It created a free trade area encompassing the United States, Mexico, and Canada. Following ratification of the Maastricht Treaty in 1993, most of Western Europe became economically united under terms of that treaty.

As trading blocs formed, competition among countries intensified. Countries within trading blocs increased trade with members and decreased trade with nonmembers. As more trading blocs were created, the pressure to seek new trading partners increased. At the time of the summit, the Western Hemisphere had twenty-three trading blocs. The creation of the Free Trade Area of the Americas would result in the merger of all these blocs through a process of negotiation.

The United States pursued increased trade with Latin America because of its trade deficit, which exceeded $100 billion. Latin America was the fastest-growing market for U.S. products. Another key to the U.S. strategy was the need to gain trade concessions in Latin America before Europe and Japan did. At the same time, the United States was concerned about the impact of exports from Latin America to the United States, particularly in the flower, garment, steel, orange juice, and shoe markets. The United States was interested in increased trade that would expand export opportunities, but it was cautious about the prospect of increased imports from Latin America. This concern caused the United States to urge a schedule for accomplishment of the agreement that was slower than most Latin American countries wanted.

Latin American countries for the most part were eager for free trade. Many countries were interested in being admitted into NAFTA. Smaller countries in Central America and the Caribbean were more cautious about free trade because of concerns about the impact it might have on their domestic economies. Latin American countries generally were experiencing significant trade deficits, particularly with the United States. To reduce those deficits, they wanted to increase exports to the United States and reduce their imports.

Consequences

In order to meet the free trade agreement deadline of 2005, it was necessary that Congress grant the president "fast track" status. This would mean that Congress would vote yes or no on the entire treaty, without changing individual provisions. The ability of U.S. Trade Representative Mickey Kantor to negotiate would be severely hampered if trade meeting agreements were overturned by Congress. Another significant early step would be the acceptance of Chile into NAFTA, which would require ratification by the governments of the United States, Mexico, and Canada.

Alene Staley

Reformers Realign Japan's Political Parties

Japanese political reformers opposing corruption and bureaucracy moved Japan toward a more responsive democracy.

What: National politics
When: December 10, 1994
Where: Japan
Who:
Ichiro Ozawa (1945-), founder of Japan's Shinshinto Party
Morihiro Hosokawa (1938-), prime minister from 1993 to 1994
Kunio Hatoyama (1946-), opposition leader
Tomiichi Murayama (1924-), prime minister from 1994 to 1996
Shin Kanemaru (1915-1996), behind-the-scenes leader of the Liberal Democratic Party

Toward Coalition Government

On December 10, 1994, Ichiro Ozawa launched Japan's latest opposition party, the Shinshinto Party (literally, the New New Japan Party). Ozawa, a former prime minister as well as a one-time Liberal Democratic Party (LDP) member, abandoned the LDP in 1993. Thereafter he earned recognition as Japan's most powerful opponent of the political organization that had governed the country without serious opposition—except from rebellious factions within its own ranks—for thirty-eight years.

The Shinshinto Party, in English called the New Frontier Party, fell in line with other aspiring nationally organized political enemies of the LDP such as the Japan New Party, the (Buddhist) Clean Government Party, the Japan Renewal Party, and the New Party Harbinger, along with the traditional, if ineffectual, LDP opposition, the Social Democratic Party (SDP) and the Japan Communist Party. A persistent critic of the unresponsive bureaucracies that managed Japanese life irrespective of parties, politicians, and popular sentiments, Ozawa also deplored the LDP's institutionalized corruption. Japan's big businesses provided deep pockets from which to satisfy the personal and political financial desires of LDP politicians. Cozy with their country's giant business and financial institutions, politicians in turn spent lavishly wooing, subsidizing, and often bribing government ministers and members of their own local constituencies. The result was a series of major LDP scandals.

Ozawa knew that the hastily organized New Frontier Party represented a weak, last-minute effort to mount fresh opposition to the LDP. Working behind the scenes, therefore, he had masterminded formation of the New Progressive Party, a nine-party coalition capable of capturing a majority of seats in the lower house of Japan's Diet, the country's legislature. The nine-party union embodied in the New Progressive Party (NPP) also was publicly announced on December 10, 1994.

Rule by the LDP

In August, 1989, for the first time in its thirty-four-year rule, the LDP lost its majority in the Diet's upper house. Several factors accounted for this upset. The party's lengthy dominance of Japanese political life had been marked by a series of scandals involving highly placed politicians, ministers, and businesspeople. In 1948, the Showa Denko affair rocked the Liberal and Japan Democrat parties, whose merger in 1955 created the LDP.

During the 1950's, LDP politicians were implicated in an extensive shipbuilding scandal. The Lockheed scandal played itself out from 1974 to 1978. In 1978 and 1979, the McDonnell Douglas and KDD affairs, overlapping scandals, rocked the public. The Recruit scandal, which resulted in indictments and convictions of seventy-one ministers, business leaders, and politicians, began in 1984. During the 1980's, the so-called Pachinko scandal and the indictment of top SDP officials indicated that the LDP enjoyed no monopoly over corruption, but the LDP was by no means out of the picture. In 1992, the Sagawa Kyubin scandal involving Shin Kanemaru, Ozawa's former mentor and the leader of an important LDP faction, divided his party. He faced trial for corruption and in 1993 was arrested on suspicion of evading income taxes, prompting Ozawa and other reformers to defect.

Japan's widespread prosperity shielded the LDP from the public's wrath and its electoral revenge. By the early 1990's, however, the country was mired in a tenacious recession. All economic indicators pointed downward. Major bankruptcies became common, the unemployment rate rose, and strike activity increased. Japanese voters, hitherto ready to identify the LDP as the agent of their prosperity, in large numbers refused to vote. Ozawa, Kunio Hatoyama, Morihiro Hosokawa, and other politicians violated Japan's culture of loyalty by blocking LDP legislation, by rebelling against the LDP's elderly leadership, and, after defecting, by founding several of the opposition parties that were to combine as the NPP. Hosokawa resigned as prime minister in April, 1994, amid charges of corruption.

Between March and November, 1994, these reformers succeeded in pushing through the Diet four reform measures that altered the LDP's political playing field. Multimember constituencies that had worked to the advantage of the LDP were replaced by single-member constituencies. New boundaries for electoral districts were drawn, a system of proportional representation was instituted, and restrictions were placed on campaign financing. Japan's postwar two-party system of a dominant LDP and its traditional opposition, the SDP, apparently underwent substantial realignment. That realignment was marked by creation of the Shinshinto Party and the NPP.

Consequences

Despite the reforms that accompanied them, Japan's new coalition governments proved to be weak and short-lived, and they evinced a bizarre blend of principles. Socialists who were no longer socialists governed side by side with Communists, Buddhists, and right-wing conservatives. Gone were the strong and relatively durable LDP governments of the 1980's, when Yasuhiro Nakasone and Tatsuo Tanaka held sway. Instead, during 1994 and 1995, three prime ministers came and went in a matter of months, leaving Tomiichi Murayama as a figurehead within a coalition in which the LDP remained the strongest member. The coalition was troubled. Japan's recession lingered, and opinion polls regularly showed public disgust with politicians and their business cronies. Turnouts in major elections hovered around 50 percent, historically a low figure for the Japanese.

The Murayama government conducted a trade war with the United States that further undermined the value of the yen. Bitter memories of Japan's role in World War II resurfaced as the fiftieth anniversary of the war's conclusion was celebrated among victorious allied nations. Japan faced the agony of apology and displays of new, unrepentant nationalism. Amid these vexations, the Murayama coalition and a host of lesser officials were denounced widely for their bureaucratic sloth and mismanagement of relief efforts following the tragic 1995 Kobe earthquake. The LDP dominated the coalition despite the reforms of 1994, but the New Frontier Party announced its dissolution at the end of 1997.

Clifton K. Yearley

AP/Wide World Photos

Ichiro Ozawa of the Shinshinto Party.

Russian Army Invades Chechnya

The Russian invasion of Chechnya brought destruction, death, and displacement to the people of Chechnya and severely weakened the political status of Russian president Boris Yeltsin.

What: Political aggression
When: December 11, 1994
Where: Chechnya and Russia
Who:
BORIS YELTSIN (1931-), Russian president from 1991 to 1999
DZHOKHAR DUDAYEV (1944-), president of Chechnya from 1991 to 1996
NICHOLAS I (NIKOLAY PAVLOVICH, 1796-1855), emperor of Russia from 1825 to 1855
JOSEPH STALIN (JOSEPH VISSARIONOVICH DZHUGASHVILI, 1879-1953), leader of the Soviet Union from 1927 to 1953
NIKITA S. KHRUSHCHEV (1894-1971), leader of the Soviet Union from 1953 to 1964

Russian Forces Enter Chechnya

On December 11, 1994, Russian president Boris Yeltsin ordered nearly forty thousand Russian troops into the southern region of Chechnya. They were sent because Chechnya's defiant president, Dzhokhar Dudayev, refused to accept Russian rule following the collapse of the Soviet Union in 1991. In that year, Chechnya, whose people are primarily Muslim, declared its independence. Dudayev based his political future on maintaining that independence. Three years of persuasion from Yeltsin's emissaries failed to convince the Chechens to yield to government from Moscow.

The first wave of Russian forces, representing the largest Russian military action since the invasion of Afghanistan in December, 1979, stopped short of the Chechen capital of Grozny. Grozny, located about one thousand miles south of Moscow, is Chechnya's largest city, with more than 400,000 residents. As the Chechens continued to resist and frantically built fortifications, Yeltsin ordered his troops to surround Grozny. Ground forces established operations on the edge of the city. This marked the beginning of serious fighting and bloodshed.

The resistance from the Chechens surprised the Russians and exposed weaknesses in the Russian army's leadership. Chechen rebels, using every able-bodied male over the age of eleven, refused to succumb to the Russian bombardment. Late in December, President Yeltsin responded by escalating the ground war and, after a fierce battle, succeeded in isolating and destroying the town of Argun, about ten miles east of Grozny. This was in preparation for a major assault on the Chechen capital in the event that President Dudayev refused to concede. Negotiations failed to resolve the dispute, and the attacks on Grozny began in earnest. Hundreds, perhaps thousands, died in the intense conflict.

During the following six months, President Yeltsin, despite expressions of concern from the United States, England, France, and Germany as well as opposition from some high-ranking Russian military and political officials, continued the attack on the Chechens. Grozny was completely destroyed, but the Chechens, still unwilling to concede, began moving into the southern mountains in preparation for guerrilla warfare. In June, 1995, some Chechen military units began to assault Russian villages between Grozny and Moscow.

Centuries of Conflict

The Caucasus region of which Chechnya is a part has long been a center of strife involving Russia, Persia (Iran), and Turkey. The Muslim legacy in Chechnya stems from the successful conquest of Byzantium by the Ottoman Turks in 1453. In 1762, Russian empress Catherine the Great, in an effort to gain access to the Black Sea, tried to overwhelm the Chechens and other Caucasus peoples. Her efforts left a persistent resentment and hatred. Periodic fighting continued through successive Russian regimes until Emperor Nicholas I, in the late 1820's, demanded the removal of all Muslims from Chechnya. This was also the wish of the Russian Orthodox church. Violence intensified during the next fifty years. By 1877, nearly half the Chechen population had been killed. Their villages and forests were destroyed, and 500,000 of them were forced to flee into Turkey.

In 1944, communist leader Joseph Stalin again attacked Chechen villages and disbursed thousands of people into Central Asian Soviet states. The reason for Stalin's assault was the fact that the Chechens, like the Ukrainians, helped Adolf Hitler's Nazi forces in their unsuccessful 1941 invasion of Soviet territory. This help was given on the mistaken assumption that the Nazis would liberate subject peoples from communist domination. In 1957, Soviet leader Nikita Khrushchev allowed the Chechens to return to their homeland. This was part of Khrushchev's general, if brief, policy of showing a clear break from the Stalinist years. Dudayev was one of those who, as an infant, was exiled by Stalin and who returned in 1957. He later became a high-ranking Soviet air force official.

After 1991, when Chechnya declared its independence, Dudayev confounded and infuriated Russian officials by creating ties with militant Islamic organizations such as Hezbollah (the Iranian-supported Party of God) and the Muslim Brotherhood. He also sought to strengthen connections with Muslim governments in Turkey, Jordan, and Lebanon. Moscow also was aggravated by the fact that Grozny became a popular city of settlement for many Muslim veterans of the Afghanistan war.

President Yeltsin, reviving an opinion put forward by Nicholas I, condemned the Chechens as an outlaw people known for their banditry and organized criminal activities. This description,

calculated to lessen world sympathy for the Chechens, is not entirely false. The Chechens have survived for centuries as rebels and as a group do not feel constrained by laws, especially those imposed by Russia.

Given the long history of religious, cultural, and political rivalry between Russia (including the imperial and Soviet eras) and Chechnya, it is not surprising that President Yeltsin believed it necessary to extend Moscow's control over rebellious Chechens. His failure to achieve a quick victory might have been anticipated.

Consequences

The Chechnya invasion called into question President Yeltsin's judgment, both within Russia and in the world community. Many heads of state, including U.S. president Bill Clinton, urged Yeltsin to use restraint and pursue every opportunity to achieve peace. In Russia, the invasion raised serious issues pertaining to leadership, preparation, and loyalty within the Russian army. As a result, President Yeltsin found public opinion turning against him as his principal political rivals, including the popular General Alexander Lebed, continued to deride his decision to invade Chechnya. Meanwhile, the historical hatred of Chechens toward Russians was renewed and strengthened, and the conflict continued, intermittently, through the rest of the decade, outlasting Yeltsin's presidency.

Ronald K. Huch

Berlusconi Resigns as Prime Minister of Italy

Silvio Berlusconi resigned as Italy's prime minister in the face of upcoming votes of no confidence, plunging the nation into turmoil.

What: National politics
When: December 22, 1994
Where: Rome, Italy
Who:
Silvio Berlusconi (1936-   ), prime minister of Italy from 1994 to 1995 and from 2001

Berlusconi Resigns

On December 22, 1994, Prime Minister Silvio Berlusconi of Italy resigned as important groups within his coalition announced their intention to leave the government. Berlusconi handed in his resignation after opposition led by the left wing announced that it had the support of 325 members of the 630-seat lower house of Parliament, including some former government supporters, for a motion of no confidence.

Although he was elected in May, 1994, with a mandate to end the corruption rampant within the Italian government, Prime Minister Berlusconi quickly found himself the center of a multitude of damaging charges. The week before his resignation, he was personally questioned by Milan magistrates probing into payments his Fininvest business empire had made to tax inspectors in return for favorable audits.

Nor were the corruption charges limited to the business practices of Berlusconi's media-based kingdom. The government leader's brother, Paolo Berlusconi, was convicted by a Milan court of making nearly $100,000 in illegal contributions to the Christian Democratic Party and issued a five-month suspended prison sentence.

These developments were particularly devastating for the Italian prime minister because he had been elected on the promise to end precisely this sort of corruption, which had been endemic among the old political elite running the government since World War II. Another broken promise that paved the road toward Berlusconi's resignation was his failure to give up control of his many television stations even while he, as prime minister, made numerous decisions about the state-owned television operation and antitrust regulations that directly affected his businesses. All of this led to a widespread belief that Berlusconi was more interested in engaging in corruption than in eliminating it.

Berlusconi campaigned on a platform advocating free-market capitalism and stood firmly against the former communists while working with the neo-fascist party. He had been hailed as a man who could be trusted to clean up Italian politics. Within a matter of months, he found himself increasingly viewed as no less corrupt than previous leaders, only more skillful in using slick television advertising to promote himself to the public.

Collapse of Traditional Parties

When Berlusconi was elected prime minister in May, 1994, his new regime was the fifty-third government of Italy since the end of World War II. Given the highly politicized and fractionalized nature of Italy's multiparty system, changes in government were both frequent and predictable. Despite the outward appearance of instability, there was a surprising continuity among these frequently changing governments. Most previous governments were coalitions dominated by the Christian Democratic Party, which typically won about a third of the vote.

As governments came and went, the same politicians and parties tended to remain and dominate the governmental apparatus. These coalitions held together primarily because of their joint opposition to Italy's large Communist Party and their own self-interest in the spoils of office. This resulted in an entrenched political elite that used its control of public office to enrich itself and its friends. Dishonesty may exist in all governments, but the level of corruption in Italy had reached crisis proportions by the 1990's.

With the end of the Cold War and the fall of the Soviet Union, neither the Christian Democratic-led right nor the Communist Party-led left had any clear ideological justification for its existence. Combined with a growing public weariness with political corruption, this led to the collapse of the old political parties that had dominated Italy since 1945. The Communist Party split into the reform-oriented Democratic Party of the Left and the still-hardline Communist Refounding. The Christian Democratic Party fell apart with only the small Popular Party as successor.

Out of this chaos, Berlusconi built a new party called Forza Italia that retained the old conservative ideas of the Christian Democrats but was to be honest and efficient. Pledging lower taxes and an end to corruption, and capitalizing on Berlusconi's ownership of powerful television outlets, the new party quickly captured the imagination of many disgruntled conservatives. Although Forza Italia received only about one-fourth of the vote, it was able to form an alliance with the Northern League, a regional party that opposed redistribution of tax revenue from the prosperous north to the south, and the National Alliance, a party with a core made up of neo-fascists and led by Benito Mussolini's granddaughter.

Berlusconi was a popular figure in part because he was seen by many as a political outsider, someone who was not part of the old corrupt political elite. He was also seen as an honest and efficient manager who would be able to reshape the Italian political system. This perception was combined with a clever U.S.-style media campaign that was long on image and short on specifics. The result was the development of a new movement that was able to become the center of a new right-wing coalition government in May, 1994.

AP/Wide World Photos

Silvio Berlusconi.

Consequences

The promise of a rapid transformation of the Italian political system into more stable, efficient, and honest institutions appeared to be doomed for the foreseeable future. Although many changes took place, the cloud of corruption continued to hover over many major politicians, including the man who was to clean up the mess, Berlusconi. Berlusconi was far from finished in politics, however. With a 65 percent share of the television advertising market and successes in referendums in June, 1995, he had the power and desire to fight to regain his position even as Milan prosecutors charged him with fiscal fraud in connection with his purchase of a $3,000,000 villa for $346,000. Indeed, although he was convicted of corruption in 1998, he made a remarkable comeback three years later and regained the premiership in June, 2001.

William A. Pelz

Ebola Epidemic Breaks Out in Zaire

National and international health organizations reacted quickly to contain an outbreak of the deadly Ebola virus in Kikwit, Zaire.

What: Medicine; Health
When: January-June, 1995
Where: Zaire (now Democratic Republic of Congo)
Who:
MOBUTU SESE SEKO (JOSEPH MOBUTU, 1930-1997), president of Zaire from 1965 to 1997
HIROSHI NAKAJIMA (1928-   ), director-general of the World Health Organization (WHO)
JEAN-JACQUES MUYEMBE TAMFUN, virologist at the University of Kinshasa
EBRAHIM SAMBA (1932-   ), director of the WHO West Africa office

The Epidemic

In April of 1995, cases of infection with the Ebola virus surfaced in Kikwit, Zaire, a major city of 500,000 people located in south central Zaire (later renamed Congo), a few hundred miles from the capital city of Kinshasa. The first victim to be documented died within days of reporting to the Kikwit hospital. His death was followed in subsequent weeks by those of several hospital staff members and patients. Later investigations traced the earliest cases in the outbreak to late January of 1995, although it took until May, 1995, for health workers to identify the deadly virus.

The symptoms of the disease, which include severely debilitating diarrhea and fever, mimic at first those of dysentery, but Ebola's mortality rate is much higher. The virus kills 50 to 90 percent of those infected within days. Its symptoms also are much more severe than those of dysentery. Victims experience severe bruising and deterioration of the skin; hemorrhaging from the eyes, ears, nose, and mouth; and eventually the collapse and disintegration of internal organs. These frightening symptoms cause panic among the stricken population, as they did in Kikwit. Many people fled to neighboring areas, thus potentially exposing many others to the disease.

Local and international responses to halt the spread of the disease were essential. Health officials in Zaire, fearing the onset of a terrible epidemic, notified the World Health Organization (WHO), headquartered in Geneva, Switzerland. This United Nations specialized agency, headed by Hiroshi Nakajima of Japan, is charged with coordinating international health policy, particularly international responses to epidemics. WHO put together a team of its own specialists, along with doctors, virologists, microbiologists, and experts in tropical medicine from the U.S. Centers for Disease Control (CDC) and Prevention and other agencies. These specialists flew into Kikwit in order to track down the source of the epidemic and to identify the contaminating agent.

This team directed its initial investigation at the hospital where the first known victim died. Blood samples from victims were rushed to CDC headquarters in Atlanta, Georgia. It was determined that the epidemic was the result of the deadly Ebola virus. In the meantime, Mobutu Sese Seko's government cooperated with international agencies, and local government officials declared a short-lived quarantine to prevent the local population from fleeing Kikwit and the surrounding region. The Red Cross of the Republic of Zaire undertook burial of the dead and sanitization of their lodgings. A number of Red Cross volunteers contracted the disease before adequate supplies were made available to ensure the safety of the burial details and sanitation teams.

Fortunately, the Ebola virus, which is spread through contact with the bodily secretions of its victims, can be controlled through the careful use of sanitary precautions. There is, however, no known cure. Through prompt action, local and international agencies were able to halt the spread of the disease, which kills its hosts so quickly that not many people are exposed to infection, excepting family members and health workers, who can thwart the disease by taking appropriate hygienic precautions. The death toll in this particular outbreak was about 150.

Increased Risk

The Ebola virus was discovered in 1976 and has reappeared several times, including a major outbreak in 1979 in Zaire and western Sudan. Virologists, including Zaire's Jean-Jacques Muyembe Tamfun, who is an expert on Ebola, are still seeking to determine its normal hosts and discover how it is transferred to humans. In the meantime, WHO-coordinated efforts to track down, isolate, and prevent the spread of such viruses often are successful—as they have been in every outbreak of Ebola—despite the fact that the teams sent to contain the spread of these deadly killers usually operate under difficult conditions and in a climate of pervasive fear.

Human contact with viruses is not merely a function of biology. It is also a function of social and economic conditions. With the explosion in the size of the global population and with the advent of transcontinental travel, coupled with increases in international trade and tourism, many more opportunities exist to contract and transmit exotic viruses.

The squalid social and economic conditions in which many people live, especially in underdeveloped countries of the Third World, create situations in which diseases can spread more readily, particularly along highways and trade routes. As more people move into previously uninhabited areas in pursuit of economic opportunities, they make contact with viruses that normally do not have access to human hosts. At the same time, however, medical science has made strides in learning more about viruses. Cures for viral infections remain elusive, however, so outbreaks like those of the Ebola virus are a source of fear and international concern.

Consequences

Once health officials determined what they were dealing with, the 1995 outbreak of Ebola was contained within a matter of weeks. Because the medical community knew so little about how the virus is first contracted by humans and where it survived between outbreaks, medical personnel could only brace for its reappearance, hoping that it would not recur in a heavily populated area, and be prepared to act quickly—as they were when new outbreaks of ebola broke out in central Uganda in late 2000 and along the Gabon-Congo (Brazzaville) border region at the end of 2001.

Rapid action in the Kikwit outbreak was vital in stopping the disease from spreading. In a world of widespread global intercourse, international cooperation in scientific and health matters, such as through the WHO and various national disease control and prevention centers, remained essential in combating the spread of this and other dreaded viral diseases.

Robert F. Gorman

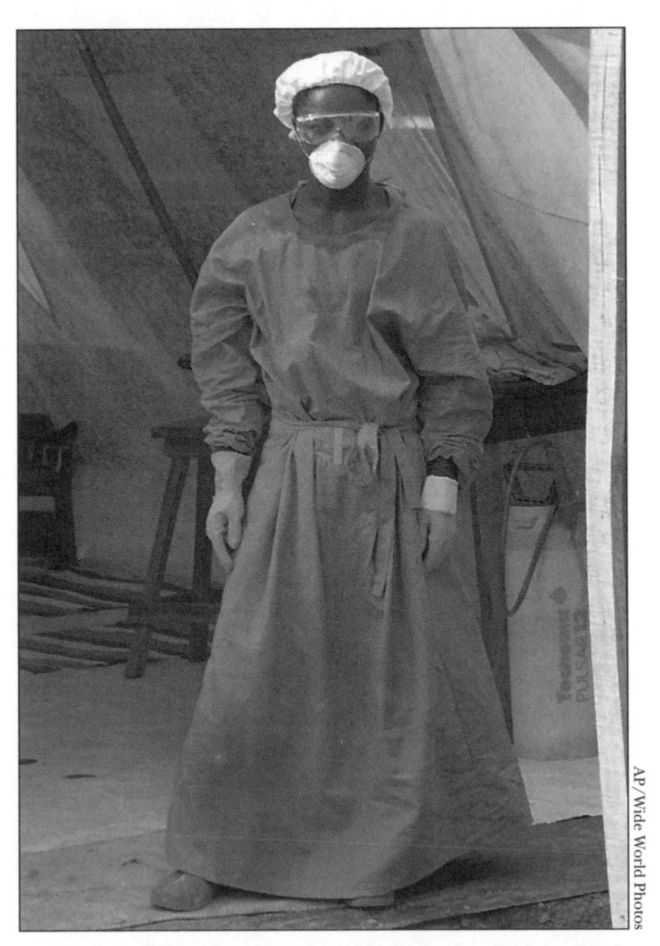
AP/Wide World Photos

A nurse in full protective gear stands in front of an Ebola isolation ward at the hospital in Gozon, Ivory Coast.

Gray Wolves Are Returned to Yellowstone National Park

After sixty years' absence, wolves were restored to Yellowstone National Park in the western United States under a provision of the Endangered Species Act.

What: Biology; Environment
When: January 13, 1995
Where: Yellowstone National Park, Wyoming
Who:
BRUCE BABBITT (1938-), U.S. secretary of the interior
DOUGLAS SMITH, chief biologist of Yellowstone Wolf Restoration Project
ALDO LEOPOLD (1886-1948), forest ranger and author, first biologist to call for restoration of wolves to Yellowstone

The Decline of the Wolf

Humans have always competed with wolves for top-predator status on the food chain. Although Native Americans respected the wolf as a skillful hunter and kindred spirit, European settlers viewed them as a threat. Plains and mountains teemed with wildlife when ranchers brought cattle to the West in the nineteenth century. As the great bison herds disappeared, wolves, deprived of their natural prey, turned to the cattle. Ranchers promoted laws favoring the elimination of wolves and paid mountain men bounties for killing them. From the 1850's through the 1930's, wolves were hunted and poisoned everywhere in the American West. The last wolves in Yellowstone, two cubs, were trapped and killed in 1926.

Just as these practices were bringing wolves near to extinction, biologists began to study ecology, the way entire ecosystems work. They found that removing a species' top predator seldom benefits the prey species. Wolves usually take down weak or sick deer or elk. Without predators, herds grow until they run out of food; then animals starve and their numbers plummet. Other species who feed on the carcasses left by wolves, from ravens to coyotes, are affected, as is the vegetation. Aldo Leopold, a forest ranger, described his change of heart about wolf killing in a powerful essay, "Thinking Like a Mountain," in *A Sand County Almanac.* Published in 1944, it called for wolves to be restored to Yellowstone Park, the greatest wildlife preserve in the United States.

In the 1960's and 1970's, ecology came to public attention, spurred by books such as Rachel Carson's *Silent Spring* (1962), which showed the damage done by heedless treatment of the natural world. The Endangered Species Act, aimed at preventing loss of more species, was passed by Congress and signed by President Richard M. Nixon in 1973. The act made it possible to restore some endangered species to their original habitats. Study groups explored the potential of bringing back wolves to Western states. Researchers had to make sure no free-ranging wolves still lived in the park and to design the best way to reintroduce and protect the new packs. Public education and political support were essential to the process. Many ranchers still believed "the only good wolf is a dead wolf," and the villainous wolf images from fairy tales lingered in popular culture. It took more than twenty years from the first official efforts before wolves actually returned to the park.

Wolves Return

Early on the morning of January 13, 1995, eight wolves from Canada were released into two large pens in Yellowstone. Tranquilized for the flight, they had arrived the previous evening but were kept caged until a court ruled on their release. Interior Secretary Bruce Babbitt, who had

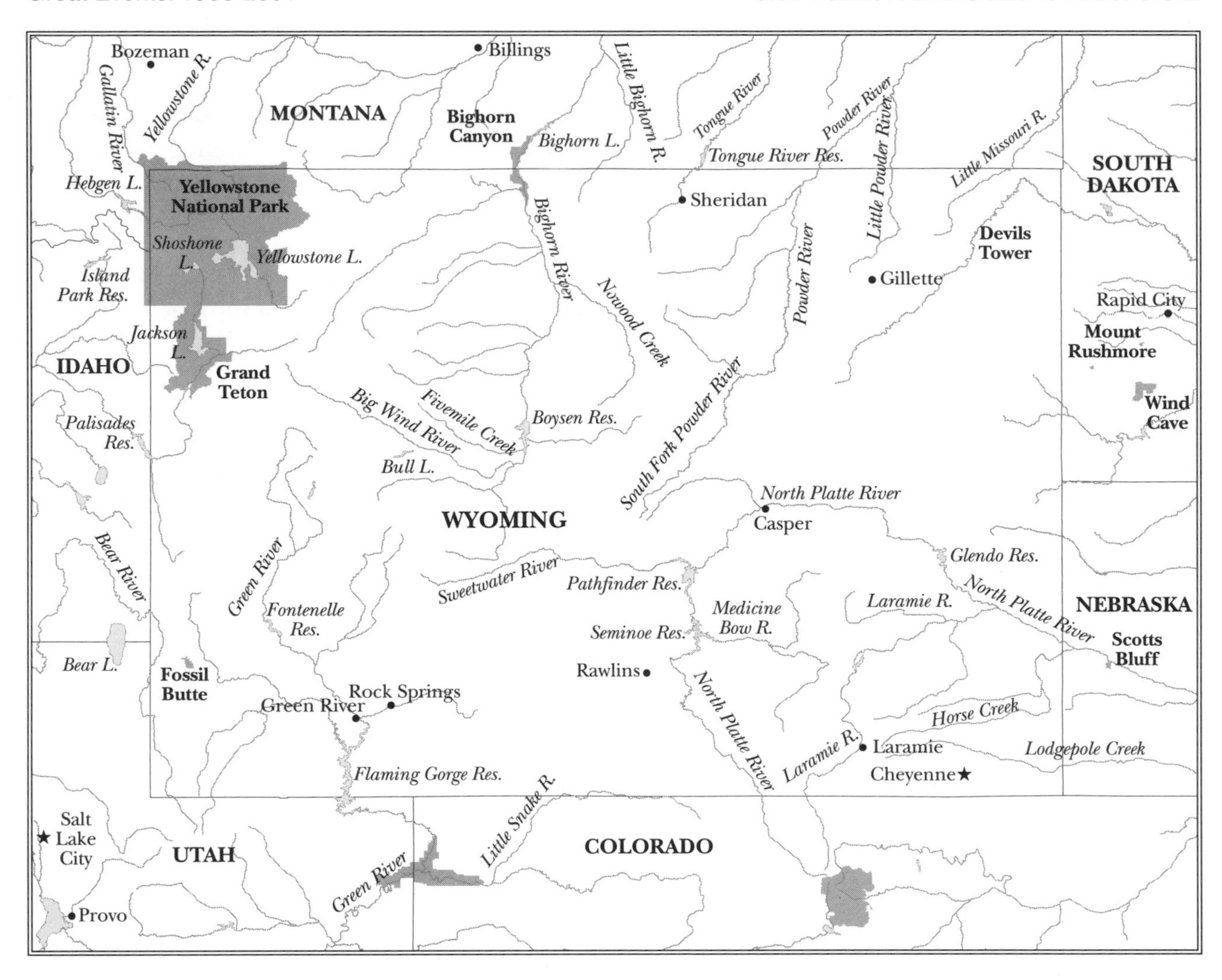

worked tirelessly for wolf reintroduction despite the political problems, carried the first wolf into the reclamation pen. Seven more wolves arrived on January 20.

Because transplanted wolves tend to head back to their home range, the new Yellowstone wolves were kept in enclosures for ten weeks. This gave them time to acclimate to the new area and to form pack bonds. When the gate was opened, they refused to go through it but exited through a hole cut in the far side of the pen. The three different packs initially stayed close to their release sites. After about three weeks, each group set out to explore. Traveling about twenty-five miles a day, the packs eventually found new ranges. In 1996, more wolves were brought from Canada, acclimated in pens, and then released.

The wolves' return showed that restoring a wilderness area to its original ecology is possible, given scientific work and public support. The Yellowstone wolves have been intensely studied. Many were fitted with radio collars, enabling park biologists to track them from a distance. New litters and pack movements are monitored and documented. Sometimes wolves wandering into settled areas outside the park are brought back. When the first shipment of wolves arrived, people gathered at the park's entrance, greeting them from a distance. Native Americans prayed and drummed to honor the wolves, and children were dismissed from school to watch the convoy. As for the wolves, despite being moved more than five hundred miles, medicated, studied, and photographed, they adapted marvelously to their new territory.

Consequences

The full impact of restoring wolves to Yellowstone will not be known for many years, but some effects on the park's ecology are clear. The wolves' prey is 87 percent elk, 4 percent bison, and 2 percent moose. The elk population has not decreased noticeably, although wolf presence has prompted some elk migration, which lets streamside vegetation in meadows flourish, giving songbirds more habitat. Wolves have killed far fewer livestock than expected; those who do are given "aversive training" before being returned to their home range. Coyote numbers are down by half, opening up spaces for foxes and badgers, who compete with coyotes for small game.

Not all the results have been good. Some wolves have died from natural causes or car accidents. A few have been illegally shot. The American Farm Bureau's suit to remove the wolves, which probably would have meant their death, won in a district court but was reversed on appeal. Some environmentalists object to the reintroduced wolves' classification as an "experimental population," which allows ranchers or farmers to kill a wolf caught in the act of attacking livestock.

Overall, reintroduction has worked even better than expected. New packs have formed, and many puppies have been born. By summer, 2000, ten wolf packs and approximately three hundred wolves lived in the Yellowstone area. Douglas Smith, the biologist in charge of the wolf restoration, says he expects only a modest increase in their numbers in the future.

Wolves have replaced bears as the most popular wildlife sight with the park's visitors. The restoration program's success, and the popular support it drew have inspired plans to reintroduce wolves and other native species to more American wilderness areas.

Emily Alward

Powerful Earthquake Hits Kobe, Japan

The Kobe, Japan, earthquake of January 17, 1995, killed 5,500 people, injured 37,000, and did damage exceeding $50 billion, one of the most costly natural disasters on record.

What: Disasters; Earth science
When: January 17, 1995
Where: Kobe, Japan
Who:
NOBUO ISHIHARA (1926-), deputy chief cabinet secretary
TOMIICHI MURAYAMA (1924 -), prime minister of Japan from 1994 to 1996

Japan's Geology

Kobe's population of 1.4 million lives along a narrow coastal plain on the southern coast of Japan's main island, Honshu. The city is Japan's second largest seaport and an important industrial center. Honshu and the other islands that make up Japan are located along the Ring of Fire that rims the Pacific Ocean, where the collision of great plates of Earth's crust causes frequent earthquakes, occasional great earthquakes, and volcanism.

The Earthquake

Just before dawn on January 17, 1995, the Kobe area was struck by an earthquake, the most devastating temblor in Japan since the 1923 Tokyo earthquake and probably the most expensive natural disaster in world history. The epicenter—the location directly over the rock slippage, or fault, where the shaking starts—was just 20 miles (32 kilometers) southwest of downtown Kobe and at a shallow depth of 12 miles (19 kilometers). The sudden slippage was about 6 feet (9.7 kilometers) along the ruptured fault that extended about 20 to 30 miles (32 to 48 kilometers) below the surface.

The magnitude (a measure of the energy released) of the temblor was 6.9, which is less than truly "great" earthquakes, which exceed magnitude 7.5. However, the temblor's closeness to a densely populated urban area meant that the effects were particularly severe. Strong vibrations up and down and back and forth lasted for twenty seconds. There were also numerous aftershocks (earthquakes of smaller magnitude following and directly related to the first event).

Although some modern buildings of good construction withstood the shaking, many others tilted, collapsed, or suffered other severe damage. Those built on top of loose materials such as soil, sediment, or landfill rather than solid rock were more severely damaged because of greater

AP/Wide World Photos

Part of the Hanshin Expressway in Nishinomiya lies on its side after a powerful earthquake hit the Kobe area.

ground motions. Along the port's waterfront, land heaved and subsided, and a 700-yard-long (630-meter-long) section of elevated highway toppled over. The elevated track of a high-speed train line, constructed to be almost indestructible, was snapped in eight places. The earthquake vibrations ruptured gas and electricity lines, resulting in more than three hundred fires that raged unchecked because of the disabled water mains and blocked roads. With the loss of utilities, residents had no heat for the cold January weather and no water for drinking or hygiene.

The rescue efforts and distribution of emergency relief materials—food, water, fuel, blankets—were hampered by an initially slow response by local government authorities and disorganization. Relief efforts were also slowed by the impassable roadways and rubble on the streets in congested urban areas. Roads that could have been cleared for emergency vehicles—fire, police, and search and rescue—were not cordoned off and were therefore clogged with residents with their vehicles and possessions. The local officials also delayed in calling in the national armed forces for assistance.

The government of Prime Minister Tomiichi Murayama came under fire for not providing aid sooner, although the central government tended to shift the blame to local authorities, who were slow to ask for help. Three days after the quake, Nobuo Ishihara, deputy chief cabinet secretary and coordinator of the emergency response, was quoted as saying that the response had been "less than optimal." Although foreign organizations such as the International Rescue Corps, a British-based search-and-rescue group, immediately offered assistance, Japan initially refused their help. For example, the IRC was not asked to help until January 21 and did not arrive in Japan until January 23, five days after the earthquake.

The lack of civic preparedness for the earthquake disaster was surprising, considering the generally high awareness in Japan of the prospect of such an event. Many people have an earthquake emergency kit in their homes. Every September 1, the anniversary of the destructive earthquake that hit Tokyo and Yokohama in 1923, communities nationwide review protective measures and participate in drills to practice disaster response, evacuations, and rescues. Ironically, Kobe became a busy port and international trading city after the 1923 earthquake partly because foreign merchants relocated there from the devastated port city of Yokohama. Some Japanese who had experienced the 1923 earthquake were living in Kobe in January, 1995, and thus experienced the two most devastating earthquakes in Japan in the twentieth century.

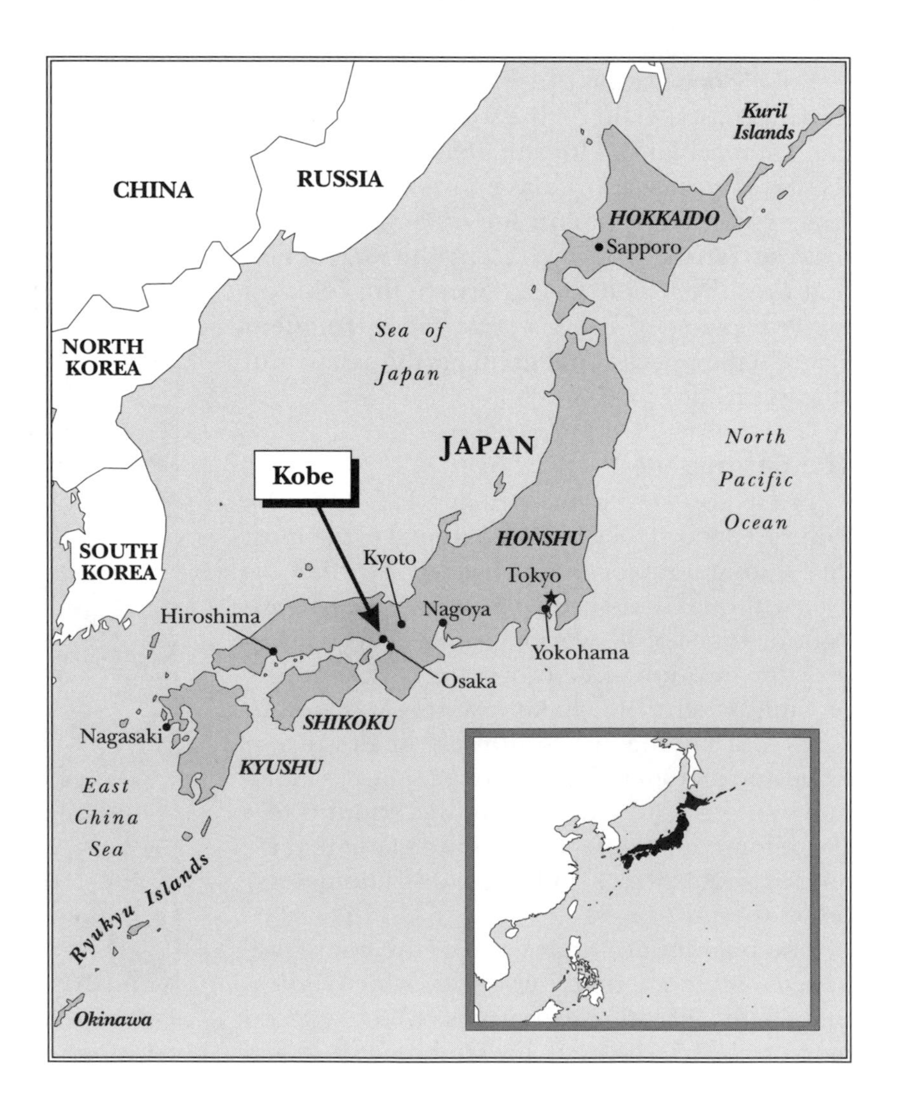

Kobe was less well prepared, psychologically and organizationally, for a major earthquake than Tokyo or other areas with more recent seismic activity. It is some distance from the seismically active zone associated with the ocean trenches off Japan's southern and eastern coasts and was therefore believed to have less potential of suffering a major shock. In addition, many people believed that modern engineering and design of buildings, freeways, and other structures had made them less susceptible to being damaged and disabled by an earthquake. This temblor, a shallow, nearby event with fairly large magnitude, demonstrated the vulnerability of the urban infrastructure.

Consequences

The casualties and destruction in the Kobe area were staggering: 5,502 people were killed, mostly from immediate crushing or from being trapped in the rubble, and 37,000 were injured. The death toll included 28 people who were killed in a landslide in a nearby town. More than 300,000 people had to be evacuated to temporary shelters, including school gyms and city offices. Initially, many had to camp out in the freezing January weather.

Despite the destruction, which caused many homes and stores to be temporarily abandoned, virtually no looting or public disturbances occurred. The Japanese virtues of order and discipline, civility, and mutual support were evident, and Kobe residents focused their attention on the hard tasks of survival and the reconstruction of their lives, homes, and workplaces.

About 200,000 buildings were destroyed or damaged. Many older wooden structures were consumed in the fires that followed the temblor. Japan is a nation with a high cost of living and contains modern cities with elaborate infrastructure—buildings, transportation lines, utilities, and equipment. Rebuilding costs, public and private, have been estimated to be from $40 billion to $100 billion, exceeding the cost of any previous natural disaster in the world.

Robert S. Carmichael

Clinton Announces U.S. Bailout of Mexico

In the face of strong opposition from Congress, President Bill Clinton announced an international $50 billion emergency plan to help bail out Mexico.

What: International relations; Economics
When: January 31, 1995
Where: Washington, D.C.
Who:
Bill Clinton (1946-), president of the United States from 1993 to 2001
Ernesto Zedillo Ponce de León (1951-), president of Mexico from 1994 to 2000
Carlos Salinas de Gortari (1948-), president of Mexico from1988 to 1994

Keeping Mexico Solvent

In one of the most decisive acts of his presidency, President Bill Clinton defied the wishes of many members of Congress and hastily put together an emergency $50 billion aid package to stabilize the declining Mexico peso. Although the president and spokespersons for the administration refused to characterize the aid package as a "bailout," few analysts doubted that the purpose of the package was to prevent the collapse of the Mexican economy. The package included a $20 billion line of credit from the U.S. Treasury's Exchange Stabilization Fund, $17.8 billion in loans from the International Monetary Fund (IMF), and $10 billion in loans from the central banks of the leading industrial countries (to be distributed through the Bank of International Settlements or BIS).

Clinton acted swiftly to circumvent Congress after his original aid proposal—a $40 billion loan guarantee package—bogged down on Capitol Hill. Arguing that a delay in responding to the Mexican crisis would mean the loss of thousands of American jobs and the risk of an increase in illegal immigration, the Clinton administration swiftly developed a cooperative rescue plan with international lending agencies and its allies. Some members of Congress complained that their legislative authority over fiscal matters was usurped by the president's actions.

For Mexico, the bailout was a bitter fiscal pill to swallow. It came scarcely one year after the hard-fought ratification of the North American Free Trade Agreement (NAFTA), which appeared to offer Mexico a more prosperous future. The economic treaty, which united Mexico, the United States, and Canada in a regional common market, was hailed as one of many prudent fiscal and economic measures taken by President Carlos Salinas de Gortari to improve the chances for Mexico's future economic success. Mexico's financial collapse in late 1994 and early 1995 and Washington's swift action to support the peso raised a number of significant issues regarding whether Clinton's actions had damaged his future working relationship with Congress, whether Mexico would ever have the resources to repay the loans, and the long-term effects of the loan package on Mexico's economic future and the future of U.S.-Mexican relations.

Reversal of Mexico's Economic Miracle

Despite their unhappiness with Clinton's bailout package, even critics recognized Mexico's dire economic situation. President Ernesto Zedillo Ponce de León decided to permit the overvalued peso to "float" against the dollar in late December, 1994, meaning that free markets would determine its value, rather than the government buying and selling pesos to stabilize their value relative to the dollar. The peso lost half of its value in a matter of weeks.

Mexican investors abandoned their stock market, selling stocks and putting the money into more secure investments abroad. Foreign investors, especially stock funds specializing in emerging markets and growth stocks, pulled their capital out of the country, seeking to cut their losses in one of the worst disasters in mutual fund history. The value of Mexican stocks fell as much as 70 percent in dollar terms. Large multinational corporations halted development plans, and interest rates skyrocketed. Bank failures were expected. Mexican workers watched as prices increased, cutting the value of their wages by a third. Saddled with a $160 billion foreign debt, Mexico would, even after the bailout, need to find a way to pay debt service of at least $40 billion per year.

Only one year earlier, many reputable financial analysts considered Mexico to be a textbook illustration of how a developing country should pursue economic growth. North American investors had reasoned that the combination of a favorable investment climate, a growing Mexican middle class, the privatization of hundreds of state-owned industries, the lessening of protectionist restrictions, and NAFTA offered good prospects for the future.

Mexico's problems appear to be more political than economic. The Partido Revolucionario Institucional (PRI) ruled Mexico for more than seventy years, and it was only with the rise of Vicente Fox in 2000 that a different party came into power. Although one-party rule spared Mexico the kind of political instability and revolutionary movements that plagued much of Latin America, the transition of power from one PRI president to the next every six years was fraught with vicious political infighting that, in turn, led to a crisis of public confidence. It is not surprising that the 1994-1995 economic collapse came during an electoral year, 1994, during the transition from President Carlos Salinas to his handpicked successor, Ernesto Zedillo. Earlier in the campaigning, Luis Donaldo Colosio, the original PRI candidate, had been assassinated.

Consequences

The White House's actions not only upset con-

AP/Wide World Photos

Mexican stockbrokers react to the news that U.S. president Bill Clinton's original $40 billion aid package has failed to win approval.

gressional critics but also elicited a strong reaction from several European allies. Those countries believed that the United States had not consulted them sufficiently before securing pledges of funds from the IMF and the BIS. Those funds would come from contributions from countries all over the world, but the United States had taken a leading role in pressuring the agencies to commit funds to Mexico. Senate Banking Committee chair Alfonse D'Amato, a Republican from New York, held hearings that were sharply critical of the Clinton plan. In accepting the loans, the Mexican government agreed to implement a number of economic reforms, including privatization of state assets and imposition of wage and price controls. Committee members voiced concern about Mexican compliance. On February 15, 1995, D'Amato joined several other senators in introducing a joint resolution that would require the White House to provide monthly reports detailing Mexico's compliance with the terms of the bailout. Public opinion ran three to one against the bailout, but Congress, after raising criticisms about the plan, did not take steps to derail the aid package.

The economic collapse made it clear that meaningful economic growth was unlikely to occur until Mexico's political system is reformed. Zedillo announced an unpopular austerity plan that included increased taxes, reduced government spending, and increases in fuel and electricity prices. The potential for political instability cannot be discounted unless political liberalization accompanies these measures. The rescue package also stipulated that Mexico obtain permission from Washington before making decisions on a variety of economic matters.

Allen Wells

U.S. State Department Criticizes Human Rights in China

The U.S. State Department issued a report on human rights stating that China made no progress in that area during the previous year.

What: International relations; Human rights
When: February 1, 1995
Where: Washington, D.C.
Who:
Bill Clinton (1946-), president of the United States from 1993 to 2001
Mickey Kantor (1939-), U.S. trade representative from 1993 to 1996
James O'Dea (1935-), Washington, D.C., director of Amnesty International
Holly Burkhalter, Washington, D.C., director of Human Rights Watch

No Progress on Rights

A U.S. State Department report on human rights concluded that China had made no progress on human rights during 1994. The report dealt a blow to President Bill Clinton's policy of not making trade linked to or contingent on improvements in human rights concerns. The president had argued that economic changes would bring about human rights reforms and, therefore, that trade sanctions or other measures were not necessary to bring about such change. However, the report contradicted his contentions, stating that China had violated "internationally accepted norms" of human rights. Violations stemmed from its failure to tolerate dissent and to safeguard freedom of speech, press, association, and religion. Moreover, the State Department called China "an authoritarian state" and attacked its "arbitrary and lengthy incommunicado detention," citing its torture and general mistreatment of prisoners.

The report also criticized China for detaining thousands of "prisoners of conscience," failing to provide an adequate account for those people detained after the 1989 uprisings, and denying dissidents fair trials and arresting those who traveled outside China. The report further attacked China for failing to allow the International Committee of the Red Cross access to political prisoners and for jamming the Voice of America. Other practices condemned by the report were forced abortions and sterilizations as well as the mandatory donation by prisoners of their organs for transplant.

China's treatment of Tibet drew further criticism. Conditions deteriorated in Tibet in 1994. According to the report, there were "widespread human rights abuses." These abuses included torture, arbitrary arrest, prosecution of nuns and monks who opposed government policies, and restrictions on freedom of expression. China refused to allow the displaying of any image of the Dalai Lama, Tibet's spiritual leader, even privately in people's homes. Officials removed his photographs from sale in bazaar shops. These actions defied the Clinton administration's insistence that China begin discussions with the Dalai Lama, who was living in exile in India.

The State Department grouped China with Iraq, Iran, Burma, North Korea, and Cuba as countries in which "flagrant and systematic abuses of basic human rights continued at the level of authoritarian regimes." This ranking demonstrated that contrary to the administration's hopes, economic progress in China was not leading to improvements in its human rights record. The one bright spot mentioned in the report, China's passage of a new law allowing citizens to sue the government for improper treatment, was darkened by the fact that the law was never implemented.

Activists React

Few were surprised that the State Department found human rights violations in China; even ardent supporters of China acknowledged that the country had a problem in this area. Human rights activists, enraged by China's failure to begin negotiations with the Dalai Lama, praised the State Department's strong statements. However, that praise was tempered by doubt that the administration would actually do anything to bring about human rights reform. For example, James O'Dea, Washington, D.C., director of Amnesty International, noted that the Clinton administration would not "participate in reducing the abuses because of special relationships." It was a sentiment echoed by Holly Burkhalter, Washington, D.C., director of Human Rights Watch. Human rights advocates would have preferred to see the administration move toward trade restrictions and military realignment in the region to pressure China to end its abuses and move toward reform.

Consequences

In light of the apparent failure of trade to effect reform, the Clinton administration began to rethink its China policy and to get tough on China's pirating of U.S. videos, films, books, and other copyrighted material. It threatened to impose more than a billion dollars in punitive tariffs on China. Mickey Kantor, U.S. trade representative, advocated the imposition of tariff penalties on Chinese imports. However, few believed that the administration would go much beyond tough talk in its punishment of China.

China reacted strongly to the report, rejecting it as nonsense and blatant interference in the internal affairs of a sovereign nation. Chinese officials pointed out failures in the United States' own record of human rights and dismissed the report, terming it hypocritical and imperialistic. China argued that it had to deal with internal threats to its survival in its own way. The report on human rights brought to the forefront the problems in Chinese-American relations, which had been minimized under the Clinton administration.

Frank A. Salamone

Peru and Ecuador Agree to a Cease-Fire in Their Border Dispute

An agreement signed in Brazil ended the third major conflict between Ecuador and Peru over disputed territory near their common border in the Amazon region. The agreement led to further negotiations and the final settlement of a border that had been disputed for more than a century.

What: International relations; Military; War
When: February 17, 1995
Where: Brasilia, Brazil
Who:
MARCELO FERNANDEZ DE CORDOBA, Ecuadorian vice foreign minister
JOSE PONCE VIVANCO, Peruvian vice foreign minister
FERNANDO HENRIQUE CARDOSO (1931-), Brazilian president from 1995

Conflicting Claims

Disputes between Peru and Ecuador regarding their common border originate in conflicting claims dating from the period when both countries were colonies of Spain. Peru was a viceroyalty of Spain until it gained independence in 1821. Ecuador was one of three countries to emerge from the collapse of Gran Colombia in 1830. For more than a century following independence, the two countries claimed overlapping territory in the Amazon Basin, based on two different colonial declarations. Ecuador cited the 1563 Audience of Quito, in which Spain established Ecuador with territory extending as far north as the Amazon River and the 1830 Pedemonte-Mosquera Protocol, which established the boundary at the Marañon-Amazon River. Conversely, Peru cited boundaries under its administrative control as a viceroyalty, as established by the Royal Seal of 1802.

In 1936, representatives of Peru and Ecuador met in Washington, D.C., to negotiate an end to more than a century of dispute over the border region. The Status Quo Line of 1936 was established as a beginning point for these negotiations, and an agreement was signed in Washington. Negotiations held from 1936 to 1940 failed to confirm this line in a treaty.

Between June and August, 1941, 15,000 Peruvian troops, led by General Eloy Ureta, engaged 3,000 poorly prepared Ecuadorian troops in a brief but intense effort to finalize Peru's claim on the region. The actions of the Peruvian army were a challenge to the civilian leadership of President Manuel Prado. Prado had authorized General Ureta to secure the border area, but he had not authorized activity on the Ecuadorian side of the border. Ecuador's military defeat likewise had domestic political repercussions that influenced its future actions in the region.

In February, 1942, the conflict was formally ended by an agreement signed in Rio de Janeiro, Brazil. The United States, Argentina, Brazil, and Chile served as guarantors of the Rio Protocol of Peace, Friendship, and Boundaries. The protocol identified the watershed between the Zamora and Santiago Rivers as a boundary, resulting in Ecuador's loss of only about 5,000 square miles relative to the 1936 Status Quo Line, but approximately 78,000 square miles in comparison with Ecuador's original claim.

Beginning in 1961, Ecuador began to pursue a policy of nullifying the 1942 Rio Protocol. The Cordillera del Condor served as a de facto boundary, but a 48-mile (77-kilometer) stretch of the border in this area had not been demarcated on the ground. Ecuador argued that the agreement could not be carried out because mapping carried out by the U.S. Air Force between 1943 and 1946 reveals that the Cordillera del Condor could not be the watershed described in the Rio Protocol.

The Event

A two-day conflict erupted in the region in 1981, and tensions were elevated on an almost annual basis leading up to the Rio Protocol anniversary. On January 26, 1995, new fighting emerged in the Upper Amazon border area. As many as 3,000 Ecuadorian and 2,000 Peruvian soldiers were involved in three weeks of fighting, in which Ecuador benefited from air superiority, short lines of communication, and dominant artillery positions. Dozens of soldiers from both sides were killed in the conflict.

The guarantors of the Rio Protocol, concerned that the conflict could lead to broader war, worked closely with the two sides to end the fighting. On February 17, 1995, representatives of Peru and Ecuador met in Brasilia and concluded a cease-fire agreement, known as the Itamaraty Peace Declaration. The guarantor nations formed the Military Observer Mission Ecuador-Peru (MOMEP) to enforce the cease-fire and to supervise the withdrawal of 5,000 troops from the Cenepa valley and the demobilization of 140,000 troops.

Consequences

No direct combat took place following the cease-fire of February 17, 1995, but the diplomatic struggle continued, as did allegations of occasional cross-border infiltrations. Ecuador and Peru entered into a series of negotiations over the next several years. The United States, Brazil, Argentina, and Chile helped Peru and Ecuador negotiate issues related to the concerns that continued following the cease-fire. Four commissions were established to address these concerns, and they began their work simultaneously on February 17, 1998, the third anniversary of the cease-fire. The four commissions addressed commerce and navigation, border integration, the common land boundary, and mutual confidence and security. Peru and Ecuador provided representatives to the commissions, each of which met in the capital city of one of the four Rio Protocol guarantor nations. Additionally, a commission met in March, 1998, to address issues associated with the Zarumilla Canal, which makes up the binational border on the Pacific Coast.

Following the work of these commissions, a comprehensive peace accord was signed on October 26, 1998, and was unanimously ratified in both national legislatures. Demarcation of the agreed-upon border with Ecuador was completed on May 13, 1999. MOMEP was formally disbanded in a ceremony in Patuca, Ecuador, on June 17, 1999, with all guarantor troops withdrawing by the end of that month.

James Hayes-Bohanan

Illegal Trading Destroys Barings, Britain's Oldest Bank

Barings Bank, the oldest merchant bank in Great Britain doing business continuously since 1762, collapsed under losses suffered because of one of its traders.

What: Business; Economics
When: February 26, 1995
Where: Singapore
Who:
NICHOLAS W. LEESON (1967-), a futures trader in the Singapore office of Barings Bank

A Bank Falls

The story of the collapse of Barings Bank is largely the story of one man. Great Britain's oldest commercial bank, operating continuously since 1762, was forced to close because of unwise and unregulated trades.

Twenty-eight years old at the time of the bank's collapse, Nicholas W. Leeson had been hired by Barings in 1989, after his graduation from high school, as a back-office clerk in the London office, settling and accounting for daily trades and payments. By 1992, he was a roving troubleshooter, assigned to a team investigating allegations of fraud within the bank.

Barings sought to establish a presence in burgeoning Far East financial markets. Leeson was sent to Singapore as a settlement officer and eventually allowed to execute trades. He gained a reputation for making money even if the Asian markets sagged, making small but steady profits by detecting the minute differences in the price of financial instruments between the Osaka, Japan, and Singapore financial markets, and buying low in one market and immediately selling high in the other market.

In the fall of 1994, Leeson began making riskier investments in derivatives and futures. The investments were tied to the Nikkei 225, the Japanese stock index equivalent to the Dow Jones Industrial Average, and required that the index stay at or above a certain number to realize a profit. Only a small percentage of the value of the futures, usually 6 to 10 percent, was needed as a down payment. It was easy to lose many times the original cash investment. Leeson began reporting his successes to the London office and hiding his losses in fictitious accounts.

Disaster

By the beginning of January, 1995, Leeson had the equivalent of US$80 million in losses. He might have eventually recovered, but Kobe, Japan, was devastated by an earthquake on January 17, 1995. The Nikkei lost 7 percent in a week. By January 26, Leeson had stopped selling and was buying only as prices fell in the hopes of recouping losses. Asian financial markets required daily payment on contracts—the difference between the price of the contract and its value at the close of each trading session. Barings had placed the trades and, therefore, was responsible.

Leeson had "bet" that the Nikkei would not drop below 19,000 points on March 10, 1995. After January 23, 1995, it continued to drop below 18,500. He had purchased so much in the way of futures that each point the Nikkei fell below 18,500 caused Barings to lose US$200,000.

On February 23, Leeson and his wife left Singapore. They were detained at the Frankfurt airport a week later, on March 2. According to Leeson, he was attempting to return to London where he expected to receive a fairer hearing than he believed possible in Singapore.

His wife was permitted to continue to London, but Leeson remained in a German prison, fighting extradition to Singapore, for nine months. On November 22, 1995, a Frankfurt court authorized his extradition on eleven of twelve charges of fraud, forgery, and breach of trust. The day before, he had pleaded guilty in a Frankfurt court in the hope of ameliorating his eventual sentence, which could be fourteen years in prison.

Early in December, Judge Richard Magnus in Singapore sentenced Leeson to six and one-half years on two counts of fraud. After two years, Leeson applied for early release, citing poor health. He returned to Britain in July, 1999, after spending four and one-half years in prison. His wife had divorced him and married another securities trader. All assets he had in England had been frozen because it was believed that he had hidden assets. He had colon cancer and a 30 percent chance of dying in four years.

Consequences

By February 26, three days after Leeson left the country, the full extent of Barings' losses became known. When internal auditors examined his transactions, they found the amount of credit Leeson extended to cover his positions exceeded the bank's capital. The bank had lost US$1.38 billion.

AP/Wide World Photos

Futures trader Nicholas W. Leeson, who had fled Singapore, is taken into custody at Frankfurt's airport.

The Bank of England tried but was unable to save Barings. The Dutch ING Group acquired all the businesses, assets, and liabilities of the Barings Group, including Baring Brothers Bank, Baring Securities, and Baring Asset Management for one pound sterling (about US$2.30). Barings PLC, the holding company for the Barings group, was not purchased.

World markets faltered and then stabilized following the Barings collapse—and tried to understand how it could have occurred. Lack of oversight and control of trading, along with a lack of understanding of derivatives trading that prevented proper monitoring of activities, were determined to be the main causes. Leeson traded and did his own settlements, in essence auditing himself, which removed one of the automatic checks on his activities. No one was aware how much money was involved or how much was being lost (or made). Because he was accountable to no one, Leeson was able to hide his losses in dummy accounts, make fraudulent transfers, and claim that the trades were done at the behest of an important client and thus did not involve Barings' money. Barings continued to supply capital for Leeson's trades, believing that the customer's money would soon be forthcoming.

Later, suspicion arose that Barings officials had some idea of what was going on but hoped that Leeson's record would enable them to weather any problems. The ING group eventually removed twenty-one executives, either by resignation or termination, who were involved with the Singapore operations. They believed that at least some of the blame rested with management.

The fallout caused by Barings' collapse caused other organizations to examine their trading practices and to analyze the uses and dangers of derivatives. Although many organizations were aware of the rises and how to minimize these and audit trades, before the Barings debacle, procedures were largely ignored. Most organizations have now acknowledged the need for an independent auditor to monitor all trades and investigate suspicious or unusual activities of traders.

Elizabeth Algren Shaw

Top Quark Is Tracked at Fermilab

Two teams of physicists announced the discovery of the top quark, the last of six such subatomic particles predicted by scientific theory.

What: Physics
When: March 2, 1995
Where: Batavia, Illinois
Who:
John Peoples (1933-), the Fermilab director during the search for the top quark
Leon M. Lederman (1922-), the Fermilab director who led the search for the bottom quark
Murray Gell-Mann (1929-), a California Institute of Technology physicist who proposed the quark theory in 1964
Sheldon L. Glashow (1932-), a Harvard physicist who predicted the existence of a new quark in 1970

Units of Matter

From the time of the earliest Greek philosophers until the end of the nineteenth century, atoms were viewed as indivisible units of matter. When the English physicist Joseph John Thomson identified electrons as negatively charged particles in 1897, however, he recognized that they must be subatomic particles that combine with positive matter to make up an electrically neutral atom.

This new understanding of the divisible atom was expanded by the English physicist Ernest Rutherford, who in 1911 showed that the atom consists of a very small positive nucleus surrounded by orbiting electrons. By 1932, Rutherford and his students had shown that the nucleus was made up of two other subatomic particles: protons, which have a positive electrical charge equal and opposite to that of the electron, and neutrons, which have about the same mass as protons but no electrical charge. Also in 1932, Ernest Lawrence of the University of California developed the cyclotron, in which charged particles could be accelerated to high energies as a means of probing the structure of matter. After World War II, large accelerators began to produce energies high enough that colliding particles could create many new kinds of particles by converting energy into mass as implied by Albert Einstein's theory of mass-energy equivalence ($E = mc^2$).

By 1960, several hundred new subatomic particles had been studied and classified into three families: leptons (light particles such as electrons, muons, and neutrinos), mesons (intermediate mass particles that transmitted nuclear forces), and baryons (heavy particles such as protons and neutrons). In 1964, California Institute of Technology physicist Murray Gell-Mann developed a theory that successfully accounted for the properties of all mesons and baryons by assuming that they were composed of smaller particles he called "quarks," held together by "gluons." His theory required three quarks, nicknamed "up," "down," and "strange," and three "antiquarks." All baryons consisted of three quarks in differing combinations, and all mesons consisted of quark-antiquark pairs.

From the Bottom to the Top

The three-quark theory predicted several new particles, and it was confirmed by observations made at Stanford University in 1968. New particles discovered in 1974, however, required recognition of a fourth quark, the "charmed" quark, to account for the properties of these particles. At the Fermi National Accelerator Laboratory in Batavia, Illinois, in 1977, a group led by Leon Lederman discovered evidence of the existence of a fifth quark, the "bottom" quark. Fermilab operates the largest proton accelerator in the world, with a main ring that is four miles in circumference. By smashing protons with energies of 500 GeV (billion electron-volts) into a fixed nuclear target, they were able to identify a new meson, the upsilon particle, with properties matching a bottom quark-antiquark pair.

Quark theory suggested that the bottom quark should be related to a sixth quark, the "top" quark. Since even higher energies would be required to produce the top quark, Fermilab began the construction of another four-mile beam tube in which counter-rotating antiprotons could be accelerated for head-on collisions with protons from the main ring. By 1986, these proton-antiproton collisions reached energies of 1.8 TeV (trillion electron-volts) in the collision detector (CDF). This energy was enough to produce top quarks, but the observers could only detect a few of the decay reactions expected from top quarks. Under Fermilab director John Peoples, a new detector that was more sensitive to other top-quark decay reactions, the DZero, was completed in 1992.

The top-quark search was conducted with these detectors by two international teams, each consisting of about 450 experimenters. On March 2, 1995, both teams announced their independent discovery of the top quark. The CDF group reported detection of thirty-three top-quark "events" under conditions in which only eight such events would be expected without a top-quark decay. The DZero group reported seventeen top-quark events with an expected background of only four such events. The probability of so many events occurring without a top quark decay was about one in a million.

Consequences

The discovery of the top quark strongly supports the revised quark theory, which identifies six quarks matched with six leptons as the basis of all matter. At Harvard University in 1970, Sheldon Glashow and his associates predicted the existence of the charmed quark to explain why certain expected particle reactions never occur. Their theory suggested a paired symmetry of quarks and leptons, with the up and down quarks paired with the electron and its neutrino, and the strange and charmed quarks paired with the muon and its neutrino. The muon neutrino was distinguished from the electron neutrino in 1962 by Lederman and his associates, then at Columbia University, and the 1974 discovery of the charmed quark confirmed Glashow's revised quark theory. The discovery at Stanford in 1975 of the tau lepton, heavier than the muon, suggested the possibility of a tau neutrino and implied the existence of a third quark pair. The discoveries at Fermilab of the bottom and top quarks completed this symmetry.

In 1983, David Schramm and his associates at the University of Chicago used evidence from astronomy concerning the density of matter in the universe to indicate that no more than three types of neutrinos can exist.

Joseph L. Spradley

Terrorists Release Toxic Gas in Tokyo Subway

In an attack blamed on members of a religious sect, nerve gas released in Tokyo subway trains killed twelve people and injured more than five thousand.

What: Civil strife; Terrorism; Religion; Crime
When: March 20, 1995
Where: Tokyo, Japan
Who:
Shoko Asahara (Chizuo Matsumoto, 1955-), leader of the Aum Shinrikyo sect
Hideo Murai (1959-1995), head scientist for Aum Shinrikyo
Tomiichi Murayama (1924-), prime minister of Japan from 1994 to 1996

Death on the Subway

On March 20, 1995, hidden containers of poison gas were placed on five subway trains in Tokyo, Japan. The gas was released during the morning rush hour as the trains made their way along three subway lines leading to the center of the city. Twelve people died from exposure to the gas, and more than five thousand were injured. The poison was identified as sarin, a potent nerve gas.

Prime Minister Tomiichi Murayama held an emergency meeting with members of his cabinet and ordered an immediate, full-scale investigation. The Japanese military mobilized its chemical warfare unit, and the Transportation Ministry increased security at all railway stations, airports, and seaports.

Police soon targeted the Aum Shinrikyo religious sect for investigation. The sect previously had been associated with less serious incidents involving sarin, but no charges had been brought against it. On March 22, more than two thousand police officers raided twenty-five buildings across Japan owned by the sect. They found large amounts of chemicals that could be used to make sarin. The next day, more than one thousand police officers raided the sect's headquarters in the village of Kamikuishiki, about six hundred miles west of Tokyo. More chemicals were discovered. The sect denied all responsibility for the March 20 incident.

Police gathered evidence against the sect by arresting members on minor charges and then questioning them about the release of the sarin. Meanwhile, violent incidents possibly related to the attack occurred in Tokyo and the nearby seaport city of Yokohama. On March 30, an unidentified gunman shot Takaji Kunimatsu, director of the National Police Agency. He survived the attack but was wounded severely. On April 19, a chlorine-based poison gas was released in a subway station in Yokohama, injuring more than five hundred people. A similar attack on April 21 injured twenty-four people in a Yokohama department store. On May 6, two bags containing enough cyanide to kill ten thousand people were discovered in a Tokyo subway.

The most dramatic incident occurred on April 24, when Hideo Murai, the sect's senior scientist, was stabbed to death in front of television camera crews. Police arrested Hiroyuki Jo, a South Korean citizen living in Japan, for the murder. It was unclear if the motive was revenge for the subway attack or to prevent Murai from giving evidence against the sect. On May 16, police arrested Shoko Asahara, leader of the sect, on charges of organizing the attack.

Shoko Asahara and Aum Shinrikyo

Shoko Asahara, the founder and leader of the Aum Shinrikyo religious sect, was born Chizuo Matsumoto in 1955. Born with limited eyesight, he attended a school for the blind. After gradua-

AP/Wide World Photos

Shoko Asahara, leader of Aum Shinrikyo sect, is driven to the Tokyo metropolitan police headquarters.

tion, he worked as an acupuncturist, then opened an apothecary selling traditional Chinese medications. In 1982, he was arrested and fined for selling fake cures. In 1984, he began a yoga school that developed into Aum Shinrikyo in 1987.

Aum Shinrikyo takes its name from "Aum," a sacred word in Hinduism, and Japanese words meaning supreme truth. Its teachings combine Buddhism, Hinduism, and Asahara's own mystical beliefs. At the time of his arrest, the sect claimed about ten thousand members in Japan and perhaps as many as thirty thousand in Russia. In 1990, the sect ran two dozen candidates for office in Japan. The overwhelming defeat of all these candidates may have led Asahara to hold more apocalyptic beliefs. He announced that the world would end in 1997, 1999, or 2000. In his book *Disaster Approaches the Land of the Rising Sun* (1995), he predicted that a cloud of poison gas from the United States would kill almost everyone on Earth.

The sect's association with criminal activities began in November, 1989, when Tsutsumi Sakamoto, a lawyer fighting the sect, disappeared along with his wife and infant son. A similar event occurred in February, 1995, when lawyer Kiyoshi Kariya vanished. The sect was suspected of kidnapping in both cases, but no arrests were made.

Aum Shinrikyo was linked to an incident involving sarin in June, 1994. Seven people died and about two hundred were injured when sarin was released in the city of Matsumoto, about one hundred miles west of Tokyo. Three of those injured were judges about to rule on a land dispute between the sect and the city. No arrests were made. One month later, traces of chemicals related to sarin were detected near the sect's headquarters in Kamikuishiki after residents of the village complained of foul-smelling, irritating fumes coming from the sect's compound.

On March 5, 1995, eleven passengers on a Yokohama subway car were hospitalized for dizziness and eye pain caused by an unidentified gas. On March 15, three attaché cases containing battery-operated fans and tanks filled with an unknown liquid were discovered in a Tokyo subway station. These two incidents may have been rehearsals for the March 20 attack.

Consequences

The March 20 attack stunned the Japanese, who pride themselves on having a relatively crime-free society, with no major terrorist incidents since the 1970's. Tensions increased as violence continued following Asahara's arrest. On June 21, 1995, sect member Saburo Kobayashi hijacked an airplane carrying 365 passengers and demanded Asahara's release. On July 2, an unknown gas in a Yokohama subway sent thirty-six people to the hospital. On July 4 and 5, containers of cyanide were discovered in Tokyo subways, and three passengers were injured by unidentified fumes.

The suspected involvement of Aum Shinrikyo in the March 20 attack led the Japanese to question the power of their nearly 184,000 registered religious organizations. On October 30, 1995, Tokyo District Court Presiding Judge Seishi Kanetsuki removed Aum Shinrikyo's tax-exempt status, the first time such a judgment had been made against any sect. On December 8, the first change made in the Religious Corporation Law in more than forty years gave the Japanese government more authority over religious groups.

Rose Secrest

Terrorists Bomb Oklahoma City's Federal Building

In one of the most shocking terrorist attacks in American history, the Murrah Federal Building in Oklahoma City was damaged severely by a car bomb.

What: Crime; Terrorism
When: April 19, 1995
Where: Oklahoma City, Oklahoma
Who:
TIMOTHY MCVEIGH (1968-2001), primary suspect in the bombing

The Bombing

Early in the morning on April 19, 1995, a rented Ryder truck parked on the street next to the Alfred P. Murrah Federal Building in Oklahoma City suddenly exploded. The truck contained about forty-eight hundred pounds of an explosive mixture of common ammonium nitrate fertilizer and diesel oil, packed into twenty or more blue plastic fifty-five-gallon drums. The entire front of the nine-story building was brought down by the explosion. At least 192 people were killed, and scores more were injured. Many of those killed were children in a day care center within the building.

Intense efforts to rescue and care for survivors began immediately, under the aegis of the Oklahoma City fire and police departments. Dozens of people were rescued from the rubble. The rescue effort continued for more than a week.

The explosion at first was assumed to have been the work of Middle Eastern terrorists. Initial investigations focused on people traveling to and from Middle Eastern countries. Investigators had a bit of luck within hours of the explosion. An agent of the Federal Bureau of Investigation (FBI) searching the streets near the blast found a piece of truck axle that still carried a legible vehicle identification number. The number was traced to a 1993 Ford truck that had been rented from a Ryder agency in Junction City, Kansas.

Although the two men who had rented the truck had used false identification papers, FBI agents were able to make composite drawings of them using information obtained from interviews with the rental agent. The owner of a nearby motel identified one of the suspects from a drawing, and it turned out that he had rented the room in his real name. He was Timothy McVeigh, a twenty-seven-year-old Army veteran.

McVeigh was an adherent of right-wing fringe causes. A nationwide search for him began immediately and soon revealed that he already was in jail, in Perry, Oklahoma. He had been arrested by an Oklahoma state trooper on a concealed weapons charge ninety minutes after the explosion in Oklahoma City. Two of McVeigh's friends, Terry and James Nichols, also were arrested within a few days, on suspicion of having assisted him and for manufacturing other bombs that they had tried out on their farm in Decker, Michigan. The search for McVeigh's co-conspirators continued, particularly John Doe number two, the person accompanying McVeigh when he rented the truck.

McVeigh was charged with the bombing, a capital offense under federal law. In 1997 he was convicted by a Colorado jury on eight counts of murder, two counts of conspiracy, one count of using a weapon of mass destruction, and was sentenced to death. In June, 2001, his sentence was carried out at the U.S. Penitentiary in Terre Haute, Indiana, where he became the first federal inmate to be executed since 1963.

Waco, Ruby Ridge, and the Far Right

In 1993, members of a religious sect known as the Branch Davidians were besieged by the federal government while inside a compound at Waco, Texas. Firearms charges had been filed against Branch Davidian leader David Koresh. On April 19, 1993, the compound was assaulted by agents of the Bureau of Alcohol, Tobacco and Firearms and the FBI. Most of the Branch Davidians, including many women and children, were killed.

In another notable confrontation between the government and dissident citizens known as the Ruby Ridge incident, FBI agents shot and killed, from ambush, the wife and son of Randy Weaver. Weaver was an Idaho separatist who had been charged with failing to appear in court to answer minor firearms charges that had been brought against him.

Most Americans saw the Waco and Ruby Ridge incidents as examples of bureaucratic incompetence and overreaction. The Department of Justice was criticized severely regarding both occurrences. In the case of the Ruby Ridge shootings, several senior FBI officials were disciplined.

Some Americans who adhere to the views of the far right believe that much of the political universe is ruled by conspiracy. The two violent incidents strengthened their view that the government of the United States is preparing to wage war on the American public or somehow betray it to vaguely described internationalists or "one-worlders."

McVeigh and the Nichols brothers consorted with such extremist groups and evidently shared their views. Apparently it was not an accident that the Oklahoma City bombing took place on the second anniversary of the attack on the Branch Davidians.

As soon as McVeigh was arrested, it became clear that the bombing was the work of domestic rather than foreign terrorists. Debate began in the United States about whether additional legal measures should be taken to control the possible activities of right-wing groups. Until an organization violates some specific statute, there is little that can be done to hinder its actions. Any restrictions on politically motivated meetings or right-wing rhetoric would violate First Amendment

AP/Wide World Photos

The Alfred P. Murrah Federal Building in Oklahoma City, after the bombing.

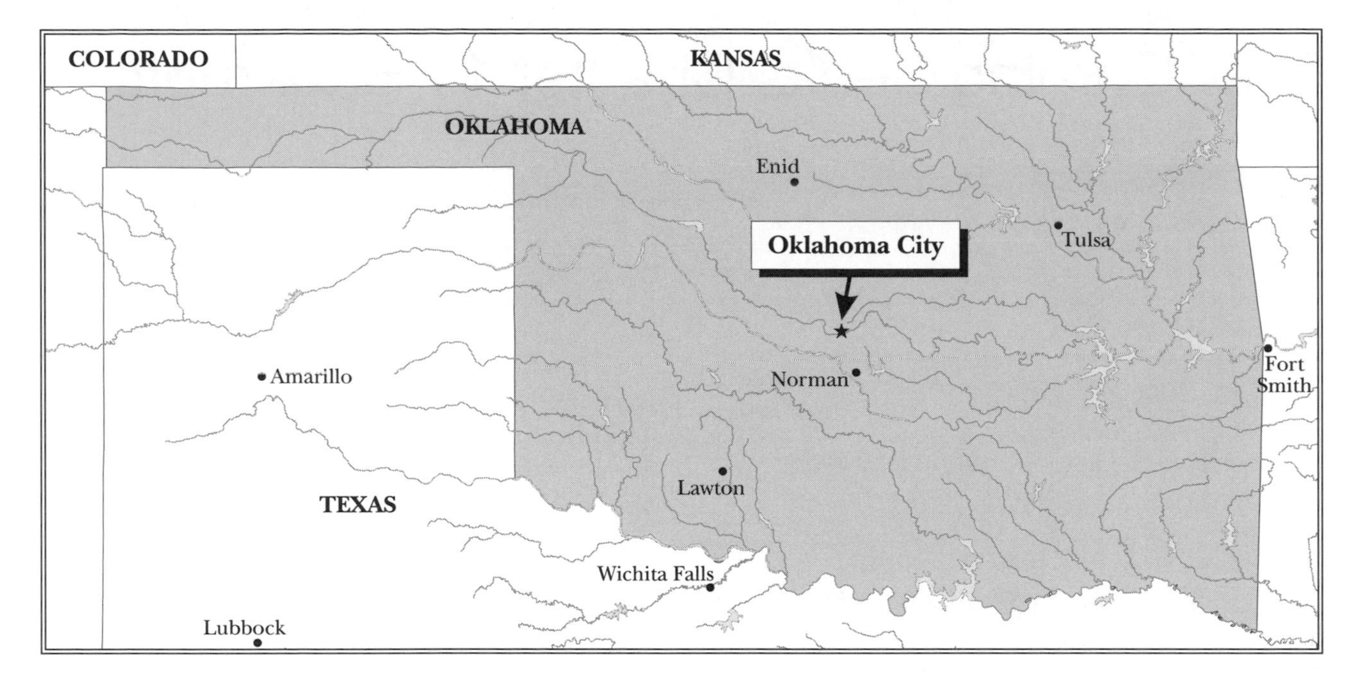

guarantees directly and probably would be held unconstitutional by the courts. State governments already have the power to prevent the formation of "armed companies," and the citizen militias formed by some right-wing groups would fall under that term. Laws banning private militias do not violate the First Amendment. By late 1995, eight states had such statutes on their books, and at least ten others were considering enacting similar laws.

Consequences

Terrorism succeeds, by its nature, in altering the system at which it is aimed. Public buildings and public institutions must be more heavily guarded, and liberties may be curtailed. The sense of public unity, safety, and dignity essential to successful democratic government is injured. The security organs of the state penetrate and infiltrate groups that conceivably may be dangerous, infringing on the rights of both innocent and malevolent parties.

In the months following the Oklahoma City bombing, no new antiterrorist statutes were passed, but many of the other general consequences of terrorism had occurred. FBI surveillance of right-wing groups intensified, and more government buildings and institutions required that visitors pass through metal detectors or be subjected to search of their parcels or baggage. This scrutiny intensified on a national level after the crisis of September 11, 2001, when hijacked passenger flights crashed into the World Trade Center towers in New York and the Pentagon in Washington, D.C.

Robert Jacobs

Dow Corning Seeks Chapter 11 Bankruptcy

Dow Corning, leading producer of breast implants, filed for bankruptcy because of lawsuits involving its breast implants. By filing for bankruptcy, Dow froze all lawsuits against it.

What: Business; Law
When: May 15, 1995
Where: U.S. District Court in Bay City, Michigan
Who:
Richard Hazleton (1941-), chairman and chief executive officer of Dow Corning
Sam Pointer (1934-), U.S. district judge

A Problem and a Settlement

Breast implants were used by women to replace breasts that had been removed for medical reasons such as cancer or to enhance their appearance through breast augmentation or shape alteration. Dow Corning, the manufacturer of a silicone gel breast implant, along with several other breast implant manufacturers, was sued by hundreds of thousands of women who claimed that these implants had damaged their health. These women attributed a number of ailments to the implants, including lupus and hardening of the breasts. About 500,000 women asserted that the implants had hardened or ruptured and that this caused excruciating pain, injury, or other health problems. Many had multiple surgeries to replace damaged or ruptured implants. These serious claims led the Food and Drug Administration to ban silicone breast implants for purely cosmetic reasons in 1992.

In September, 1993, a settlement was reached between the women and the breast implant manufacturers that awarded the plaintiffs an overall sum of $4.2 billion, of which Dow Corning would provide $2 billion. U.S. District Judge Sam Pointer approved the settlement in Birmingham, Alabama. Approximately 400,000 women were involved in the settlement, and about 11,000 additional women opted out of the settlement and filed their own lawsuits. The payments to individual women who were part of the settlement were to range from $140,000 to $1.4 million each. Each woman could be released from the agreement and gain the option to file an individual lawsuit if she agreed to a reduced payment.

Banking on Bankruptcy

This settlement seemed likely to undergo global expansion, which would involve a larger contribution from Dow Corning and the other manufacturers, including Bristol-Myers Squibb, 3M Company, and Baxter International. In the face of the additional lawsuits by individual women and the possibility of having to ante up more money for the global settlement, Dow Corning, an equal joint venture of Dow Chemical and Corning, took action.

On May 15, 1995, Dow Corning, the nation's leading manufacturer of silicone breast implants, filed for Chapter 11 bankruptcy protection, thereby freezing all lawsuits against the company. Although the company filed for Chapter 11 protection, its spokesperson insisted that the venture's underlying business remained strong. The Chapter 11 filing allowed Dow Corning to reorganize its finances and postpone payment of its debts until completion of that reorganization.

Dow Corning hoped that bankruptcy filing, by giving it time to reorganize its finances, would enable it to take part in the overall settlement of claims. It insisted, however, that the implants were safe and that it would not stop making them. Richard Hazelton, chairman and chief executive officer of Dow Corning, stated that the company would "continue to supply products to our customers, and compensate our suppliers and employees."

Consequences

By September, 1995, the $4.2 billion settlement had collapsed. Many women accused Dow Corning of seeking to dilute bad publicity by stalling the lawsuits and settlement process and trying to force those women desperate for the money to settle quickly rather than risk a lengthy period of delay.

Dow Corning vehemently denied these charges. Hazelton argued that the delay was necessary to enable Dow Corning to compensate all the women fairly and that this case demonstrated the need for congressional reform of compensation litigation. The company reiterated its stand that no connection had been found between implants and the types of health problems the plaintiffs claimed.

In November, 1995, the other companies involved in the lawsuit, Baxter International, Bristol-Myers Squibb, and the 3M Company approved a smaller settlement resolving many of the cases against them. In February, 1998, Dow Corning proposed a settlement plan that was countered by the plaintiffs' plan the following month. In July, 1998, both sides agreed to a $3.2 billion settlement plan that would allow the company to begin compensating women and to start emerging from bankruptcy.

Frank A. Salamone

Supreme Court Voids State Term-Limit Laws

The U.S. Supreme Court declared unconstitutional an Arkansas state law that imposed term limits on its representatives and senators elected to the U.S. Congress.

What: Government; Law; National politics; Political reform
When: May 22, 1995
Where: Washington, D.C.
Who:
JOHN PAUL STEVENS (1920-), senior associate justice of the U.S. Supreme Court
CLARENCE THOMAS (1948-), associate justice of the U.S. Supreme Court

Term Limits

On May 22, 1995, the U.S. Supreme Court rendered its decision in *United States Term Limits v. Thornton.* In this case, the justices—the nine members of the Court—declared unconstitutional an Arkansas law that imposed term limits on individuals elected to the U.S. Congress from the state of Arkansas. This decision marked the first time that the Court passed judgment on whether term limits on federal legislators violated the U.S. Constitution.

The qualifications required by the Constitution for holding office in the U.S. Congress are minimal. Article I, section 2, clause 2, of the Constitution requires that members of the House of Representatives be twenty-five years of age, citizens of the United States for seven years, and residents of the state from which they are elected. Article I, section 3, clause 3, of the Constitution requires that members of the Senate be thirty years of age, citizens of the United States for nine years, and residents of the state from which they are elected. These are the only qualifications for holding office in the U.S. Congress specifically mentioned in the Constitution.

Term limits are limitations on the number of terms a person can hold a particular office. Article I of the U.S. Constitution places no limits on the number of terms a member of Congress may serve. In fact, the individuals who drafted the Constitution specifically rejected any term limits for members of Congress out of fear that talented members of the legislative branch would be forced from office, thus rendering the institution less effective. Nevertheless, term limits resurfaced as a political issue in the 1990's, in part because of widespread distrust of politicians. Between 1990 and 1995, twenty-two states enacted term limits on their members of the U.S. Congress. Many of these proposals limited the number of total years of service to twelve.

In 1992, voters in Arkansas approved an amendment to the state constitution to impose term limits on certain elected officials. The proposed Term Limitation Amendment prohibited the name of an otherwise-eligible candidate for Congress from appearing on the general election ballot if that candidate had previously served three terms in the House of Representatives or two terms in the Senate. Following its adoption, a state lower court declared the amendment to be in violation of Article I of the Constitution. The Arkansas supreme court agreed, noting that states possessed no authority "to change, add to, or diminish" the requirements for congressional service enumerated in the qualifications clauses of Article I. U.S. Term Limits, an organization favoring term limits for elected officials, appealed that decision to the U.S. Supreme Court.

The Case Before the Court

The question presented to the Supreme Court in *United States Term Limits v. Thornton* was straightforward: Does the Constitution prohibit states from adding to or changing the qualifications specifically stated in Article I of the Constitution? The Court answered that question affirmatively. The justices were divided five to four; five justices agreed that the states were prohibited from altering the qualifications specifically mentioned, and the other four justices did not believe that the states were prohibited from doing so.

Justice John Paul Stevens, the senior associate justice, wrote the opinion for the Court. (Typically when the Court renders a decision, a justice describes the reasons for that decision, which are listed in what becomes known as the Court's "opinion.") Associate Justices Anthony M. Kennedy, David H. Souter, Ruth Bader Ginsburg, and Stephen G. Breyer joined the Court's opinion. The majority reasoned that the Framers—the people who wrote the Constitution—decided that the qualifications for service in Congress should "be fixed in the Constitution and be uniform throughout the Nation." In other words, the Framers intended the Constitution to be the exclusive source of qualifications for members of Congress.

Allowing the states to require additional qualifications, the opinion stated, would violate the basic principle of a representative democracy—that voters should be able to choose whomever they pleased to govern them. As such, the states had no power whatsoever to add qualifications. The Court's opinion also made clear that the Constitution prohibited Congress from adopting term limits by way of federal legislation.

Justice Clarence Thomas, an associate justice, authored a dissenting opinion. Chief Justice William H. Rehnquist and Associate Justices Sandra Day O'Connor and Antonin Scalia joined Justice Thomas's dissent. The minority maintained that the qualifications clauses of Article I should be read simply as straightforward recitations of the minimum eligibility requirements that the Framers thought necessary for every member of Congress. Article I of the Constitution was silent on the matter of additional qualifications. Because the Constitution was silent, the states were not prohibited from requiring additional qualifications beyond those specified in the Constitution for age, citizenship, and residency for their members in Congress.

Consequences

As a result of the Court's decision in *United States Term Limits v. Thornton*, states do not have the constitutional power to impose term limits on their members of the U.S. Congress. Moreover, if term limits were to be adopted via legislation by the Congress, those limits would, by the same rationale, be viewed as unconstitutional. In short, if term limits for members of Congress are to be imposed, either the Constitution must be amended or the U.S. Supreme Court must reverse its decision.

Richard A. Glenn

Supreme Court Limits Affirmative Action

In Adarand Constructors v. Pena, *the Supreme Court struck down a complex web of federal laws that gave racially based preferences in federal contracting to minority-owned firms.*

What: Civil rights and liberties; Law
When: June 12, 1995
Where: Washington, D.C.
Who:
Sandra Day O'Connor (1930-), associate justice of the United States Supreme Court
Federico Pena (1947-), secretary of transportation in the Bill Clinton administration
John Paul Stevens (1920-), associate justice of the United States Supreme Court

The Decision to Limit Affirmative Action

In 1989, a division of the United States Department of Transportation awarded a highway construction project to Mountain Gravel, a Colorado firm. Two smaller firms, Adarand Constructors and Gonzales Construction Company, bid for the subcontract to build guardrails. Although Adarand submitted the low bid, the contract was awarded to Gonzales because under the federal law governing the project, the prime contractor received a bonus if subcontracts were awarded to firms certified as owned by members of minority groups. Gonzales Construction qualified under that provision. Had the minority preference not been a factor, Adarand Constructors would have received the contract. Adarand sued the Department of Transportation, claiming that the race-based presumptions of the law violated its right to the equal protection of the laws.

In deciding the case in favor of Adarand Constructors, the Supreme Court held 5 to 4 that the strictest possible judicial scrutiny of federal race-based programs is required. Justice Sandra Day O'Connor, who wrote the Court's opinion, said that the constitutional guarantee that no person will be deprived of "the equal protection of the laws" confers an individual right with which the government may not interfere. The only racial classifications that are constitutional are those that "serve a compelling governmental interest." Remedying past racial discrimination by the government might be one such governmental interest, but race-based programs must be very narrowly tailored to an acceptable purpose. The case was remanded to the lower courts for a decision on whether the federal program challenged by Adarand Constructors could meet the requirements of strict scrutiny.

Justice John Paul Stevens wrote a lengthy dissenting opinion. His opinion supported affirmative action programs. In his view, the Court can distinguish between invidious and benign racial discrimination. Programs that disadvantage the majority to remedy injuries suffered by the minority are acceptable under the Constitution. Stevens was joined in dissent by justices Ruth Bader Ginsburg, Stephen Breyer, and David Souter.

Benign Racial Discrimination?

Invidious racial discrimination was held constitutional in *Plessy v. Ferguson* in 1896. Until *Plessy* was overruled by *Brown v. Board of Education* in 1954, both the federal and state governments were allowed to establish racially segregated schools and other institutions. A new consciousness of civil rights and racial justice arose in the early 1960's. Affirmative action programs and minority set-aside provisions—which amount to official racial discrimination for "benign" or remedial purposes—were first established at the federal level in the Civil Rights Act of 1964.

Affirmative action was held constitutional in close decisions of the Supreme Court in *University of California v. Bakke* (1978) and *Fullilove v. Klutznick* (1980). In those cases, the justices failed to agree on the level of judicial "scrutiny" that race-conscious remedial programs should receive. In subsequent cases, the Court vacillated between the strictest possible examination of such programs and a lesser degree of constitutional skepticism called intermediate scrutiny. The test for strict scrutiny is so rigorous that it is almost impossible for race-conscious government programs to pass. If strict scrutiny persists as the constitutional rule, it may signal the end of affirmative action in the United States. The result would be "color-blind" policies that many constitutional scholars believe are consistent with the true intent of the equal protection clause.

Other Supreme Court decisions of the October, 1994, term gave indications that the Court of the 1990's favors such an outcome. In *Miller v. Johnson*, decided in June, 1995, a week after *Adarand Constructors v. Pena*, the Court struck down a Georgia congressional redistricting plan that deliberately established three voting districts with black majorities. Georgia had drawn what the court called "bizarre" district lines with the sole purpose of segregating voters on the basis of race. The facts showed that "race was the predominant overriding factor behind the . . . drawing [of the lines]." Because of that, the majority on the Court applied the strict scrutiny called for by *Adarand Constructors v. Pena*. Georgia's districting plan did not survive the Court's analysis, as there was no compelling governmental purpose. In *Miller v. Johnson*, the justices were divided exactly as they had been in *Adarand Constructors v. Pena*.

Consequences

As is often the case, what seem to be technical interpretive rules have massive practical and constitutional consequences. The choices before the Court amounted to whether the Constitution should be "color-blind." At issue is the constitutionality of race-conscious affirmative action programs designed to remedy the economic and educational disadvantages suffered by racial minorities in America. There is an irony to a constitutional rule that suddenly imposes a color-blind Constitution: It forbids the remediation of conditions created under a Constitution in which discrimination could take place.

Although it is said that affirmative action discriminates only against the majority, the reality is that individuals always are the victims and are deprived of the constitutional right to the equal protection of the laws. Although such programs do in some sense make amends, they are likely to engender continuing racial hostility as individuals compete for various benefits partly on the basis of race.

Robert Jacobs

AP/Wide World Photos

In 1995, University of California, Los Angeles, students protest the regents' decision to end affirmative action policies at the campus in the Westwood section of Los Angeles.

French Plans for Nuclear Tests Draw Protests

The French government announced that it would resume testing of nuclear weapons, provoking vigorous condemnation from governments, environmental groups, and peace organizations.

What: International relations; Environment; Weapons technology; Military capability
When: June 13, 1995
Where: Paris, France
Who:
JACQUES CHIRAC (1932-), president of France from 1995
JAMES BOLGER (1935-), prime minister of New Zealand from 1990 to 1997
PAUL KEATING (1944-), prime minister of Australia from 1991 to 1996

The Announcement to Resume Testing

On June 13, 1995, Jacques Chirac, the newly elected president of France, announced that his country would conduct a series of eight underground nuclear tests in the South Pacific between September, 1995, and May, 1996. As justification, he stated that these tests were necessary to ensure the safety and reliability of France's nuclear weapons and to enable the country's scientists to calibrate computer instruments properly so that future tests could be done by laboratory simulation. In the following weeks, French officials stressed that once the tests were completed, the government would sign both the Comprehensive Test Ban Treaty and the Nuclear Non-Proliferation Treaty.

Although the announcement was not completely unexpected, it resulted in furious worldwide protests. Because the testing site was in French Polynesia, 750 miles southeast of Tahiti, it was understandable that the reaction in Australasia was strongest. On June 17, Paul Keating, the prime minister of Australia, called the French action "environmental vandalism"; the prime minister of New Zealand, James Bolger, called it the arrogant action of a European colonial power. The same day, the French consulate in Perth, Australia, was firebombed. Later in June, Gareth Evans, the Australian foreign minister, led delegates of South Pacific nations to Paris to protest against the planned tests. The environmental organization Greenpeace sent its ship *Rainbow Warrior II* into the testing area, drawing unfavorable attention to the tests.

A number of European nations as well as the United States expressed regret over the French decision; only Great Britain tacitly supported the French. When Chirac addressed the European Parliament in Strasbourg, he was jeered. One deputy called him a "neo-Gaullist Rambo." In November, the United Nations voted ninety-five to two to condemn the French action. Chirac canceled scheduled meetings with Italian and Belgian leaders because of the strength of criticism emanating from those two countries. Even in France, public opinion polls showed a majority against the decision.

The Post-Cold War Era

The two central questions in this drama were why Chirac ordered the tests and why reaction was so bitter. Aside from Chirac's stated reasons, there were other factors influencing him. First, there was his political philosophy. He believed that he was a worthy inheritor of the Gaullist tradition in France. Chirac had refounded the Gaullist movement after the death of former president Charles de Gaulle, creating the modern Rassemblement pour la République party and becoming its chief in 1976. The Gaullist tradition decreed that French presidents must be strong and decisive, that France should not rely on the United States for national security, and that the country must be a major player on the international stage and pursue its destiny, irrespective of world opinion.

Second, Chirac was squeezed by an impending deadline. If France was to sign the Comprehensive Test Ban Treaty in 1996, then any tests would have to be undertaken almost immediately. Third, Chirac argued that these tests were done in the name of Europe. During the Cold War, France had been the only independent Western European power possessing such weapons, Great Britain being essentially under America's nuclear umbrella. Without the French deterrent, there was the danger that in the future, Europe would have no credible independent response to the blackmail of another nuclear power.

As to why the reaction was so hostile, it was a case of different forces, interests, and groups converging. In the post-Cold War era, nuclear weapons had lost their legitimacy. A strong antinuclear mentality had set in, and in response the major nuclear powers, with the exception of China, had instituted a moratorium on tests in 1992. France not only violated this popular moratorium but also did so near the time when the world solemnly commemorated the fiftieth anniversary of the dropping of the atomic bomb on Hiroshima. After the Cold War, nuclear weapons appeared to be inappropriate. Military theorists placed greater emphasis on creating rapid-response mobile forces using conventional weaponry in order to fight regional conflicts or carry out United Nations peacekeeping missions.

The tests also aroused the ire of environmentalists, a well-funded and influential lobby in many countries. To neutralize criticism, the French government pointed out that recent independent and respected scientific studies had showed that radiation levels at the test site posed no serious health risk. Environmentalists were concerned about the long-term ecological impact, especially possible seepage of radioactive materials into the ocean. They also questioned why, if there were no environmental dangers, France conducted these tests ten thousand miles away.

France also was accused of acting as a colonial power in an age when European imperialism no longer could be justified morally. To test weapons of such awesome power and jeopardize the ecology of the South Pacific was tantamount to saying that the peoples of these areas were less worthy of human respect than were Europeans. The argument was made that French Polynesia should be given its independence so that it no longer could be used as a testing ground. To some fervid anticolonialists, French indifference to the sensitivities of Polynesians conjured up parallels with the American decision during World War II to drop atomic bombs on people considered to be racially inferior.

Consequences

There is little doubt that Chirac's popularity took a severe blow within France as a result of these tests and that the testing issue reduced his credibility and authority in tackling the country's massive domestic problems. On the international stage, France's image suffered badly, not only in the South Pacific but also in Europe, where France believed its mission was to be a guiding force in the European Union, particularly in defense matters. After having conducted four underground tests, the French government made the dramatic announcement on December 6, 1995, that it would end its testing program by the end of February in the following year, not in May as had been planned. World opinion played a large role in this decision.

David C. Lukowitz

Shuttle *Atlantis* Docks with Space Station Mir

U.S. astronauts aboard the shuttle Atlantis *docked with the space station Mir on a mission that set the stage for future rendezvous and construction of an international space station.*

What: Technology; International relations
When: June 27-July 7, 1995
Where: Earth orbit, United States, and Russia
Who:
Robert Gibson (1946-), space shuttle commander
Charlie Precourt (1955-), space shuttle pilot
Ellen Baker (1953-), mission specialist-1
Greg Harbaugh (1956-), mission specialist-2
Bonnie Dunbar (1949-), mission specialist-3
Norm Thagard (1943-), American astronaut on Mir
Vladimir Dezhurov (1962-), Mir-18 commander
Gennady Strekalov (1940-), Mir-18 flight engineer
Anatoly Soloyvev (1948-), Mir-19 commander
Nikolai Budarin (1953-), Mir-19 flight engineer

Atlantis Docks with Mir

The space shuttle *Atlantis* lifted off on the STS-71 mission on June 27, 1995, with seven astronauts and two replacement cosmonauts (the crew of the Mir-19 mission). Two days later, shuttle commander Robert Gibson positioned *Atlantis* below space station Mir and guided the shuttle into the docking mechanism of the Russian space station. This unique approach was designed to reduce the number of firings of the shuttle's engines to avoid any damage to Mir's solar panels. The two spacecraft sailed in Earth orbit, 245 miles above Central Asia, at a speed of about 17,500 miles per hour. The 100-ton *Atlantis* and 123-ton Mir together formed the largest human-made object to orbit Earth to that time.

Atlantis was modified to carry a docking system compatible with the Russian space station. Two hours after the linkup, Commander Gibson and cosmonauts Anatoly Soloyvev and Nikolai Budarin traveled through a three-foot-long pressurized tunnel, known as the Androgynous Peripheral Docking System (APDS), to Mir. The orbiter docking system, weighing about thirty-five hundred pounds, consisted of an airlock, a supporting truss structure, a docking base, and the APDS; it cost $95.2 million to build.

Gibson and Mir commander Vladimir Dezhurov met in the tunnel and shook hands as a symbol of friendship between American and Russian people. All the astronauts and cosmonauts later gathered in Mir for a group photograph. The two spacecraft circled Earth together seventy-seven times during the five-day docking.

Physical examinations were conducted on American astronaut Norm Thagard and Russian cosmonauts Vladimir Dezhurov and Gennady Strekalov of the Mir-18 mission. They had been in space for more than one hundred days. The medical tests were intended to determine the effects of weightlessness on the cardiovascular system, bones and muscles, and the immune and cardiopulmonary systems. Scientists from the National Aeronautics and Space Administration (NASA) also wanted to study the astronauts' mental health to see how well they had adapted to living in a highly contained environment for so long.

Thagard and the crew of Mir-18 transferred to *Atlantis* and joined the crew of STS-71 for the return to Earth. The crew of Mir-19 remained on board the Russian space station. The space shuttle returned to Earth and landed at the Kennedy Space Center in Florida on July 7, 1995.

The Space Race

A competition for military dominance in space began in 1957 with the launch of the Soviet satellite Sputnik 1. The United States formed NASA, and astronauts were selected from the United States military services. Astronauts in the United States and cosmonauts in the Soviet Union frantically trained during the 1960's to be the first to reach the moon. After the successful Apollo 11 moon landing, the Soviet Union withdrew from attempts to make a moon mission and focused its efforts on working in Earth orbit.

In the early 1970's, the United States redirected its human space missions to research in Earth orbit. Skylab, the first American space station, was launched into orbit in 1973. Modified from the third stage of a Saturn V rocket, Skylab included living and working quarters. Astronauts conducted experiments using the microgravity environment and proved that people could work successfully for long periods in space. Skylab fell to Earth on July 11, 1979.

In an effort to promote long-term cooperation between the United States and the Soviet Union, astronauts aboard an Apollo capsule and cosmonauts in a Soyuz capsule docked in Earth orbit on July 17, 1975, in the highly publicized Apollo-Soyuz Test Project (ASTP). The American spacecraft was a modified version of the Command and Service Module flown during the first several lunar landing missions. The Soviet Soyuz spacecraft was modified to use a compatible rendezvous and docking system that NASA and Soviet engineers designed. A major publicized goal of the mission was the testing of jointly designed docking equipment and techniques for conducting international scientific missions and emergency rescue operations. It was planned that subsequent to ASTP, all American and Russian spacecraft would be fitted with the docking equipment to allow the craft to dock in the future. The mission's main purpose was political, and it did not unify the two countries' space programs.

AP/Wide World Photos

Atlantis *commander Robert Gibson (right) shakes hands with Mir commander Vladimir Dezhurov after opening the hatch connecting the spacecrafts.*

During the 1970's, NASA began tests of a reusable space vehicle, the space shuttle. The first orbital test flight of the space shuttle, STS-1, was launched on April 12, 1981. Since then, astronauts have used the space shuttle to conduct experiments and launch such satellites as the Hubble Space Telescope.

The Soviet Union launched the core module of its space station, Mir (the Russian word for peace), into orbit in 1986. Robotic resupply vehicles took food, propellant, and supplies to people in Mir.

Consequences

After the loss of Skylab, NASA developed plans for a new American space station. Economic pressures from Congress threatened to cancel the U.S. space station completely in the 1990's. Relations between the United States and the Soviet Union improved, and following the disintegration of the Soviet Union, President Bill Clinton signed an agreement to pay Russia to work on constructing an international space station. The strategy was for both countries to work together and combine their engineering skills to build a new international space station in orbit around Earth.

The docking of *Atlantis* and Mir marked an end of competition for primacy in space and the start of a cooperative space effort involving the United States and Russia. The STS-71 mission was the first of several Space Shuttle/Mir linkups designed to give American and Russian crews experience in working together in space. The first module of the International Space Station was launched in 2001.

Noreen A. Grice

United States Establishes Full Diplomatic Relations with Vietnam

U.S. normalization of diplomatic relations with the Socialist Republic of Vietnam was achieved two decades after U.S. forces admitted defeat in the Vietnam War and typified a post-Cold War emphasis on the global economy.

What: International relations
When: July 11, 1995
Where: Vietnam
Who:
Bill Clinton (1946-   ), president of the United States from 1993 to 2001
Vo Van Kiet (1922-   ), prime minister of the Socialist Republic of Vietnam from 1992 to 1997

Twenty Years of Controversy

On April 30, 1975, the U.S.-backed government in South Vietnam, called the Republic of Vietnam, fell to the forces of the Hanoi-based Democratic Republic of Vietnam, and the country was subsequently reunified as the Socialist Republic of Vietnam, a government with which the United States never had diplomatic relations. The United States imposed a trade embargo on the reunified nation of Vietnam that remained in effect for nineteen years.

In the 1970's, Vietnam had called for the establishment of diplomatic relations, requesting that the United States fulfill the provisions of the Paris Peace Agreement of 1973, under which the United States pledged postwar aid for Vietnam's reconstruction. President Gerald R. Ford rejected this request, saying there could be no normalization of relations without a full accounting of American prisoners of war (POWs) or soldier missing in action (MIAs). It is interesting to note, however, that no full accounting of the 8,177 U.S. MIAs from the Korean War or the 78,751 U.S. MIAs from World War II or the 200,000 to 300,000 Vietnamese MIAs has ever been made. President Bill Clinton acknowledged in his July 11, 1995, speech that "Never before in the history of warfare has such an extensive effort been made to resolve the fate of soldiers who did not return."

The United States vetoed Vietnam's application for membership in the United Nations three times in 1975-1976; Vietnam was admitted in 1977, during the administration of President Jimmy Carter. Although Carter proposed establishing diplomatic relations, Congress opposed any movement until Vietnam agreed to aid in accounting for MIAs, and Vietnam continued to demand reparations. No further progress was then made. The administration of President Ronald Reagan opposed normalization of relations unless Vietnamese forces withdrew from Cambodia. In 1990, the administration of President George Bush ended U.S. support for the Cambodian coalition government (which included the Khmer Rouge) and began peace negotiations with the Vietnamese-backed government in Pnomh Penh.

The Road to Normalization

Vietnam made efforts to comply with the United States' demand on POW/MIAs. Between 1974 and 1992, it repatriated the remains of three hundred Americans. In April, 1991, the Bush administration laid out a four-phase "road map" toward normalization of relations. An office to handle POW/MIA affairs was established in Hanoi, and the United States sent $1 million of humanitarian aid, mostly in the form of prosthetics, to Vietnam. Travel restrictions to Vietnam were eased. In 1993, the Clinton administration endorsed the Bush "road map." On Me-

AP/Wide World Photos

After announcing the restoration of diplomatic relations with Vietnam, President Bill Clinton meets with Senator John McCain, a former prisoner of war in Vietnam who led the effort for the normalization of relations.

morial Day, 1993, a congressional delegation led by Senator John Kerry, a veteran of the war, traveled to Vietnam; upon their return, they advocated further steps toward normalization.

International pressure had been brought to bear on the United States by France, Japan, and other nations to ease the trade embargo. On July 2, 1993, Clinton announced that the United States would not oppose financial aid to Vietnam from other nations. In December, 1993, he further eased the embargo by allowing U.S. companies to bid on development projects sponsored by international financial institutions. Finally, in February, 1994, Clinton lifted the embargo. After the congressional vote of confirmation in April, 1994, the United States and Vietnam established liaison offices in Washington, D.C., and Hanoi.

On July 11, 1995, a little more than twenty years after the end of the war, Vietnam and the United States established full diplomatic relations.

Consequences

The decision was controversial in the United States. Senator Bob Dole, a celebrated veteran of World War II, denounced the announcement on the grounds that the question of whether Vietnam was withholding information concerning U.S. MIAs was unresolved. The decision was applauded by others, including Senator Bob Kerrey, a veteran who had lost a leg and won a Medal of Honor, and Senator John McCain, a former Vietnam POW.

U.S. businesses were eager for Vietnam to be granted most-favored-nation status, thus reducing tariffs. Some Vietnamese Americans disapproved of U.S. recognition of Vietnam's Communist government; others welcomed the move as long overdue.

On May 23, 1995, President Clinton announced the appointment of Congressman Pete Peterson, a retired U.S. Air Force colonel and former POW, as ambassador to Vietnam. After 1995, cooperation between the United States and Vietnam expanded in U.S. aid in combating health threats in Vietnam, cooperation in science and technology, disaster relief, and labor protection. The United States has been engaging in the humanitarian project of removing or deactivating land mines in Vietnam. A bilateral trade agreement was signed on July 13, 2000; in the announcement of this agreement, the U.S. government noted that exports to Vietnam had increased from $4 million in 1992 to $291 million in 1999.

Renny Christopher

Disney Company Announces Plans to Acquire Capital Cities/ABC

In a surprise announcement, Disney declared that it would spend $19 billion to purchase Capital Cities/ABC to gain television access for Disney products.

What: Business; Entertainment
When: July 31, 1995
Where: Burbank, California
Who:
Michael Eisner (1942-), chairman/ chief executive officer Disney
Thomas Murphy (1925-), chairman/ chief executive officer Capital Cities/ ABC
Stephen Bollenbach (1942-), senior executive vice president/chief financial officer Disney
Sanford Litvack (1936-), senior executive vice president/chief operating officer Disney

The Mouse that Roared

The company that Walter E. Disney started in 1923 grew to be one of the entertainment industry's giants. The debut of Mickey Mouse, Disney's animated character, in 1928 had a dramatic effect on the growth of the fledgling movie studio. The Walt Disney Company, headquartered in Burbank, California, is known worldwide for producing family entertainment in a number of venues. It employs more than 65,000 people and engages in film production, theme park management, character merchandising, and consumer product marketing. In 1994, the company had record revenues of $10.1 billion, of which almost half came from film production and distribution.

Much of the credit for the expansion of Disney is given to Michael Eisner, who joined Disney in 1984 as chairman/CEO (chief executive officer). Eisner had held a variety of executive positions at ABC (American Broadcasting Company) over a ten-year period before moving on to Paramount and finally to Disney in 1984. During his tenure at Disney, revenues had grown sevenfold and net income had increased elevenfold.

Disney's expansion occurred both within the company's areas of expertise as well as in new areas of the entertainment industry. Walt Disney Studios expanded. The company created Buena Vista Pictures and Touchstone Pictures, as well as Hollywood and Miramax Pictures. In addition to its theme parks in California and Florida, the company became involved in new entertainment-oriented ventures that included Disney Cruises, Disney Stores, several record companies, a book publishing company, and several professional sports teams.

Despite all this, what Disney did not have was television access. The growth in the number of households subscribing to cable and satellite services had dramatically increased the market for cable and made-for-television movies. However, Disney could not guarantee access to this market for its products because it had no control over whether television stations purchased or aired its movies and programs. Regarding this issue, Eisner said, "We have to guarantee access for the mouse."

Capital Cities/ABC was a corporate giant in the television industry. It owned not only ABC but also had television access through its 225 affiliated broadcast stations. Capital Cities/ABC also owned a number of cable distribution stations, including Eastern Sports Programming Network (ESPN), Entertainment (E!), Arts and Entertainment (A&E), the History Channel, and the Lifetime Channel. Disney's purchase of this group would guarantee access for its products on television not only at the broadcast level but also through cable distributions.

A Friendly Transition

Thomas Murphy, the first head of Capital Cities, played an important role in negotiating the purchase of his company by Disney. After a period of very secret discussions, the two principals agreed to a deal. Eisner and Murphy agreed that Disney would pay $19 billion for Capital Cities/ABC. Because an agreement on the cash or stock portion of the deal could not be reached, the two sides agreed to a compromise: Disney would pay Capital Cities/ABC shareholders sixty-five dollars per share, or about $10 billion in cash, for their stock. Each stockholder would also be given one share of Disney stock to replace each of his or her shares of Capital Cities stock, for a total of about $9 billion.

Aside from Eisner, other important members of the Disney team included Stephen F. Bollenbach, senior executive vice president and CFO (chief financial officer), and Sanford M. Litvack, senior executive vice president and COO (chief operating officer). Bollenbach had joined Disney only months before the merger after leaving Host Marriott where he had been CEO. Litvack had been at Disney since 1991 as general counsel (attorney).

Following the purchase of Capital Cities/ABC by Disney, it was clear that Murphy would step down and Eisner, Bollenbach, and Litvack would run the organization. However, before the sale, many people were surprised that Disney did not pursue either Columbia Broadcasting System (CBS) or National Broadcasting Company (NBC). Both of these companies would have given Disney television access at a fraction of the price paid for ABC.

Consequences

The purchase by Disney of Capital Cities/ABC created a very large and powerful force in the entertainment industry. Disney had always espoused "family values" in its movies and had been a source of quality entertainment in its theme parks and other venues. One of the first indications of the incorporation of ABC into the Disney family was the return of *The Wonderful*

AP/Wide World Photos

Disney's Michael Eisner (left) shakes hands with Thomas Murphy of Capital Cities/ABC.

World of Disney to ABC on Sunday evenings. A second indication was the creation of the Disney Channel, a cable station on which movies and television shows of Disney origin aired continuously.

Apparently, however, the crown jewel of the purchase was not on the entertainment side but rather on the sports side of the house. ABC sports continued to air *Monday Night Football* as well as *ABC's Wide World of Sports.* At the same time, ESPN quickly expanded into ESPN2, ESPN Classic, and the Classic Sports Channel. ABC and ESPN and its derivatives combined to create a powerful force in worldwide sports programming. Capital Cities estimated that the ESPN network was seen in 66.3 million American households and 95 million households around the world.

ABC and ESPN advertised programming on each other's stations and brought unlikely sports such as yacht racing and ice curling to the mainstream of American households. The hub of ESPN turned out to be Sports Center, a nightly hour-long sports news, scores, and highlights show. Because of its news-show format, Sports Center covers all the day's sports, even events not covered by ABC and ESPN, using video tape obtained from other stations. Through this program, ESPN avoids incurring the costs of reporting an event but still provides coverage of it.

ABC television stations have access to more than ninety-three million television-equipped homes in the United States, thus giving Disney's programming, television movies, and animated films a guaranteed market.

Robert J. Stewart

INS Frees Thai Workers Held Captive in Clothes Factory

Officers of the U.S. Immigration and Naturalization Service raided a factory in El Monte, California, where seventy-two Thai workers were held in slavelike conditions by their employers.

What: Human rights; Law
When: August 2, 1995
Where: El Monte, California, about 12 miles east of downtown Los Angeles
Who:
Robert B. Reich (1946-), U.S. secretary of labor from 1993 to 1997
Virginia Bradshaw, California labor commissioner
Rojana Cheunchujit, a Thai garment worker, spokesperson for the other workers

Garment Workers Freed

Before dawn on August 2, 1995, U.S. Immigration and Naturalization Service (INS) officials staged a raid on a garment factory in El Monte, California. The factory, surrounded by barbed wire, held seventy-two workers from Thailand, some of whom had lived and worked in the factory for years. When the workers first arrived in the United States, their employers took them from the airport directly to the factory. Each night, the workers, who sewed clothing that was sold under major brand names, were locked up and guarded. They worked from 7:00 A.M. to midnight every day, for $1.60 per hour, with no extra pay for working more than forty hours a week. Most of this pay was withheld by their employers as repayment for transportation costs to the United States. Factory owners often held children of the workers hostage to force the adults to keep working. The employers also threatened to beat workers who tried to escape.

Held Captive

Neighbors thought that the high walls and barbed wire surrounding the El Monte factory had been put in place to keep criminals out, not to keep workers in. Immigration officers had been suspicious for a long time, however, and in 1992, the INS had sought a warrant (legal permission to conduct a search) to search the building. On that first occasion, federal prosecutors refused to grant the warrant, saying that the evidence of wrong-doing was not sufficient.

By the time INS officers gained legal permission to stage their raid on the factory, some of the workers had been imprisoned for as long as seven years. The operation began in the late 1980's when the Manasurangkun brothers from Thailand, Wirachai, Phanasak, and Surachai, together with their mother, Suni, joined with three other Thai people to recruit poor women in their native land. By bringing these women to the United States, the Manasurangkuns and their partners could get inexpensive labor to sew clothing for name-brand manufacturers. Over time, the treatment of the workers grew increasingly harsh, and the Manasurangkuns hired guards to keep them from escaping. According to Rojana Cheunchujit, a worker who spoke English and came to serve as a spokesperson for the others, the Thai women had to work sixteen hours a day and sleep on a dirty floor with cockroaches and mice. Two women who tried to escape were beaten and sent back to Thailand.

Consequences

The case of the Thai workers in El Monte helped call attention to the plight of garment workers in the United States. Since the 1960's, the sewing of clothing has moved away from large factories and toward small producers who supply large retail stores with a variety of clothes designed to appeal to consumers with varied tastes. These large retail stores are relatively few in number and control much of the American market. To make a profit, clothing manufacturers have had to keep their costs down because the retail stores want to supply customers with fairly inexpensive clothes. The clothing manufacturers compete with each other to make garments as cheaply as possible, and the manufacturers therefore try to find the cheapest workers they can. Because immigrants, especially those in the country illegally, will work for lower wages than other people in the United States, by the 1990's, a majority of garment workers were immigrant women.

The slavelike conditions found at the El Monte factory are rare in the United States. Still, many garment workers labor in difficult and often illegal circumstances. For example, a 1994 investigation by California labor officials looked into the operations of sixty-nine randomly selected manufacturers. All but two of these manufacturers were found to be breaking federal or state laws or both. Half of them were violating minimum wage laws, 68 percent were violating laws regarding overtime, and 93 percent were violating health and safety regulations.

The publicity created by the raid at El Monte led to an investigation of the clothing industry by the U.S. Labor Department within two weeks after the incident. The Labor Department warned more than a dozen of the largest U.S. retail merchants that they may have received goods made by the Thai workers. Labor Secretary Robert B. Reich called a meeting with the retailers to discuss ways to avoid selling goods made by enslaved workers.

In California, within two weeks of the raid, the California Labor Department demanded business records from sixteen garment makers believed to have had connections with the El Monte factory. California labor commissioner Virginia Bradshaw found that many of the manufacturers who did business with the El Monte factory were themselves engaging in illegal activities and the

AP/Wide World Photos

Razor wire helped keep Thai immigrants inside the apartment complex where they lived as slave laborers.

California Labor Department fined several of them $35,000 each for failing to register their operations with the state.

In late September, 1999, the California State Assembly passed Assembly Bill 633, a law designed to crack down on clothing sweatshops, businesses employing workers to make clothes under unfair and illegal conditions. Cheunchujit testified before the assembly when it was considering the law.

The workers also sued the companies that hired the El Monte factory to make clothes. In July, 1999, their attorneys agreed with these companies that the workers would be paid $1.2 million for back wages and damages. Under the agreement, the workers would receive $10,000 to $80,000 each, depending on how many years they had been forced to work in the factory. The Manasurangkuns pleaded guilty to charges of smuggling the workers into the United States and keeping them in slavelike conditions. The four family members and three other Thai people who worked with them were sentenced to prison terms.

Carl L. Bankston III

Principal *Roe v. Wade* Figure Takes Stand Against Abortion

Norma McCorvey, identified as "Jane Roe" in the landmark Roe v. Wade *1973 abortion lawsuit, announced that a newfound faith in Christianity had caused her to oppose abortions.*

What: Gender issues; Human rights; Social reform
When: August 10, 1995
Where: Dallas, Texas
Who:
NORMA MCCORVEY (1947-), the original plaintiff in the *Roe v. Wade* 1973 abortion lawsuit

Choices

Norma McCorvey, the "Jane Roe" of the 1973 *Roe v. Wade* U.S. Supreme Court case that resulted in the legalization of abortion in the United States, made history again on August 10, 1995, when she publicly stated that her newfound faith in Christianity would no longer allow her to support legalized abortion. McCorvey, who was baptized two days before the announcement, initially stated that she still supported first trimester abortions under certain conditions, but within three weeks, she changed her position to "100 percent prolife, no exceptions." McCorvey's involvement in the abortion issue began in 1970 when she was the lead plaintiff in a class-action lawsuit filed to challenge the strict antiabortion laws in Texas.

The case was appealed to the U.S. Supreme Court, which handed down its controversial decision on January 22, 1973, by a vote of 7-2. The decision of the Court legalized a woman's right to an abortion under certain conditions in all fifty states and has been a controversial sociopolitical issue since that time. As mandated by the Court, during the first three months (trimester) of pregnancy, the decision to abort lies with the woman and her doctor. During the second trimester, the right to an abortion still exists, but each state may regulate the procedure in ways reasonably related to protecting the woman's health (such as requiring that it be done at a hospital). During the last three months of pregnancy, if the fetus is judged capable of surviving if born (the concept of viability), any state may regulate or prohibit abortion except when the abortion is necessary to preserve the life or health of the mother. If a pregnancy were terminated during the third trimester, a viable fetus would not be allowed to die.

When the case was filed, McCorvey was twenty-one years old and pregnant for the third time. She did not have an abortion, nor had she had abortions in her first two pregnancies. All three of the children she delivered were given up for adoption. In her two books, McCorvey has described herself as the child of an impoverished Louisiana home who suffered emotional and physical abuse as a child, a runaway at age ten, a student at reform schools, a ninth-grade dropout, and a rape victim in her teens. She was beaten by a man she married at age sixteen and had abused alcohol and other drugs for most of her life.

Conversion

McCorvey's life and beliefs about abortion were changed while working at Choice for Women, a Dallas, Texas, medical clinic that performed abortions. An antiabortion group called Operation Rescue moved into an office next door. Because of the close proximity of the offices and the controversial purpose for which each stood, there were daily crowds, confusion, picketing, and demonstrations. McCorvey had gone public with her identity in the 1980's, published her first book, and had made many appearances with the media and at proabortion gatherings. Therefore, she was often recognized when she went outside the clinic to smoke a cigarette and sometimes participated in rude verbal exchanges with opponents. The office staff next door made a special effort to be polite to McCorvey, and over time, she began to communicate with the staff and form friendships. Most notably, she became friends with the Reverend Phillip Benham, Operation Rescue's national director, and the seven-year-old daughter of Operation Rescue's office manager. One day at work, McCorvey accepted an invitation from the girl to attend the Hillcrest Christian Church and that same night at the service converted to Christianity. On August 8, 1995, McCorvey was baptized by Reverend Benham in a swimming pool at a Dallas home, and a film of the event was shared with media and news organizations worldwide over the next several days.

During the 1995 news event, McCorvey explained that her philosophical views of abortion had gradually changed over the years as a result of much contemplation and a few specific factors, which she identified. One was that in the 1980's when she wrote her first book and went public with her identity, she became a celebrity of sorts, being asked to make personal appearances at proabortion rallies and being sought by the media for interviews. On some of these occasions, she felt that she was treated disrespectfully by intellectuals, feminist leaders, and activists of women's rights organizations. She believed the poor treatment was because of her lack of education and poor socioeconomic status.

Another reason McCorvey gave for her conversion was that she believed that the doctor she worked for at the Dallas clinic and other physicians were unethical in many ways. Some, she said, placed financial gain ahead of the welfare of the women patients. Specifically, McCorvey also accused some of approving abortions in violation of the *Roe v. Wade* and states' legal standards, performing fake abortions on women who were not pregnant, ignoring followup medical care standards, and providing limited pre- and post-abortion counseling of the women patients.

McCorvey also said that because she had worked at several abortion clinics over the years but never aborted any of her own pregnancies that she could judge more objectively than most the emotional scars of women before and after abortion procedures. She stated her belief that many women are coerced and threatened into having abortions by boyfriends and husbands. She also said that she believed that the post-traumatic stress that women experience is greater than professional medical and psychological associations acknowledge.

Consequences

McCorvey became the head of her own ministry called Roe No More, whose primary mission is counseling for women who have had abortions in the Dallas, Texas, area. On August 17, 1998, she was formally received into the Roman Catholic Church.

Through 2001, the January 22, 1973, Supreme Court decision legalizing a woman's right to an abortion under specific conditions had not been reversed or otherwise nullified, although the abortion issue remained controversial.

Alan P. Peterson

Microsoft Releases Windows 95

At 12:01 A.M. on August 24, 1995, the first copy of Microsoft Windows 95 was sold. Windows 95, which made using an Intel personal computer easy and intuitive, became the operating system of choice for personal computers and one of the most successful software products ever developed.

What: Business; Computer science; Technology
When: August 24, 1995
Where: Microsoft Corporation, Redmond, Washington
Who:
Bill Gates (1955-), Microsoft cofounder and chief executive officer
Steve Ballmer (1956-), Microsoft vice president in charge of marketing

Making Computers Easy to Use

The earliest computers were little more than electronic calculators. Data and the instructions to manipulate the computers were entered by toggling switches on the front panel. The output of a program was obtained by viewing lights on the front panel. Clearly, these early computers were not easy to use. Hardware and software for computers improved over the years. One of the major improvements in software was the development of the operating system—the software that controls how one inputs data, gets output, starts programs, and manages files.

In the 1980's, some of the mainframe computers and all of the new personal computers introduced operating systems that supported interactive use. Data was entered into the computer by using a keyboard and the results of a program were viewed on a video display screen. Microsoft developed an interactive operating system, called the Disk Operating System (DOS), which became the most popular personal computer operating system of the 1980's. Operating systems research at the Xerox Palo Alto labs during this period resulted in the development of a new computer, the Xerox Star, with a Graphical User Interface (GUI). The Star displayed its output in overlapping windows on the computer screen and got its input from a mouse as well as a keyboard. The Star was the inspiration for both the Apple Macintosh and Microsoft Windows.

Windows 3.1 popularized the GUI for both home and business users. Windows 3.1 provided a program manager that made it easy to keep track of all applications and a file manager that supported moving, copying, and deleting files. Windows 3.1 made it reasonably easy for someone to use a computer. In spite of the success of Windows 3.1 as an operating system, there were a number of problems. For example, Windows 3.1 had been added to the earlier DOS operating system and had numerous problems properly integrating with DOS. Another problem was its lack of support for networking. Vendors created multiple networking applications for Windows 3.1, but these applications often failed to work. Rather than simply trying to fix Windows 3.1, Microsoft decided to develop a completely new operating system, Windows 95, which would be reliable and easy to use.

The Introduction of Windows 95

Initially called Chicago, the successor to Windows 3.1 was named Windows 95 as the product moved to market. Microsoft spent millions of dollars on the development of Windows 95. Much of the basic technology of Windows 95 existed in other operating systems, but the integration of this technology into a single working product was a major technical achievement. The most obvious feature of Windows 95 is its graphical user interface. To start a program, you simply click the start button and select the application you want from the program menu. If you use an application regularly, you can drop an icon representing the application on your desktop. If you want to create a folder to hold some of your word processing files, you launch the Windows Explorer and a few mouse clicks later type in the name of your new folder. By selecting the appropriate object in the Windows Explorer and clicking the mouse a few times, you can format a floppy disk or delete a file.

AP/Wide World Photos

Microsoft chairman Bill Gates sits on the stage during an event to launch Windows 95.

In addition to the friendly GUI of Windows 95, there were a number of substantial changes made to the operating system itself. The kernel—the basic process, memory, and input/output management systems of an operating system—of Windows 95 was much more stable and efficient than that of Windows 3.1. Better support was provided for networking, printing, and multimedia. Support for reading and writing to compact discs (CDs) was provided as a part of the operating system. A number of useful utilities and games, including a pinball game, were provided with Windows 95. One of the most important additions to Windows 95 was that of Microsoft's Internet browser called Internet Explorer.

Consequences

Windows 95 is a much more reliable and stable operating system than Windows 3.1. Most agree that Windows 95 is the most successful software product ever sold. By the first decade of the twenty-first century, more than 90 percent of the desktop computer operating systems were versions of Windows with a GUI based on Windows 95 and its successors. This domination of the desktop personal computer market, combined with Microsoft's close working relationship with other software developers, had resulted in the creation of an enormous amount of high-quality, low-cost software. It is impossible to overstate the importance of Windows 95 in setting the standard for GUIs for personal computers. Windows 95 has made personal computers easy to use.

With the development of Windows 95, Microsoft adopted the position that new technologies should be incorporated in the operating system rather than added by other vendors. In Windows 3.1, third-party vendors created key parts of the memory management subsystem, but Windows 95 included all of the technology needed for memory management. Windows 95 also incorporated a full set of networking utilities, eliminating the need for most third-party networking products. Although some companies were forced to close because their products were no longer needed, others were created to add new software for the user-friendly Windows 95 environment. The decision by Microsoft to aggressively incorporate new technologies into its Windows operating systems continues to have a major impact on the computing industry.

George Martin Whitson III

Beijing Hosts U.N. Fourth World Conference on Women

About thirty-five thousand delegates from more than 180 countries attended the fourth in a series of United Nations governmental conferences on the economic, social, and political condition of women.

What: Human rights; International relations
When: September 4-15, 1995
Where: Beijing and Huairou, China
Who:
Hillary Rodham Clinton (1947-   ), first lady of the United States from 1993 to 2001
Deng Xiaoping (1904-1997), dominant figure in Chinese politics since 1978
Benazir Bhutto (1953-   ), prime minister of Pakistan from 1988 to 1990 and from 1993 to 1996

Defining Women's Rights

In the largest international assembly to date, thirty-five thousand delegates gathered in China at two parallel United Nations (U.N.) conferences. A governmental conference took center stage in Beijing, while a parallel conference of nongovernmental organizations (NGOs) was held fifty miles away in Huairou. Both assemblies met to discuss the social, economic, and political condition of women around the world. The governmental conference developed a *Platform for Action* that reached 150 pages in length.

Debate at the governmental conference resulted in several fissures among the delegations. On matters of sexuality and family law, delegations backed by religious fundamentalists (including Protestants, Orthodox, Catholics, Muslims, and Jews) lined up against more secular-minded governments, principally but not exclusively from the West. Delegations from Iran, Sudan, and other fundamentalist Muslim regimes worked closely with the Vatican and Catholic-minded governments, including those of Argentina, Honduras, Guatemala, and Ecuador. Together, they sought to block resolutions put forward by Western delegations and more liberal Third World allies on issues of sexual freedom, contraception, abortion, and definition of the role of women both in and outside the family.

For example, a compromise substituted the word "equity" for "equality" in regard to women's status in relation to men. This appeased Muslim governments in Iran and Sudan, which argued that the Koran states a moral obligation to treat women unequally on issues such as inheritance, the value of court testimony, divorce, and employment law. Not all Muslim countries took this position, however; Tunisia and Morocco usually sided with the liberal bloc.

Another major fissure was between the Vatican and the sixteen nations of the European Union, concerning the definition of family rights and the issue of sexual orientation. Finally, most Western governments balked at language implying that they would have to pay to promote women's social and economic advance globally. Many Third World delegations pushed for such language on financial obligations even as they resisted real change in their own social policies.

As is often the case at large international conferences, most issues were settled by achieving paper agreement on vague generalizations that did not imply any real commitment to change. Where irreconcilable differences remained, either no agreement was reached or individual states announced formal reservations on the obligation to promote the goals and ideals set out in the *Platform for Action.* That is, they agreed to sign but would not actually implement its provisions.

The NGO delegates released the "Beijing Declaration," which rang with radical phrases about the supposed special evil of market economics and the putative need for massive transfers of wealth from developed to less developed countries. Its principles were even less likely than those of the *Platform for Action* to change government policies or the lives of women.

The Power and Powerlessness of Words

Three earlier World Conferences on Women, held ten years apart, had sought to establish international standards for government policy, with little expectation that most states would apply these standards in domestic legislation or practice. The fourth conference followed the much publicized, but hardly practical, U.N. proclamation of a "Decade for Women."

With this background in mind, some questioned the need for and usefulness of a fourth U.N. conference. In all countries, statistics show that women are gravely disadvantaged relative to men in their working opportunities, earning power, and ownership of property. In many societies, they also suffer disproportionately from illiteracy, domestic violence, and legal discrimination concerning their political and economic rights. The U.N. world conferences were designed to focus attention on the search for practical ways to raise the status and standards of living of women.

Even the locale of the Fourth World Conference generated controversy. Some observers objected to the choice of Beijing as host city because of China's generally poor record on human rights and its especially poor performance on women's rights, particularly concerning forced abortion and sterilization practices stemming from harsh enforcement of a law allowing only one child per family. Outside criticism did not prevent Chinese authorities from conducting a preconference roundup of dissidents and likely protesters, from censoring material imported by the delegates, or from breaking up public demonstrations during the conferences. The government of Deng Xiaoping was criticized for blocking the participation of NGOs that it opposed on political grounds.

The Chinese delegation and some others were denounced for doing little more than state government policies, rather than addressing the situation of women and devising possible reforms to improve their social, economic, and political condition. Delegations from Taiwan and Tibet were denied visas because of issues unrelated to women, and individuals from other countries were denied rights by Chinese authorities.

The primary criticism of China's behavior as host arose over its unilateral decision to physically isolate the NGO conference, in which about thirty thousand women participated, from the more compact and tightly controlled governmental conference. NGO delegates were sent to the provincial city of Huairou, which had fewer

AP/Wide World Photos

First Lady Hillary Rodham Clinton addresses the Beijing conference on women.

and poorer facilities and was largely inaccessible to the international press.

Consequences

Some view the results of the four U.N. conferences on women as negligible. They argue that the setting of international standards has merely raised the level of hypocrisy among states. Rhetorical and legal commitment to women's rights has increased, but denial of these rights continues.

Others believe that it is better to have standards for state behavior that are disregarded than to have no standards at all. They maintain that the information exchanges, consciousness raising, and networking among women's groups that occurred will lead to true reform over time. The conferences at least have focused international attention on abuses unique to women or suffered disproportionately by them. Perhaps most important, they raised awareness that without increased involvement of women in economic and political life, no nation can achieve its potential for prosperity and modernization.

Cathal J. Nolan

Washington Post Prints Unabomber Manifesto

The Washington Post *published a lengthy article written by the Unabomber, who promised to end his terrorism upon publication of the article.*

What: Social reform; Crime; Terrorism
When: September 19, 1995
Where: Washington, D.C.
Who:
The Unabomber, an unknown terrorist who mails and delivers bombs
LEONARD DOWNIE, JR. (1942-), executive editor of *The Washington Post*
THEODORE KACZYNSKI (1942-), suspected of being the Unabomber

The Unabomber Gets a Public Forum

On September 19, 1995, *The Washington Post* printed an article written by a terrorist known as the Unabomber. The article, about thirty-five thousand words long and titled "Industrial Society and Its Future," condemned modern society and spoke out against the kind of life brought about by the Industrial Revolution. The article declared that the products of modern technology are "a disaster for the human race."

The manuscript was mailed June 24, 1995, to designated editors at *The Washington Post*, *The New York Times*, and *Penthouse* magazine. The Unabomber called for the manuscript to be published within three months of receipt and demanded that space be made available for three annual articles. If these conditions were met, he promised to stop his terrorist activities, although he reserved his right to bomb any property.

The editor of *Penthouse* already had agreed to publish the article and let its author have a column in his magazine. The Unabomber, however, declared that if his writings appeared only in *Penthouse*, he reserved the right to kill one more person.

Opposed to Science and Technology

On May 25, 1978, a security guard at Northwestern University was injured when he opened a package that exploded. Slightly less than one year later, on May 9, 1979, a student at the same school was injured after opening a package left in the university's Technological Institute. On November 15 of that same year, a bomb exploded aboard American Airlines Flight 444 as it flew from Chicago to Washington, D.C., injuring twelve people. These three explosions have been identified as the first of seventeen Unabomber attacks or threats made between May, 1978, and June, 1995. A total of three people were killed in the attacks, and twenty-three were injured. Nine of the seventeen attacks were made against universities or university instructors, staff, or students. Four incidents were directed at airlines, two at computer stores, one at an advertising executive, and one at a forestry association lobbyist.

The investigative search for what federal authorities believed was only one man resulted in a widely distributed profile of the suspect. First given the code name Junkyard Bomber, he was later renamed "Unabom" (university/airline bomber) because the early attacks were against property or individuals associated with universities or airlines. It was believed that the Unabomber carried a grudge against a college or university, possibly because of a bad experience he may have had while a student. His actions and writings also displayed a belief that modern technology is turning people into technocrats with no regard for nature.

His first bombs were crude, made of pipe, lamp cords, scrap metal, and wood. By 1985, the Unabomber's skills had become far more sophisticated, and his bombs were far more powerful and deadly. Instead of match heads and gunpowder, he had begun using ammonium nitrate and aluminum powder. Investigators believe that this development indicated a self-taught individual with a background or training in metalwork or carpentry.

The bombs were all made by hand and appear to have been constructed meticulously, almost to perfection. Made from many parts, the bombs included several tiny levers carved from wood. Authorities believe that all the bomb parts, including the tiny pins and screws, were made by the Unabomber and that the construction of each bomb took many hours of work. Each was polished and handled many times, and most were finally placed inside a homemade wooden box. Officials searched for fingerprints, hair, or even saliva on the labels or stamps that could be tested for DNA identification. No clues of any kind were found on the bombs, the wood boxes, or the packaging in which they were mailed or delivered.

One investigator called the Unabomber "one of the most creative and elusive bombers ever encountered." The only sighting believed to have been made of him was in 1987, in Salt Lake City, Utah. A woman looking out a window saw a white man in a hooded sweatshirt, wearing sunglasses, place a laundry bag filled with wooden boards in a parking space beside a computer store. Less than an hour later, the computer store owner was injured when he knelt beside the laundry bag.

Many of the bombs were engraved with the letters "FC." The engraving was etched in such a place as to ensure survival after the explosion. For years, FC was unexplained but was interpreted by investigators as a vulgar term, "f—— computers." In 1985, a letter sent to the *San Francisco Examiner* said that these letters signify the Freedom Club, a terrorist group that is anticommunist, antisocialist, and antileftist.

Four days after mailing his manifesto to the two newspapers and the magazine, the Unabomber threatened the international airport in Los Angeles (LAX). The LAX bomb threat was canceled by the Unabomber, who declared it to be merely "one last prank." Authorities found no bomb and speculated that the threat to the Los Angeles airport may have been related to a major terrorist bombing in Oklahoma City, Oklahoma. Some criminal psychologists suggest that the threat was made to retake the spotlight from the bombing in Oklahoma City.

Consequences

As *The Washington Post* and *The New York Times* contemplated their options, public and private opinions varied on whether publication would be capitulation to a terrorist or instead would be a public service. Working with investigators from the United States Postal Service, the Federal Bureau of Investigation (FBI), and the Bureau of Alcohol, Tobacco and Firearms, *The Washington Post* published the manifesto with the hope that it would save lives. Publication was, according to executive editor Leonard Downie, Jr., an agonizing decision. It may have done some good: No Unabomber attacks were reported in the months following publication.

In April, 1996, FBI agents raided a one-room shack owned by Theodore Kaczynski. He was charged initially with possession of an unregistered explosive device, a bomb found in the shack. Evidence provided by Kaczynski's brother, David, prompted the raid. Kaczynski was jailed in Helena, Montana, as the FBI accumulated evidence against him and law enforcement officials worked out the crimes for which he would be charged.

Kay Hively

Turner Broadcasting System Agrees to Merge with Time Warner

The Time Warner company officially acquired the Turner Broadcasting System in 1996, creating what was, at that time, the world's largest media and entertainment company, with revenues of more than $18 billion a year.

What: Business; Communications
When: September 20, 1995
Where: New York City
Who:
Robert Pitofsky (1929-), chairman of the Federal Trade Commission
Gerald Levin (1939-), chairman of Time Warner
Ted Turner (1938-), president of Turner Broadcasting System

A Giant Media Company

On September 20, 1995, Time Warner, a multimedia conglomerate, and Turner Broadcasting System (TBS), a company known for its entertainment television, popular cable news station, and outspoken founder, Ted Turner, announced plans to merge their operations and create the world's largest media and entertainment company. The plan called for Time Warner, based in New York City, to acquire TBS, based in Atlanta, Georgia, and take control of all its television stations, cable stations, sports teams, and other properties. The buyout was valued at $7.5 billion.

Time Warner already had businesses involving cable networks, entertainment, and publishing. Its properties included all Home Box Office networks, Cinemax, record labels, music companies, entertainment programming companies, and dozens of magazines, including *Time*, *People*, *Sports Illustrated*, *Fortune*, *Entertainment Weekly*, *Parenting*, and *Coastal Living*. The extensive list also included partial ownership of a number of other properties.

TBS's Turner founded the first twenty-four-hour news channel, Cable News Network (CNN), in 1980. He quickly expanded his company with ownership of the TBS Superstation, the Cartoon Network, Turner Network Television (TNT), Turner Classic Movies, the Atlanta Hawks basketball team, and the Atlanta Braves baseball team. Turner was known for being an aggressive, outspoken businessperson who created an international company from humble beginnings.

Gerald Levin, chairman of Time Warner, said the merger would be "the dream deal," because of the potential of uniting two companies with expansive film libraries and cable programming assets. Even before the business deal, it would have been difficult for someone to turn on a television without seeing something produced or owned by one of these companies. Together, the two companies accounted for about 40 percent of all cable programming in the United States.

The Acquisition

Some large business deals, such as this one, require the approval of the Federal Trade Commission (FTC). The FTC's job is to enforce federal antitrust and consumer protection laws. In other words, its job is to maintain fair competition in the business world and to make sure consumers have choices in the marketplace—of what to buy or what to watch on television.

The FTC announced its preliminary approval of Time Warner's buyout of TBS in a statement released on September 12, 1996. The vote was 3-2, meaning that two commissioners did not agree with the FTC's decision. The commission alleged that the deal could mean higher television cable service prices and fewer programming choices for consumers. The FTC approved the business deal only after establishing some guidelines designed to maintain competition in television programming.

Part of the concern involved the Tele-Communications (TCI) corporation. That company owned 22 percent of TBS shares and 7.5 percent of Time Warner shares. Operating out of Englewood, Colorado, TCI sold cable television programming services throughout the United States. Some of the cable networks affiliated with TCI were Starz!, Encore, Discovery Channel, The Learning Channel, E! Entertainment Television, and the Home Shopping Network.

In a widely published news release issued by the FTC, Chairman Robert Pitofsky called the merger "one of the biggest and most complicated deals that antitrust officials have reviewed." He said the settlement was designed to preserve competition and protect consumers. The two commissioners who voted against the agreement argued that the government should not have demanded any changes in the merger.

The two companies closed the deal on October 10, 1996, creating a company worth at least $95 billion. In February, 1997, the FTC formally approved the merger, imposing two major conditions designed to maintain competition: First, that TCI transfer its Time Warner shares to a separate company spun off from TCI's control, and second, that some cable systems carry a second all-news channel in addition to CNN.

Among other things, the business deal meant that content (network television shows such as *ER* and *Friends*) could be produced through Warner Brothers. Time Warner could create new animation to air on TBS's Cartoon Network, and it could use TBS programming (including a large film library) on any of its fledgling networks and promote the programming on any of the stations owned by the company. The Hollywood film companies of Warner Brothers, Castle Rock Entertainment, and New Line Cinema would all be owned by the same company.

Consequences

Competition, or the lack there of, is the main concern whenever large corporations choose to merge. Naturally, the choices for consumers become more limited when direct competitors such as these are suddenly working together. By the late 1980's, about fifty large companies owned the majority of media outlets, including television stations, radio stations, newspapers, and magazines. Critics charged that this trend toward consolidation represented the formation of a monopoly, and many protested the 1996 Time Warner-TBS transaction. However, by the year 2000, the number of companies with the majority ownership of media had shrunk to about ten multinational media conglomerates, including Disney, Sony, and Viacom. In May, 2000, Viacom and the Columbia Broadcasting System (CBS) merged under the name Viacom, creating an even larger media and entertainment company than that formed by the merger of Time Warner and TBS.

George Bennett

Ted Turner.

Companies that own television stations operate under the guidelines of the Federal Communications Commission (FCC). The FCC is an independent U.S. government agency that regulates interstate and international communications by radio, television, wire, satellite, and cable. The FCC's job is to maintain fairness in broadcasting, particularly broadcast news. With some restrictions, the FCC approved the Time Warner-TBS merger in late 1996. Some media company executives were pushing for the FCC to lift the ban prohibiting companies from owning a newspaper and television station in the same market. Loosening the limits on such ownership would most likely lead to the creation of even more media company giants.

Advocates of media company mergers say that television producers, in particular, need the financial backing of a large company to test new shows and cover the costs of critically acclaimed (but not widely watched) television shows. In other words, they say restricting company mergers could limit variety on television. In any case, after this merger, few people were able to say that they had never heard of Time Warner.

Sherri Ward Massey

Israel Agrees to Transfer Land to Palestinian National Authority

Israel has long indicated a willingness to negotiate its boundaries and turn over certain of its occupied lands to Arabs; however, Palestinians have demanded more land than Israel can cede.

What: International relations; Political independence; Military conflict
When: September 24, 1995
Where: Israel and Cis-Jordanian Palestine
Who:
Yitzhak Rabin (1922-1995), prime minister of Israel from 1974 to 1977
Shimon Peres (1923-), foreign minister and prime minister of Israel from 1995 to 1996
Yasir Arafat (1929-), chairman of the Palestinian National Authority

Origins of the Dispute

After Jordan lost its territory on the West Bank of the Jordan River to Israel in the 1967 Six-Day War, King Hussein continued to pay the salaries of Arab teachers, Muslim clergy, and members of the Jordanian parliament who resided in the captured territories. Meanwhile, although no peace treaty was ever signed, Israel and Jordan carried on trade and cooperated fully in checking terrorism. However, Hussein eventually concluded that the principle terrorist organization, the Palestine Liberation Organization (PLO) had the full support of the Soviet Union, as well as many members of the United Nations.

In 1970 Hussein drove PLO bases out of Jordan in a bloody war that left Jordan with the PLO as an enemy. In order to maintain the appearance of Arab solidarity, however, he supported the PLO in public, while quietly cooperating with Israel. Strains between Jordan and the PLO continued to worsen until July, 1988, when Hussein surrendered his claims to the entire West Bank to the PLO. Since Jordan by then did not actually possess the West Bank, he was able to buy peace with the PLO without actually giving up anything tangible. Moreover, Hussein evidently felt confident that Israel—which still occupied the West Bank—would never surrender its disputed territories to the PLO.

Creation of the Palestine Authority

While Israel's conservative Likud Party was still in office, Prime Minister Shimon Peres and his devoted aide, Yossi Beilin, opened secret discussions with the PLO, although Israeli law strictly forbade negotiations with terrorists. Nevertheless, Peres was convinced that Israel could achieve peace with the Arabs only by an understanding with the PLO. He articulated his ideas two years later in a book expressing the philosophy that open borders would generate a level of prosperity which would make old hatreds based on ethnicity or religion obsolete.

When the Labor Party won Israel's election of 1992, Peres became foreign minister. Prime Minister Yitzhak Rabin was repulsed by the idea of an understanding with overt terrorists, but it was obvious that Labor had no chance of forming a coalition and governing effectively without the parliamentary support of the far left wing Meretz Party, which believed that only an agreement with the PLO could achieve lasting peace for Israel.

Although it was not until February 1993 that Israel's parliament, the Kenesset, legalized negotiations with terrorist organizations, Peres and Beilin began secret negotiations with Yasir Arafat's representatives with the active assistance of the Norwegian foreign minister and his wife. Thus was created the top-secret Oslo Agreement of 1993.

From the beginning, Peres and Beilin intended to give what came to be called the Palestine Authority virtually sovereign authority in the Arab-inhabited areas of the West Bank and Gaza. It was to be left to secret negotiations to determine future boundaries. President Bill Clinton of the United States was admitted to Israel's plans and on September 13, 1993, the text of the Oslo Agreement was revealed for the first time to the public, and the document signed at a ceremony on the White House Lawn in Washington. Arafat gave written assurance that he would abandon terrorism and amend the PLO Charter which called for the destruction of Israel. Only after those concessions had been committed to writing was the public admitted to the startling new agreement.

Consequences

The Rabin-Peres government immediately set to work fulfilling the Oslo Agreement, dividing the disputed territories into areas to be submitted to the government of the Palestinian National Authority, areas to be governed and policed jointly by Israel and the Palestine Authority, and areas to be governed by Israel alone. Seven cities and a substantial part of the surrounding countryside were soon evacuated by Israelis. One city, Hebron, was reserved for future partition, as it contains the burial places of biblical patriarchs and matriarchs that constitute Judaism's second-holiest shrine. Moreover, since Hebron had been the scene of a bloody massacre of Jews in 1929, the Israeli government wish to take special precautions to avoid repetitions of such events.

On the night of November 4, 1995, a Jewish dissident who resented the concessions made to the Palestine Authority, murdered Yitzhak Rabin, and Peres became prime minister. If peace had seemed to be possible under the Oslo agreements, it rapidly disintegrated as Arab terrorists then unleashed a campaign of suicide bombings, protesting any agreement with Jews. In a desperate gamble, Peres scheduled a new election for May 29, 1996. By a very narrow margin, the leader of the Likud Party, Benjamin Netanyahu won.

AP/Wide World Photos

Palestinian negotiator Ahmed Qreia initials the PLO-Israeli agreement in Taba, Egypt, as Chairman Yasir Arafat (seated second from left) and Israeli foreign minister Shimon Peres (second from right) observe.

Although Netanyahu headed a nationalist party opposed to surrendering land to Arabs, he went ahead with the painful task of dividing Hebron and further partitioning the country. Eventually, these concessions cost him the support of his own party loyalists. The Arab leadership, having gotten more than 95 percent of the Arab population in the disputed lands, took a new tack. Thereafter, they denied that Israel had the right to keep any of that territory. They now raised the demand that all Arab refugees who had fled the country when Israel was born, must be allowed to return. The Israeli government could not seriously consider that demand.

The Palestine Authority then demanded the partition of Jerusalem, the capital of Israel. Finally, the demand was made that the Temple Mount, Judaism's most sacred shrine, must be placed under control of the Palestinian National Authority or some other Muslim corporation. Finally, whether it was spontaneous or inspired by Chairman Arafat, violent attacks were launched against Jews in general and the residents of the disputed territories, in particular.

Arnold Blumberg

O. J. Simpson Is Acquitted

Following a legal process lasting more than a year, a former football and media star's acquittal for murder aroused nationwide controversy.

What: Crime; Law
When: October 3, 1995
Where: Los Angeles, California
Who:
ORENTHAL JAMES "O. J." SIMPSON (1947-), former college and professional football star on trial for murder
MARCIA CLARK (1954-), lead prosecutor for the Los Angeles County District Attorney's office in the case against Simpson
JOHNNIE L. COCHRAN, JR. (1937-), chief defense attorney for Simpson
LANCE A. ITO (1950-), Los Angeles superior court judge who presided over the trial
MARK FUHRMAN (1952-), police detective accused of racism

Simpson Is Tried for Murder

Shortly after 10 A.M. Pacific standard time on October 3, 1995, a defendant was found not guilty in a Los Angeles, California, court. This ordinary event was news around the world because of the defendant involved and the publicity surrounding his trial.

Early on the morning of June 13, 1994, the murdered bodies of Nicole Brown Simpson, the former wife of college and professional football star O. J. Simpson, and her friend Ronald Goldman were found on a walkway in front of Nicole Simpson's condominium. O. J. Simpson, in Chicago, Illinois, when told of the killings, became a suspect within days of the crime. He was scheduled to be arraigned by the Los Angeles County District Attorney's office on June 17. Simpson did not show up for the arraignment and eventually was found cruising down a Los Angeles freeway as a passenger in his white Ford Bronco automobile. After a low-speed chase by the police, Simpson surrendered and was taken into custody.

Tens of millions of people watched the Bronco chase, which was televised live across the nation and throughout the world. From that point on, the Simpson case was a huge media spectacle, with every move carefully scrutinized by everyone from legal experts to the mass media and their audiences.

A preliminary hearing held in June and July of 1994 introduced the evidence against Simpson and the major personalities in the case, including Simpson's houseguest, Brian "Kato" Kaelin. The state's case, presented by prosecutor Marcia Clark, was strong enough to warrant a trial. Superior Court Judge Lance A. Ito was selected to preside. Simpson's main attorney was Johnnie L. Cochran, Jr., a prominent African American defense attorney. It was revealed that Simpson's defense planned to rely heavily on allegations of racism and evidence-planting against Mark Fuhrman, one of the Los Angeles police detectives who initially investigated the crime.

The prosecution's case was extensive, with many witnesses, some of whom spent days or even weeks on the stand. The prosecution's case was interrupted often by legal dickering and by problems with the jury, many of whose members ended up being dismissed and replaced by alternates.

The prosecution presented scientific evidence linking Simpson to the crime scene. The defense presented scientific experts of its own, who tried to refute the prosecution's evidence, and audiotapes proving that Fuhrman, who previously had denied using racial epithets against African Americans, had done so many times.

AP/Wide World Photos

O. J. Simpson reacts after the jury announces its not guilty verdict in the murders of his former wife, Nicole Brown Simpson, and her friend Ronald Goldman.

A stark difference in racial perceptions of the trial emerged between whites and African Americans. African Americans were more likely to believe that Fuhrman could have conspired to frame Simpson. Cochran effectively used the history of police brutality and racism in Los Angeles in his closing arguments. Perhaps swayed by this, or merely wishing to be finished with the long ordeal of the trial (for which they had been sequestered for nearly a year), the jury finished deliberations quickly, in less than four hours, returning a verdict that Simpson was not guilty.

A Trial in the Media

Never before had a trial been covered in the media in the manner accorded the Simpson trial. Not only was there gavel-to-gavel live coverage on several cable television channels, but networks also covered the trial every night on their news shows. Tabloid magazines eagerly seized on the trial's mixture of celebrity and grisliness. Many observers suggested that the television coverage made the attorneys more self-indulgent and therefore prolonged the trial; most agreed that television certainly heightened the carnival atmosphere surrounding the trial.

Many who disagreed with the verdict and thought Simpson was guilty blamed the jury system, which required a unanimous twelve-person verdict. Some suggested that a majority verdict of ten to two be allowed in certain cases. Others averred that the jury should never have been sequestered and that being isolated from most contact with the outside world subjected jury members to undue stress.

The racial aspect of the trial demonstrated

that, thirty years after the Civil Rights movement, African Americans still felt a tremendous sense of grievance against the white-dominated legal system. A majority of the jury members were, like the defendant, black. Simpson's defense relied as much on allegations of racism and conspiracy as on refuting the prosecution's scientific evidence. In Los Angeles, the legacy of the police beating of Rodney King in 1991 and the subsequent acquittal of the accused white officers by a white-dominated jury still rankled African Americans and perhaps made some receptive to allegations of racism and misconduct against Fuhrman.

There was another, subtler, racial aspect of the Simpson trial. The judge was a Japanese American, the chief prosecutor was a white woman, and the chief defense attorney was an African American. Among the crucial witnesses were a Chinese American police criminologist and an India-born county coroner. Anyone watching the Simpson trial on television was reminded about how thoroughly multicultural a society America had become. In the background of the racial animosity raised by the Simpson trial, there remained evidence that the United States was beginning to accept its multiracial character.

Consequences

Once acquitted, Simpson found that he could not simply return to his previous life of celebrity, as most of white America still believed him guilty. Moreover, his legal problems were not over. After the conclusion of his criminal trial, the families of his former wife and Ronald Goldman filed a wrongful death lawsuit against him. In February, 1997, a civil court found him liable for their deaths and ordered to pay the families large damages. Simpson appealed the verdict the following year. Meanwhile, his mounting financial problems led him to sell his Brentwood, California house. Soon afterward, he moved to Florida, apparently to escape publicity, which was again becoming unfavorable.

In retrospect, the true celebrities of the Simpson case were the lawyers in his criminal trial, who became stars through their daily appearances on television and who commanded multimillion-dollar book deals. The Simpson trial revealed fundamental problems facing the American judicial and social order. The media coverage of the trial brought those problems to the attention of the public as well as raising issues of the role of the media in criminal proceedings.

Nicholas Birns

Million Man March Draws African American Men to Washington, D.C.

Before major mainstream newspapers were even aware of African American plans to march on Washington, D.C., nearly one million men had already registered for the event—thanks to the efficiency of the black press, black radio stations, community-based organizations, civil rights groups, and word of mouth.

What: Civil rights and liberties; Gender issues; Social reform
When: October 16, 1995
Where: Washington, D.C.
Who:
Louis Farrakhan (1933-), Nation of Islam leader, who issued the initial call for the event
Benjamin F. Chavis (1948-), the national march director

The Gathering in Washington, D.C.

On October 16, 1995, millions of viewers worldwide watched in awe as hundreds of thousands of people—mostly African American men—gathered in the Mall of Washington, D.C. They had gathered in response to a call from the Reverend Louis Farrakhan, the outspoken leader of the Nation of Islam. In the view of many media critics, political pundits, and those in attendance, Farrakhan delivered more than the public could have imagined. On that day, he revisited an old agenda, black self-help, but with more militant political implications. His call for an annual national day of atonement appeared to reflect a change in modern black politics.

Not coincidentally, Farrakhan's address came one hundred years and a month after Booker T. Washington had risen to national prominence by pronouncing another reorientation in African American ideology, strategy, and tactics. In an address he delivered at the Atlanta Exposition on September 18, 1895, Washington announced a shift in African American strategy and tactics "from politics to economics, from protest to self-help, and from rights to responsibilities."

Historical Context

Both Farrakhan's "Challenge to Black Men" and Booker T. Washington's Atlanta Exposition address were delivered at critical historical moments, during transitions from one ideological stage to another. Both eras also involved high points of reactionary racist assaults on African Americans and their hard-fought and earned civil rights.

Some theorists and scholars, however, suggest that the historical precedent for Farrakhan's Million Man March was the 1966 March Against Fear, also known as the Black Power March, and not Washington's speech. The 1966 march cleverly shifted the African American liberation movement's goals, strategies and tactics from transforming local and federal policy, long centerpieces of the Martin Luther King, Jr.-led Civil Rights movement, toward black self-help. Therefore, as an ideology, Black Power represented another change, shifting the focus from integration with whites to institutional development among African Americans.

The ideological notions of Black Power continued to have long-term implications, but during the 1960's it was chiefly seen as a transitional stage between the civil rights movement and the growing dominance of electoral politics. Some nationalists and pan-Africanists, however, viewed it as a definitive period, expressing African consciousness, unity and independent institution-building.

The dominance of the atonement theme, as well as the subdued role of women in the Million Man March, also suggested that many ideas expressed by Farrakhan and the Nation of Islam more closely parallelled those of the Promise Keepers, a conservative Christian men's movement, than those of earlier civil rights and Black Power leaders. However, both male and female leaders spoke at the march. They included Farrakhan himself; Maulana Karenga, the founder of US; Dorothy Height, president of the National Council of Negro Women; Coretta Scott King, the widow of Martin Luther King, Jr.; Betty Shabazz, the widow of Malcolm X; Rosa Parks, whose actions had triggered the Montgomery bus boycott in 1955; Kweisi Mfume, head of the National Association for the Advancement of Colored People; author Maya Angelou; activist Jesse Jackson; singer Stevie Wonder; and columnist Conrad Worrill.

There were also other clear examples of diversity at the march. For example, according to a study by the Wellington Group and Howard University, 80 percent of the male participants in the march had household incomes above $25,000, 43 percent made more than $50,000 and 18 percent more than $75,000. Also, 34 percent of the marchers identified themselves as either moderate or conservative, 31 percent as liberal, 4 percent as socialist, 11 percent as nationalist, and 21 percent as other.

Farrakhan's message of atonement also appeared to be largely successful because it appealed to personal responsibility, voluntarism, African culture, the role of religion, and black capitalism. In addition, the ongoing conservative assault on affirmative action generated a sense of crisis around which many African American men were willing to mobilize.

While Farrakhan issued the call for march, the large number of people who heeded it probably would not have come to Washington without the involvement of a broader base of civil rights and nationalists, activists, and intellectuals. Even Martin Luther King, Jr.'s organization, the Southern Christian Leadership Conference actively supported the event.

The National Million Man March and Day of Absence Organizing Committee's executive council also represented a wide sweeping collection of thoughts. Some members advocated a partnership with women of African descent, while others promoted a men's only project. Some championed capitalism, others opposed it; some practiced an open democratic style of leadership, while others displayed an authoritarian approach.

AP/Wide World Photos

Louis Farrakhan at the Million Man March.

Consequences

"One million Black men will not be ignored," Farrakhan said on October 16, 1995. "We must rise up in this time and seize the hour, seize the moment, because this moment can never be again." Some black scholars and activists have defined the march as having the potential to be the genesis of twenty-first century African liberation. On the other hand, others have suggested that the event could be the genesis of the demise of the African liberation movement if practical results are not seen in the near future.

The unofficial count for the Million Man March varied from 850 thousand to over two million participants. Whatever the precise number, the participants completely filled the Washington Mall—from the Lincoln Memorial to the Capitol building. Many of its participants returned to their communities and began the process of nation building and reempowerment.

Keith Orlando Hilton

Quebec Voters Narrowly Reject Independence from Canada

In 1995, voters in Quebec, a largely French-speaking area, narrowly defeated a referendum that would have given the province the right to pursue independence from Canada.

What: Government; National politics; Political independence
When: October 30, 1995
Where: Province of Quebec, Canada
Who:
René Lévesque (1922-1987), founder of the Parti Québécois (PQ) and former premier of Quebec
Lucien Bouchard (1938-   ), leader of the proindependence campaign
Pierre Elliott Trudeau (1919-2000), Quebec-born prime minister of Canada

Historical Background

Although most of Quebec's more than seven million people are French-speaking Catholics, many of whom seem greater autonomy for the province, about 18 percent are predominantly Protestant English speakers, who are adamantly opposed to independence from Canada for Quebec. Additionally, a growing number (about 2 percent) of those living in Quebec are immigrants. Their political loyalties lie with the federal government in Ottawa, as do those of Quebec's sprinkling of Native Canadians.

Canada's division into two large communities—English speakers who identify with the federal government in Ottawa and the French speakers living in Quebec whose allegiance is to the provincial government—dates from 1534, when Jacques Cartier claimed the St. Lawrence Valley for France. By 1670, when England granted the Hudson Bay Company a charter to develop the adjacent territory, the French presence was firmly entrenched in what is now Quebec. By 1759, when Britain conquered New France, the French culture was firmly established in the area—a fact London recognized in the Quebec Act, which formalized British rule over the region by preserving Quebec's French civil codes and the freedom of the Catholic Church.

Although Canada's French- and English-speaking communities often clashed politically during the nineteenth and early twentieth century over such issues as Quebec's opposition to conscription in both world wars, the origins of French-speaking Quebec's modern quest for independence lie in the 1950's. During this decade, Quebec's gradual transformation from a conservative, rural province into a modern industrialized region led to a perception by the Québécois (French speakers) that the English-speaking government in the provincial capital had colonized Quebec, relegating French speakers to second-class economic status in their own province. Almost immediately, Quebec's old political guard was ousted, and in the 1960, the Quiet Revolution began with the election of a Quebec Liberal Party government committed to making the Québécois the masters of their own house.

Although the Liberals' efforts to modernize Quebec departed substantially from the previous government's agenda, they did not go far enough for the more ardent Quebec nationalists. For them, a sovereign-association arrangement of self-rule coupled with continued economic association with English-speaking Canada was the only way to guarantee that Quebec would be for the Québécois, and in 1968, one of them, René Lévesque, founded the Parti Québécois (PQ), committed to achieving sovereignty for Quebec. Eight years later, the PQ won control of the provincial parliament. Arguing that its victory signaled Quebec's desire to pursue the sovereignty-association arrangement, the PQ in

1980 held Quebec's first referendum on whether to negotiate with the central government on independence for Quebec.

The Immediate Background, Campaign, and Vote

The outcome of the 1980 vote was a profound setback for the nationalists. The referendum failed overwhelmingly in the province as a whole, where 59.56 percent voted against it. Furthermore, exit polls indicated that even within the Francophone community, a majority voted against the sovereignty-association option. The PQ fell into disarray.

Meanwhile, Canadian prime minister Pierre Trudeau, who had promised Quebec a new form of federal association with special status if it rejected the referendum, began to pursue his lifelong dream of negotiating with Britain the full independence of Canada, still officially governed by the queen at the time. Doing so meant writing a new constitution. Trudeau succeeded but at the cost of alienating Quebec, which felt several constitutional provisions undermined provincial authority. Ratification occurred when the English-speaking provinces agreed to the document, and Quebec found itself a part of a political process whose constitution it had yet to sign.

The situation deteriorated further between 1987 and 1992 when the central government of Canada twice tried to coax Quebec into signing the constitution by amending it in such a way as to give Quebec special standing among Canada's provinces. The first time, the proposals, the Lake Meech Accord of 1987, failed when the assemblies of two English-speaking provinces refused to ratify them; the second time, the proposals, the Charlottetown Accord of 1992, sank on a province-by-province referendum basis when six of Canada's ten provinces, including Quebec, rejected them.

Suddenly the PQ, voted out of office in 1986, was reinvigorated. Also, in Ottawa, a minister in the federal cabinet, Lucien Bouchard, bolted from his party to form the Bloc Québécois (BQ), a proindependent Quebec party. By 1994, the BQ held virtually all Quebec's seats in the federal parliament, and the PQ, under the leadership of Jacques Parizeau, was again in power in Quebec province. The time seemed right to schedule a new referendum on whether to negotiate with the federal government on the sovereignty issue.

From the very beginning, the polls showed that the vote in this 1995 referendum would be extremely close. The campaign was consequently intense. Bouchard led the campaign in favor of the referendum, emphasizing Quebec's capacity for self-government and the degree to which English-speaking Canada had betrayed Quebec by rejecting the Lake Meech and Charlottetown Accords. In the end, the referendum was narrowly defeated, 50.56 percent to 49.44 percent, in Quebec as a whole; however, unlike in 1980, a solid majority of French-speaking Canadians voted in favor of pursuing the sovereignty-association option.

Consequences

In its immediate aftermath, the referendum raised tensions between English- and French-speaking Canadians. Parizeau blamed the outcome on money and the ethnic vote, meaning the central government's financing of the opposition's campaign and the province's immigrant and English-speaking communities. His opponents responded by accusing the PQ of yet again representing only French-speaking Quebec, not the province as a whole.

The long-term impact of the referendum on Canadian politics was more ambiguous. The fact that most of the French-speaking residents of Quebec voted in favor of the measure kept the issue of a sovereign Quebec alive. Bouchard promised to hold a new referendum when winning conditions prevail and when Canadian leaders have openly discussed under what conditions Ottawa would recognize a "yes" vote and negotiate the breakup of Canada. However, voters also apparently developed referendum fatigue, and public opinion polls between 1996 and 2000 persistently found a majority of Quebec's Francophones opposed to holding a new referendum. Thus, the twentieth century ended without Bouchard's winning conditions anywhere on the horizon.

Joseph R. Rudolph, Jr.

Israeli Premier Rabin Is Assassinated

Yitzhak Rabin, prime minister of Israel and recipient of the Nobel Peace Prize, was assassinated by an Israeli extremist following a peace rally.

What: Assassination; International relations
When: November 4, 1995
Where: Tel Aviv, Israel
Who:
YITZHAK RABIN (1922-1995), prime minister of Israel and recipient of the Nobel Peace Prize in 1994
YIGAL AMIR (1970-   ), right-wing Israeli extremist who carried out the assassination
SHIMON PERES (1923-   ), former Israeli minister of defense, vice prime minister under Rabin, and Rabin's successor
HAGAI AMIR (1968-   ), brother of Yigal Amir, charged as co-conspirator in the assassination

Rabin Is Assassinated

On November 4, 1995, Prime Minister Yitzhak Rabin addressed more than 100,000 persons in Kikar Malkhei Israel (Kings of Israel) Square, in the middle of Tel Aviv. Among the persons in the audience were representatives from Jordan and Morocco, countries that until recently had nominally been at war with Israel, and Egypt, with which a peace agreement had been signed and a tenuous peace maintained.

Rabin spoke of his faith in the peace process, a series of events in which he had played a central role. He said he thought that most Israelis believed as he did, that risks they were prepared to take would in the end provide both peace and security for the next generations in all Middle Eastern countries. He rejected the violence that was becoming prevalent in Israeli society, particularly among those who opposed the peace process. Rabin finished his address by singing *Shir Ha-Shalom* (song for peace) along with the large crowd.

With his wife, Leah, the prime minister left the podium, descended a stairway, and walked toward his armored Cadillac. Vice Prime Minister Shimon Peres left separately. At 9:40 P.M. (Israeli time), as Rabin was getting into the car, Yigal Amir walked up to him and at close range fired two hollow-point bullets from a .22-caliber pistol. Rabin collapsed on the driveway, mortally wounded, as bodyguards grabbed Amir. Rabin's security guards placed him in the car and rushed him to Ichilov Hospital. The hospital was nearby, but in the confusion, no clearance was made for the driver, and nobody notified the hospital staff that the car and Rabin were on their way. In any event, it probably was too late, because one bullet had severed Rabin's spinal cord. At 11:40 P.M., Eitan Haber, Rabin's speechwriter and close aide, announced to the crowd waiting outside the hospital that Rabin was dead. Peres, at one time a rival of Rabin, was sworn in as Rabin's successor.

Rabin had spent much of his life fighting the Arabs. He was the first prime minister born in what would become the state of Israel. In a sense, he was one of the few leaders the Israeli people would trust as the peace process proceeded. Rabin's role had changed from protagonist to peacemaker. His death left a vacuum.

A Political Life

Rabin was born in Jerusalem in 1922. At the age of fifteen, he entered Kadoorie Agricultural High School to study agronomy. He thought that the best way to serve his country was to become a farmer. At the beginning of World War II, Rabin joined the Haganah, the Jewish underground army in Palestine. He became part of Palmach, an elite strike force. He developed a reputation as a brilliant military tactician that would be maintained for nearly five decades. In 1964, Rabin became chief of staff of the Israeli army. In the Six-Day War in 1967, he was the architect of victory as the army captured both the West Bank territory and the Old City of Jerusalem.

Shortly afterward, Rabin entered politics. In 1968, he became ambassador to the United States, and in that role he aggressively pushed for stronger ties. In 1974, Rabin became Israel's youngest prime minister. Forced to resign in 1977, he returned as defense minister in 1984 and for a second term as prime minister in 1992.

Israeli settlements began to spring up in the West Bank territory. Some of the settlers maintained a fundamentalist viewpoint that the land was part of Israel's biblical birthright, to be retained by violence if necessary. For a time, these Jewish extremists were led by Meir Kahane, founder of the political movement Kach, which strongly opposed any peace movement. Kahane was assassinated in 1990, but his successors have been even more radical. Baruch Goldstein, a follower of Kahane, murdered twenty-nine people in a Hebron mosque in 1994.

Rabin recognized that a state of war could not be maintained forever. Following secret negotiations with Yasir Arafat, head of the Palestine Liberation Organization (PLO) and a longtime enemy of Israel, Rabin signed a self-rule agreement in 1993 that would end much of the Israeli occupation of the West Bank. In turn, the PLO would recognize Israel's right to exist. For their work in forging these accords, Arafat, Peres, and Rabin were jointly awarded the 1994 Nobel Peace Prize.

Radical groups within Israel refused to accept the agreements. Some Israelis perceived Rabin and Peres as traitors. Amir, a law student at Bar

AP/Wide World Photos

Security personnel lift a wounded Yitzhak Rabin (legs visible at bottom center) into a car.

Ilan University and member of Eyal, a radical group, decided to block the peace process by murdering both Rabin and Peres.

Consequences

The long-term consequences of Rabin's death were at first unclear. Israel's moderate right wing was as shocked by the murder as were Rabin's supporters. Benjamin Netanyahu, head of the opposing Likud Party and former parliamentary antagonist of Rabin in the Knesset, called for national unity. It was expected that Peres would continue, or perhaps even accelerate, the peace process as Israeli troops continued their withdrawal from the West Bank. However, after only six months in office as prime minister, Peres called for a national election, in which his party lost to the Likud Party, and he was replaced as prime minister by Netanyahu.

Richard Adler

Nigeria Hangs Writer Saro-Wiwa and Other Rights Advocates

General Sani Abacha, military leader of Nigeria, ordered the executions of nine outspoken critics of his government, including playwright Kenule Saro-Wiwa.

What: Civil strife; Government; International relations
When: November 10, 1995
Where: Port Harcourt, Nigeria
Who:
KENULE SARO-WIWA (1941-1995), an internationally known playwright and campaigner for minority rights and environmental protection in Nigeria
SANI ABACHA (1943-1998), Nigerian military dictator from 1993 to 1998

Decades of Repression

Home to some of the oldest and most artistically sophisticated civilizations in the world, Nigeria was a colony of Great Britain from the early twentieth century until 1960. During the colonial period, the British controlled the government and the economic system, and Nigerians were permitted only a small role in making decisions about their country. The discovery of oil in Nigeria in 1958 and the declaration of independence from Great Britain in 1960 marked the start of a new era of troubles. The oil proved to be a source of great wealth, and the promise of wealth brought with it corruption that reached to the highest levels of government in the new nation. Different ethnic groups within the country battled among themselves for control of oil profits, and a civil war in the late 1960's cost as many as one million lives. Another source of tension was religion: Many Nigerians in the north practiced Islam, as they had for hundreds of years, and many in the south practiced the Christianity that British missionaries had brought with them in the 1860's.

Between 1966 and 1985, the leadership of Nigeria changed hands six times as a result of violent *coups d'état.* In their relatively short tenures at the head of government, the different leaders amassed large bank accounts by diverting oil profits and foreign aid money into their own pockets. This political instability made it difficult for Nigeria to attend to the basic needs of its people. Nigeria experienced a time of great economic growth, fueled by the value of its huge oil reserves, but a corrupt government squandered most of the money and little of it was used to improve the life of Nigeria's poor. Even during a relatively stable period between 1985 and 1993, corruption and human rights abuses were common.

In November, 1993, Defense Minister General Sani Abacha seized power, aided by the military. Abacha soon proved to be corrupt, inefficient, and cruel, even by the standards of a country that has seen more than its share of bad leaders. Abacha used his military security forces to maintain absolute military and political power.

Prominent Protest

One of the groups working within Nigeria to reform the government was the Movement for the Survival of the Ogoni People. Its president, Kenule Saro-Wiwa, was a member of the Ogoni ethnic group, and an outspoken environmental activist. He saw that the process of obtaining and transporting oil from the Nigerian delta was polluting the area, and that oil workers and executives were living lives of luxury while the Ogoni people who depended on the polluted land were not sharing in the profits. The spoiling of the land was well documented on television and in magazines and newspapers, but nothing was done to stop the pollution, and Saro-Wiwa continued to speak out against the government.

Saro-Wiwa perhaps felt freer to speak out than many Nigerians because he was an internationally known playwright. Arresting him would bring out supporters around the world, and Nigeria depended on the goodwill of other nations. Nevertheless, he was arrested in May, 1994. As the leader of the Movement for the Survival of the Ogoni People, Saro-Wiwa, along with eight others, was accused of starting a riot that led to the killing of four Ogoni leaders. They were charged with murder. Most observers believed that the charges were false and politically motivated, and human rights groups and international writers groups pressured Abacha for fourteen months to release the activists.

While he was in prison, Saro-Wiwa was awarded the Nobel Peace Prize in recognition of his efforts to publicize Nigeria's environmental problems and to work for the betterment of the Ogoni people. However, on October 31, 1995, Saro-Wiwa and the others were found guilty of murder. Increased pressure from forces outside the country brought no relief. On November 10, the nine were hanged in a Port Harcourt prison.

Consequences

The international response to the hanging of Saro-Wiwa and the others was swift and decisive. Nigeria was expelled from membership in the British Commonwealth, a voluntary organization of fifty-two nations associated with Great Britain, and vigorously condemned by Commonwealth leaders, including British prime minister John Major and South African president Nelson Mandela. Major and Mandela stated publicly that they believed the executions were a travesty of justice. Within twelve hours of the executions, the United States and fifteen European countries withdrew their ambassadors from Nigeria, a serious diplomatic rebuke. However, Abacha and the government held firm, maintaining that Saro-Wiwa and the others had been accused of a crime, tried fairly, and punished appropriately.

Within Nigeria, many people believed that the government's interest in oil profits had led to the executions. They called on people around the world to boycott Nigerian oil, particularly the Shell Oil Company. For its part, Shell Oil issued a statement accusing environmental and human rights activists of making it impossible for the Nigerian government to release the prisoners. The company denied responsibility for environmental degradation in Nigeria. The effect of the boycott was slight—more emotional than economic.

On June 8, 1998, Abacha suddenly died of a heart attack. He was replaced by another military leader, who promised to bring Nigeria to democracy and ordered the release of several political prisoners. In February, 1999, Nigeria held its first free presidential election, won overwhelmingly by General Olusegu Obasanjo. Obasanjo appeared committed to democracy and to eradicating corruption from the government, and worldwide observers were cautiously optimistic that stability and justice were ahead.

Nigeria convened a Human Rights Commission in January, 2001, to investigate human rights abuses committed by former military regimes. Modeled after a similar commission in South Africa, it hoped to publicly expose former crimes and to begin the process of reconciliation. Abisome Wiwa, the father of Saro-Wiwa, was invited to testify, but he refused because he did not believe that the government was yet ready to provide justice for the Ogoni people.

Cynthia A. Bily

International Panel Warns of Global Warming

The second assessment report of the Intergovernmental Panel on Climate Change (IPCC) projected a rise in global mean surface temperatures. The rise would constitute the fastest rate of change since the end of the last Ice Age.

What: Environment; International relations
When: November 27-29, 1995
Where: Madrid, Spain
Who:
World Meteorological Organization
United Nations Environment Program

Political Concern Emerges Over Global Warming

In 1995, the Intergovernmental Panel on Climate Change (IPCC) confirmed in its second assessment report that evidence indicated that human beings were having a discernable influence on global climate. The report, *Climate Change 1995: The Science of Climate Change*, projected an increase of between 1.8 to 6.3 degrees Fahrenheit (1 and 3.5 degrees Celsius) in global mean surface temperatures by 2100. Such a rise in temperature would constitute the fastest rate of change since the end of the last Ice Age. The report also predicted a rise in global mean sea levels of between 5.9 and 7.5 inches (15 and 19 centimeters) by the year 2100, which would result in the flooding of many low-lying coastal areas. A rise in global temperatures would affect rainfall patterns, increasing the potential in many regions for floods, intense storms, or drought.

Natural variation has always been present in the world's climate. According to most scientists, this natural variability has begun to be overridden by rising concentrations of greenhouse gases in Earth's atmosphere. Although greenhouse gases such as water vapor, carbon dioxide, methane, nitrous oxide, and ozone can occur naturally in the atmosphere, certain human activities add to the levels of most of these gases. The increase in these gases has occurred during the last two centuries as a result of the economic and demographic growth that has taken place since the Industrial Revolution.

Climate change became a part of the political agenda in the mid-1980's as the result of growing public concern over global environmental issues caused by the emergence of scientific evidence that human activities were affecting Earth's climate. The World Meteorological Organization (WMO) and the United Nations Environment Program (UNEP) established the IPCC in 1988 in recognition of policymakers' needs for accurate and timely scientific information. The first assessment report confirming the threat of climate change was issued by the IPCC in 1990. The report brought about the involvement of the global community in an effort to address the problem, and the United Nations Framework Convention on Climate Change (UNFCCC) was adopted in 1992. The UNFCCC provides the overall policy framework for addressing climate change, and the IPCC assesses the scientific, technical, and socioeconomic information important for understanding human-induced climate change risks.

Half a decade of research on the global climate was summarized in the IPCC's second assessment report. Although the report was written, reviewed, and edited by scientists who were primarily accountable to other scientists, debate over its objectivity surfaced. Skeptics tried to discredit it, claiming findings had been reached through a political process rather than a scientific one. For example, delegates from countries who oppose reducing the use of fossil fuel, a known contributor to global warming, were accused of weakening the language in the report. However, credibility for the 1995 report was gained from the fact that 179 delegates from 96 countries, representatives from 14 nongovernmental organizations, and 28 authors participated in the formal review of the report.

The Science of Climate Change

Scientists continue to strive for an understanding of Earth's complex climate system and the timing, extent, and potential effects of climate change. The IPCC concedes that its has a limited ability to quantify with assurance the extent to which humans influence global climate. This uncertainly exists because the climate varies naturally. However, it is difficult to argue with the balance of evidence, which clearly suggests humans play a role in climate change.

Several new findings were outlined in *Climate Change 1995*. The report stated that the amounts of greenhouse gases, including carbon dioxide, methane, nitrous oxide, and sulfur hexafluoride, in the atmosphere have continued to increase. These gases are released into the atmosphere through human activities. Rising levels of greenhouse gases cause climate change because they absorb the infrared radiation that controls the flow of natural energy through the climate system.

The projected rise in temperature by about 1.8 to 6.3 degrees Fahrenheit (1 to 3.5 degrees Celsius) by the year 2100 would be the largest increase ever experienced during the last ten thousand years. This estimate is based on the assumption that no effort to limit current greenhouse gas emissions is made. Uncertainties exist about

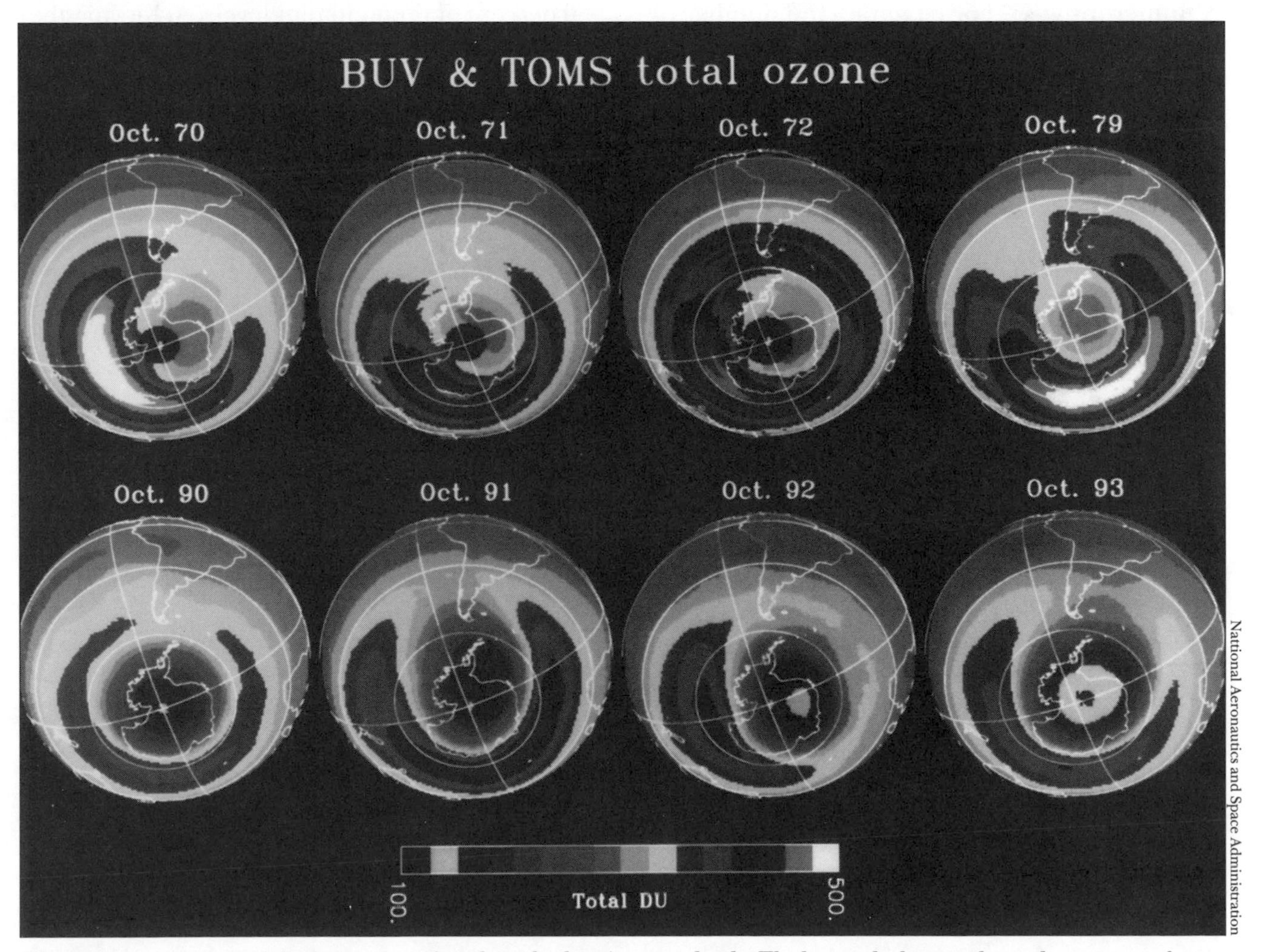

National Aeronautics and Space Administration

These images over time of the Antarctic region show the drop in ozone levels. The bar at the bottom shows the amount of ozone in Dobson units.

the precise effects and scale of climate change. For example, surface temperatures do no immediately respond to emissions of greenhouse gases because of the oceans' delaying effect. Therefore, decades after atmospheric concentrations have stabilized, climate change is likely to continue. At the same time, evidence suggests that past emissions may already be affecting the climate.

Consequences

Climate Change 1995 confirmed the first report's findings that continued accumulation of greenhouse gases would lead to climate change that would affect human and natural systems. Climate change may have an adverse effect on human health, economic activity, food security, and water resources. The faster these changes in climate occur, the greater the risk of damage.

Changes in climate will affect human health in numerous ways. For example, the number of people affected by tropical diseases such as malaria is expected to increase. In tropical areas, the geographical zone of potential transmission of malaria could grow so that the disease affects 60 percent of the population, up from about 45 percent. Falling crop yields in many regions could result in food shortages, hunger, and serious disruptions in farming. Food security at the global level, however, is unlikely to be threatened.

As global precipitation and evaporation patterns change and the sea level rises, large-scale migrations from more to less seriously affected areas could be triggered. Deserts, rangelands, and forests would face climatic stresses as ecosystems and agricultural zones could shift from midlatitude regions toward the poles. Many individual species will become extinct as a result.

Unfortunately, poor and disadvantaged people are the most vulnerable to the negative impacts of climate, and they are the least equipped to adapt to them. In order to stabilize atmospheric concentrations of greenhouse gases at safe levels, all countries will be required to limit their emissions. Emissions of carbon dioxide, for example, would eventually have to fall to about 30 percent of their current levels. Achieving that goal will require dramatic improvements in energy efficiency and fundamental changes in other economic sectors as the world economy continues to expand and populations continue to grow.

Anne Galantowicz

Chinese Government Selects Tibet's Panchen Lama

After the Dalai Lama identified the child who was to become the eleventh Panchen Lama, Beijing announced its selection of another child as the reincarnation of the tenth Panchen Lama.

What: Human rights; International relations; Religion
When: November 29, 1995
Where: Jokhang Temple, Lhasa, Tibet
Who:
TENZIN GYATSO (1935-), fourteenth Dalai Lama, Tibet's exiled spiritual leader
GEDHUN CHOEKYI NYIMA (1989-), eleventh Panchen Lama, announced by the Dalai Lama
GYAINCAIN NORBU (1990-), eleventh Panchen Lama, recognized by the Chinese government
CHADREL RINPOCHE, abbot of Tashi Lhunpo, head of the search party for the new Panchen Lama

The Dalai Lama and the Panchen Lama

As the highest religious figures in Tibetan Buddhism, the Dalai Lama and Panchen Lama have an authority that extends not only to spiritual matters but also to the temporal, political side of Tibetan life. The title "Dalai," a Mongol word meaning "ocean" (referring to the title holder's depth of wisdom), was first conferred upon Sonam Gyatso by Mongol prince Altan Khan in 1578. The title "Panchen" is a conjunction of the Sanskrit word "Pandita" meaning "scholar" and "Chenpo," a Tibetan word meaning "great." The tradition of this title began when the fifth Dalai Lama appointed Lobsang Choekyi Gyaltsen, the fourth abbot of the Tashi Lhunpo monastery, as Panchen Lama. Over the centuries, the geographically isolated Tibetans maintained a degree of autonomy, and Tibet's relationship with China was more that of a protectorate than a province. This relationship shifted from time to time, particularly when China and various other parties tried to assert control and influence.

In 1950, the People's Liberation Army invaded Tibet, and the Chinese Communist Party backed the Panchen Lama as an alternative to the Dalai Lama. Tensions increased between the atheistic leaders of China and the intensely religious Tibetans. After the Dalai Lama escaped to India in 1959, China praised the Panchen Lama for not going with him. However, the Panchen Lama criticized Mao Zedong's policies in Tibet, and he spent fourteen years in prison under house arrest as a result.

Two Panchen Lamas

When the tenth Panchen Lama died suddenly on January 28, 1989, he had just gone to Shigatse, Tibet, to inaugurate his restored monastery. Because he had recently resumed his criticism of the Chinese Communist Party, many supporters of the Tibetan resistance were convinced that he had been poisoned. According to the official Chinese News Agency, however, his sudden heart attack at age fifty-one was a result of the strain of presiding over the opening of a new Buddhist shrine at Tashi Lhunpo monastery, the traditional residence of the Panchen Lama. Two days later, Tibetans in New Delhi, India, demonstrated near the Chinese embassy. They accused the Chinese of killing the Panchen Lama, who they claimed had no history of heart problems.

In order to preempt a rival appointment from the Tibetans in exile, the Chinese authorized a search group to find their own reincarnation of the Panchen Lama. They recruited the abbot of Tashi Lhunpo monastery. The abbot contacted the exiled Dalai Lama to ask for his approval, but communications broke down, and on May 14, 1995, the Dalai Lama announced that a new Panchen Lama, a six-year-old boy named Gedhun Choekyi Nyima, had been found by the traditional method of observing omens. Although the Dalai Lama's choice was originally part of a group of candidates selected by the government-appointed committee, the Chinese government denounced this appointment, and placed the Dalai Lama's chosen boy and his parents under arrest. They also arrested the abbot Chadrel Rinpoche and others for having secretly consulted the Dalai Lama regarding the selection.

On November 12, they chose their own reincarnated Panchen Lama using another traditional method that had originally been suggested for Tibet by a former ruler of China in the eighteenth century; drawing ivory lots from a golden urn. Although this alternate technique had been used only a few times before, the political symbolism underscored a historical precedent for Chinese influence in Tibetan history, and the rival Panchen Lama chosen by this method, another six-year-old boy named Gyaincain Norbu, was announced in November, 1995.

Consequences

The selection and support of rival eleventh Panchen Lamas further strained relations between the Chinese government and the Dalai Lama's Tibetan supporters. Central to the conflict are political, governmental, and religious issues. As the second highest Tibetan Buddhist figure, the Panchen Lama is important to the Chinese government's interests in Tibet, especially because the highest Tibetan figure, the Dalai Lama, maintains a government-in-exile in India. With the long historical relationship between Panchen Lamas and Dalai Lamas, often with one helping to choose the reincarnation of the other, the Dalai Lama saw it as his religious obligation to help recognize the reincarnation of the tenth Panchen Lama. Also, the eleventh Panchen Lama is expected to help select the Dalai Lama's reincarnation and successor.

This controversy drew international attention. Amnesty International became concerned over the disappearance and house arrest of Gedhun Choekyi Nyima soon after his selection by the Dalai Lama. Later, in April, 1997, upon learning of the trial sentencing Chadrel Rinpoche, former abbot of Tashi Lhunpo monastery, to six years' imprisonment and two of his associates to prison terms, Amnesty International called for their immediate and unconditional release on the grounds that they had been imprisoned in violation of international human rights standards. In January, 2001, the European Union urged China to let an independent delegation visit Gedhun Choekyi Nyima, but the response was a report that the child was attending school and his parents did not want international publicity.

In Beijing on February 23, 2001, the Panchen Lama selected by China, Gyaincain Norbu, spoke at a ceremony unveiling the "Eleventh Panchen Lama Qoigyijabu" pictorial, on the occasion of the year of the Iron Snake in the Tibetan calendar.

Alice Myers

Galileo Spacecraft Reaches Jupiter

Six years after its deployment from the space shuttle Atlantis, *the Galileo spacecraft reached its destination, the planet Jupiter.*

What: Space and aviation
When: December 7, 1995
Where: Jupiter
Who:
William J. O'Neill, the Galileo project manager at NASA's Jet Propulsion Laboratory
Richard Young (1927-), a Galileo probe scientist at NASA's Ames Research Center
Marcia S. Smith (1951-), the Galileo probe manager at NASA's Ames Research Center
Donald Ketterer, the manager of the Galileo program
Andrew P. Ingersoll (1940-), a mission scientist at the California Institute of Technology
Tobias C. Owen (1936-), a mission scientist at the University of Hawaii

Fifth Rock from the Sun

Jupiter, the fifth planet from the sun and the largest planet in the solar system, has intrigued scientists for hundreds of years. Following the invention of the telescope in the seventeenth century, the Italian scientist Galilei Galileo's observations of the moons of Jupiter changed fundamental assumptions about the universe. In the ensuing centuries, scientists learned that Jupiter is a giant gas planet, possesses an atmosphere, has a powerful magnetic field, and emits more heat than it absorbs from the sun. Turbulence in Jupiter's outer atmosphere, such as the movement of the famous Red Spot, could be observed with telescopes, although the exact composition of the atmosphere remained unknown.

Data transmitted by U.S. Pioneer and Voyager spacecraft as they passed by Jupiter in the 1970's raised many additional questions. Voyager detected the presence of active volcanoes on Io, one of Jupiter's sixteen known satellites, as well as faint rings similar to Saturn's circling Jupiter itself. With its numerous moons and gaseous nature, Jupiter resembles a miniature solar system. Scientists believed that a close examination of the planet Jupiter could make significant contributions to understanding the early history of the solar system and could lead to advances in geochemistry, meteorology, space physics, and other fields.

Project Galileo

The National Aeronautics and Space Administration (NASA) began planning for the Galileo mission in 1977. An interdisciplinary team of scientists and engineers designed a spacecraft to be built in two sections: a probe that would separate from the main spacecraft before arrival at Jupiter and would be deployed into the planet's atmosphere, and an orbiter that would spend approximately two years moving around Jupiter observing both the planet and its satellites. NASA's Jet Propulsion Laboratory in Pasadena, California, built ten scientific instruments and two radio science experiments into the two-and-a-half-ton spacecraft, while the 746-pound atmospheric probe designed and built by the Hughes Space and Communications Company of El Segundo, California, carried an additional six instruments.

The project was designed to support research missions performed by more than one hundred scientists from six different countries. Engineers provided for two modes of radio communication with the spacecraft: a heat-tolerant low-gain antenna for use immediately following deployment from Earth orbit, and a high-gain antenna that would open after Galileo left the inner solar system. In April, 1991, the high-gain antenna failed to unfurl as designed, but NASA engineers compensated by making improvements in receiving

An artist's rendering of the probe released from the Galileo spacecraft.

equipment. NASA believed that the mission would still be able to achieve at least 70 percent of its goals.

On October 18, 1989, a crew aboard the space shuttle *Atlantis* launched Galileo on its six-year journey to the outer solar system. Although the primary focus of the mission was to study Jupiter, instruments on board the orbiter began gathering data for transmission to the Jet Propulsion Laboratory shortly after deployment. As Galileo passed through the inner solar system and the asteroid belt, it gathered new data from Venus, the Moon, and the asteroids Gaspra and Ida.

Finally, on December 7, 1995, after traveling 2.3 billion miles (3.7 billion kilometers) in a looping path through the inner solar system that passed by both Venus and Earth in gravity-assist swings, Galileo entered an orbit circling Jupiter, where it would remain making observations and transmitting data for two years. On that same day, the atmospheric probe, which had been released from the spacecraft on July 12, 1995, descended into Jupiter's atmosphere. The probe survived both intense heat and deceleration forces 230 times as great as the acceleration of gravity at Earth's surface to record information and transmit data for 57.6 minutes.

Consequences

Prior to the descent of the probe into Jupiter's atmosphere, scientists had believed that weather patterns on Jupiter would be similar to those on Earth. Scientists believed, for example, that winds would be most intense in the upper atmosphere but would weaken as the probe descended. Data obtained by the probe revealed that this theory was not correct. In fact, high-speed wind currents on Jupiter remained consistent in their intensity thousands of miles deep into the atmosphere. In addition, the probe's instruments recorded that the density and temperature of the upper atmosphere were both much higher than predicted. Scientists believe that these findings are consistent with a planet that emits more heat than it absorbs, although the processes by which Jupiter releases that heat into its atmosphere are still unknown. Some scientists speculated that the heat is emitted by volcanic activity, although the probe provided no data either to support or to disprove this theory.

Another surprising discovery was that overall visibility in Jupiter's atmosphere was much better than predicted. Observations from previous spacecraft flyby missions and from telescopes on Earth had led scientists to believe that Jupiter's atmosphere consisted of layers of thick, dense clouds. The probe, however, encountered only small concentrations of dust and haze materials along its trajectory. Instruments indicated that clouds on Jupiter, like clouds on Earth, vary in size rather than forming a consistently thick and uniform layer. Finally, the probe found very little water in the Jovian atmosphere; scientists had predicted little water would be found in the outer atmosphere but had expected to find significant amounts as the probe descended. Instead, the probe revealed that the atmosphere of Jupiter is extremely dry. Such findings led mission scientists to conclude that existing theories of how the solar system and the planets formed and evolved needed to be radically revised.

Nancy Farm Mannikko

NATO Troops Enter Balkans

NATO forces, one-third from the U.S. military, entered Bosnia in order to enforce the peace agreement drawn up by the warring sides.

What: Military; International relations
When: December 14, 1995
Where: Bosnia
Who:
Slobodan Milośevíc (1941-), president of Serbian Republic from 1990 to 1997
Franjo Tudjman (1922-1999), president of Croatia from 1990 to 1999
Alija Izetbegovic (1925-), chair of the state presidency of Bosnia from 1990 to 1996
William Lafayette Nash (1943-), U.S. general, commander of American ground troops in Bosnia
Leighton W. Smith, Jr. (1940?-), U.S. admiral, commander of NATO forces in Bosnia

Another Peace-Keeping Mission

On December 14, 1995, after being delayed by foul weather, the first troops of the United States army crossed the Sava River and entered the Republic of Bosnia. The troops were part of a North Atlantic Treaty Organization (NATO) force charged with implementing the peace treaty agreed to by Bosnia and its neighbors the previous month. NATO planned a contingent of sixty thousand military personnel, twenty thousand of them from the United States.

The first American soldiers were part of the advance group assigned to prepare the way for the remainder of the American troops. At the same time, forces from the other NATO countries entered at different locations. The NATO forces replaced the failed United Nations (U.N.) mission, which was unable to maintain peace among the communities or secure safe areas for the beleaguered Bosniaks (Bosnian Muslims). NATO sought to provide buffers separating the three ethnic communities and to ensure that the timetable of the negotiations was carried out.

Both Admiral Leighton W. Smith, Jr., the commander of the NATO operation, and General William Nash, the commander of the American contingent, emphasized that any attack on the troops would be met with a suitable response, including force. There was virtually no resistance to the entry by the warring Serbs, Croats, and Bosniaks. On the contrary, the various ethnic armies welcomed the NATO forces and promised full cooperation.

The greatest danger came from land mines strewn along the roads and fields between the points of entry and the staging areas where the troops were to encamp. Winter weather conditions also caused problems. Snow and ice jammed the river, making the crossing extremely difficult, interfering with the operation of the motorized and electronic equipment, and slowing the progress of the entry of troops as well as their travel to their destinations. Over the next days, more troops and supplies entered the country as the various contingents established their bases.

A New Initiative

In the fall of 1995, after more than three years of fighting, there appeared to be no end in sight to the war among the Serbs, Croats, and Bosniaks of Bosnia. Reports of unspeakable atrocities had aroused the moral outrage of the world. Attempts to bring peace through use of U.N. forces and NATO air raids had little effect. Political divisions kept the European nations from taking decisive action.

The final straw was the Serb attack against and conquest of Srebrenica, a Muslim safe area protected by the United Nations. The reported mas-

AP/Wide World Photos

American soldiers arrive in Sarajevo to join troops from Germany and Great Britain as part of a NATO force.

sacre of Muslim men and boys from the area by Serbs and threats to the U.N. forces drove the West to action.

It appeared to U.S. president Bill Clinton's administration that an American initiative was necessary to bring a conclusion to the conflict. Such an initiative would be complicated by divisions in the federal government: The Democratic president faced a hostile Republican Congress and a difficult election in November of 1996. President Clinton invited the presidents of Serbia, Croatia, and Bosnia to come to Dayton, Ohio, to work out a new peace plan that would establish Serbian and Croatian areas in Bosnia and allow the Bosnian government to function in the rest of the country.

The negotiations were interrupted several times by disagreements among the principals, but all sides had a commitment to see an end to the war. The talks were further complicated by a military campaign in September by the Croatian army that drove Serbian families out of the Krajina region of their country. Furthermore, the leaders of the Bosnian Serbs were not invited to take part in the negotiations, and they continued their attacks on Muslim areas.

The three parties in Dayton agreed to a peace plan including a specific timetable for withdrawal from the front lines, the exchange of prisoners, and the punishment of war criminals. The agreement called for a large NATO force, including a substantial American contingent, to enforce the treaty. The commitment of American troops caused concern in the United States because of fear of becoming embroiled in an endless conflict. The failures associated with the 1993 American operation in Somalia were still fresh in people's memories. Public opinion opposed the venture, and Congress insisted that the president get its approval before the commitment. The president realized this would not come about and insisted that as commander in chief of the U.S. military, he could commit the troops. He ordered them to Bosnia. Congress reluctantly agreed to support the troops but not the decision to send them.

Consequences

Twenty thousand American and forty thousand other troops entered Bosnia with few incidents. NATO forces, unlike those of the United Nations, responded to the occasional attacks against them. The one American casualty in the initial weeks was a soldier who was wounded by a land mine.

After the first month of the peacekeeping mission, despite some friction, the various ethnic armies cooperated with NATO and in general adhered to the timetable. Most of the difficulties concerned the investigation of Serbian war crimes, but even in that situation the Bosnian Serbian army reluctantly went along. An atmosphere of uncertainty and cautiousness nevertheless prevailed.

Many observers thought that it might take generations to sort out what had happened in the Bosnia Herzegovina war, as well as what should be done about the war's legacy. Historians and journalists have come to different conclusions. Some speculate that there is something endemic in the Balkan character that has resulted in civil wars, religious persecution, and war crimes; others dismiss the idea of blood feuds and historic animosities, preferring to analyze how Yugoslavia broke apart after its strong communist leader, Tito, died and a power vacuum opened up that former communists and nationalists of all kinds exploited.

Frederick B. Chary

France Announces End to Nuclear Bomb Tests

In the fall of 1995, France announced plans to make a series of nuclear bomb tests in the South Pacific on an atoll in French Polynesia. Worldwide diplomatic criticism and public opinion within France caused President Jacques Chirac to cancel the test series after five months.

What: International relations; Military capability; Weapons technology
When: January 29, 1996
Where: Paris, France
Who:
Jacques Chirac (1932-   ), president of France from 1995
François Mitterrand (1916-1996), president of France from 1981 to 1995
Charles de Gaulle (1890-1970), president of France from 1958 to 1969

The French Nuclear Weapons Program

In the summer of 1995, President Jacques Chirac announced plans for France to conduct a series of eight underground nuclear explosions. The test site was to be at Mururoa atoll in the South Pacific, about seven hundred miles southeast of the Fiji Islands, north of New Zealand. The purpose of the tests, according to Chirac, was to maintain the readiness of France's arsenal of nuclear weapons as a deterrence against any potential aggressor.

Chirac's announcement provoked international outrage because a worldwide moratorium on test explosions had been in existence for more than three years. The French action was a provocation that could initiate a resumption of the nuclear arms race. Nevertheless, the first explosion was carried out in early September of 1995, with a yield of twenty thousand tons (twenty kilotons) of trinitroglycerine (TNT), equal to the blast that had leveled the city of Hiroshima, Japan, in 1945.

To understand the magnitude of twenty thousand tons, visualize a freight train with two hundred box cars, each one filled to capacity with one hundred tons of explosive. If this quantity of TNT could be assembled in one place and then detonated all at one time, the released energy would equal one such nuclear explosion. Five more French test explosions of even higher intensity took place in the next several months.

An earlier international incident had taken place in 1985 when the government of France, under President François Mitterrand, was rocked by a major scandal in its nuclear weapons testing program. An antinuclear group called Greenpeace had sent its ship, *Rainbow Warrior,* to the South Pacific to protest against continuing test explosions. This ship was sunk on July 10 by two bombs attached to its hull while it was in the harbor at Auckland, New Zealand. Responsibility for the sinking was traced to an espionage mission by the French secret service. Mitterrand fired his minister of defense and halted the test series. Later in 1995, Chirac, then mayor of Paris, was elected president of France and replaced Mitterrand.

The Nuclear Arms Race

At the end of World War II in 1945, the United States was the only nation that possessed atomic bombs. However, this monopoly was broken in 1949 when scientists in the Soviet Union experienced their first successful explosion. Great Britain followed in 1952, and France exploded its first atomic bomb in 1960 in the Sahara desert of North Africa. China's first atomic explosion took place in 1963. During the Cold War of the 1950's and 1960's, the Soviet Union and the United States both developed the even more powerful hydrogen bomb. However, exploding such huge weapons created clouds of radioactive fallout contamination, provoking worldwide public demonstrations against further testing.

The Cuban Missile Crisis of 1962 brought the two superpowers to the brink of war. At that time, the U.S. government had an announced policy of "massive retaliation," in which it threatened a nuclear strike directly against the Soviet Union if vital American interests were at stake anywhere in the world. Eventually, fear of mutual annihilation led to negotiations resulting in the 1963 Limited Test Ban Treaty that prohibited nuclear tests in the atmosphere, in outer space, and under the ocean. However, underground testing was not restricted at that time.

Both the type and number of nuclear weapons built by the two superpowers grew rapidly, from about five thousand in 1957 to nearly twenty-five thousand in 1965. Also, the fleets of bombers were replaced by more sophisticated weapon-launch systems using nuclear submarines and land-based intercontinental missiles. As U.S. involvement in Vietnam escalated, the government of France under President Charles de Gaulle concluded that it could no longer rely on the United States to protect its national interests in Europe. Therefore, France decided to build its own independent nuclear arsenal, hoping to restore French prestige in international affairs. During the next twenty years, France carried out more than two hundred test explosions, mostly in the French Polynesian islands of the South Pacific.

AP/Wide World Photos

President Jacques Chirac, speaking on television, calls for an end to France's underground nuclear tests.

Consequences

The breakup of the Soviet Union in 1991 caused a major shift in public opinion about the need for nuclear weapons. In the following two years, U.S. president George Bush and Russian president Boris Yeltsin signed two Strategic Arms Reduction Treaties (START), which reduced the number of nuclear weapons on each side from twenty-five thousand to under seven thousand. Also the land-based missile systems in both countries were terminated and gradually dismantled. Against this background, the French decision to begin a new bomb-testing series in 1995 provoked international criticism.

When President Chirac announced an early end to the French testing program in January of 1996, the decision was praised by political leaders worldwide. Bringing an end to nuclear explosions was necessary to protect people and the environment from nuclear fallout and to halt the development of even more destructive weapons. Furthermore, the potential spread of nuclear arms to those countries that do not possess them has been of great concern. Later in 1996, the United Nations approved a Comprehensive Test Ban Treaty that would end all testing of nuclear weapons, including underground tests. This treaty was to go into effect as soon as it was ratified by all countries that have nuclear capability. Also, France announced its willingness to sign a treaty declaring the entire South Pacific region a nuclear-free zone.

Hans G. Graetzer

International Monetary Fund Approves Loan for Russia

An international financial organization provided substantial economic assistance to Russia in the postcommunist period in an effort to promote a free-enterprise system

What: International relations; Economics
When: February 8, 1996
Where: Moscow, Russia
Who:
MICHEL CAMDESSUS (1933-), managing director of the International Monetary Fund
BORIS YELTSIN (1931-), president of the Russian Federation from 1991 to 1999

Assisting an Economy in Crisis

The collapse of communism and the dramatic breakup of the Soviet Union in 1991 directed Russia into another phase of its long and challenging history. In the aftermath of the disintegration of the multinational Soviet state and the end of the communist regime of seven decades, the Russian people began their uneasy transition toward democracy. This included a multiparty political system, an uncensored media, expanded rights of citizens, and the dismantling of the state-owned and heavily regulated economy under the communist regime. Russian voters elected Boris Yeltsin to be the president of Russia and lead the new government. However, the parties in the Duma, the lower house of the Russian parliament, found it difficult to decide on a direction for the nation in the postcommunist era.

The efforts of Yeltsin's government to establish a free-enterprise system initially had high expectations in the euphoria after the collapse of the communist system. Unfortunately, it soon became obvious that removing existing price controls and reducing extensive government regulations created severe economic problems that could not be allowed to continue. Inflation grew rapidly and production declined. Consumers faced widespread scarcities of goods in stores. Bankruptcies threatened the nation, and unemployment increased to levels not previously seen for decades. Faced with a progressively weakening economy, in the mid-1990's, the Russian government attempted to take countermeasures. President Yeltsin and his advisers sought substantial foreign financial assistance to assist in the difficult transition to capitalism and a free-market economy.

The International Monetary Fund (IMF) offered to help. This organization, established in 1945, was created to assist governments during times of economic difficulty. One means is by providing substantial loans to governments to alleviate existing problems. In return for this financial assistance, the IMF has the authority to impose conditions related to the use of those funds. The common objective of the IMF is to help economies become more self-sufficient and reduce the need for future outside aid.

The Financial Agreement

On February 2, 1996, Michel Camdessus of the IMF announced that a $10.2 billion loan would be provided to the Russian government over a three-year period. The formal signing took place February 8. These funds could be used to provide capital for new free-enterprise efforts as well as to assist existing businesses that faced severe economic dislocation because of production decline, loss of markets, and reduced income. The IMF funds would also be used to meet other government expenses including payment of welfare benefits and back wages.

In return, the IMF required that the budget deficit must be reduced, not to exceed 4 percent of the 1996 gross domestic product (GDP)—a calculation that measures the total of production and services—with the goal to reduce the figure to 2.5 percent in 1998. The inflation rate, while still too high, was expected to decline to a more acceptable 1 percent monthly increase. Russia also agreed to abolish designated tariffs and eliminate tax advantages that especially favored large corporations. Finally, the IMF required a higher level of tax collection.

Consequences

The announcement of IMF aid provided a psychological as well as an economic boost to the struggling Russian economy. The news also came during the early period of the 1996 presidential campaign. President Yeltsin, who initially had very low support in public opinion polls, used the IMF news to assert that his leadership was leading to economic progress in the nation and that he deserved election to a second term. However, some of the economic reforms he earlier had promised or initiated were weakened with the dismissal during 1996 of several key government officials. Nonetheless, Yeltsin gradually gained popular strength by early summer and won reelection against the communist candidate.

Economically, the positive effects of the IMF plan were evident. The first installment of the loan, $3 billion, was used to help balance the 1996 budget. A portion was used to pay teachers, coal miners, and others who had not been paid for many months. The loan allowed the Russian government to meet its scheduled payments of international debt. Without these IMF funds, Russia probably would have defaulted on a number of its obligations, which would have dramatically worsened its opportunity to improve its international standing. Finally, the inflation rate fell in 1996 as compared with 1995.

On the basis of this good news, Russian economists and government officials predicted that the infusion of these international funds would continue to have a positive effect on the Russian economy. They estimated growth rates of 2 percent to 4 percent in the 1996-1998 period, possibly rising to 6 percent to 7 percent by 2000.

These optimistic predictions fell short of actuality, as economic problems revealed the unresolved challenges that still existed. The GDP rate actually fell by 6 percent in the first eleven months of 1996, indicating an economy in decline rather than one undergoing growth. The IMF delayed its July, 1996, installment of $340 million when it became obvious that Russian authorities had made little progress in increasing collection of tax revenues. IMF authorities recommended delaying the November funding of a similar amount for the same reason, although the transfers eventually were made.

By early 1998, the Russian government was forced to apply to the IMF for even larger financial loan amounts than in the 1996 agreement. The 1998 total exceeded $17 billion. In retrospect, the 1996 IMF aid was a useful but temporary transfusion to a seriously ill patient. More fundamental economic reforms would be required before the promise of Russia's uneven journey toward a democratic free-enterprise system might be realized.

Taylor Stults

Liggett Group Agrees to Help Fund Antismoking Program

Liggett breaks rank with other tobacco manufacturers and settles a class action lawsuit. It promises to pay 5 percent of its profits for a quarter of a century to help end smoking.

What: Business; Law; Health
When: March 13, 1996
Where: New Orleans, Louisiana
Who:
Mike Moore, attorney general of Missouri
David A. Kessler (1951-), Food and Drug Administration commissioner
David M. Burns (1947-), medical school professor at University of California, San Diego

A Landmark Settlement

The Liggett Group, the smallest of the five major tobacco companies, broke ranks with the other members of the industry also named as defendants and reached a settlement in a class-action lawsuit brought by millions of smokers or former smokers who claimed to be addicted to nicotine. Among the terms of the agreement, announced on March 13, was a stipulation that Liggett pay 5 percent of its pretax income each year for twenty-five years to a national smoking-cessation campaign. Liggett also promised to drop its fight against the efforts of the Food and Drug Administration (FDA) to regulate the selling and marketing of tobacco and to comply with the agency's interim rules designed to discourage underage smoking.

Missouri attorney general Mike Moore noted that Liggett's action, which marked a sharp departure for the tobacco industry, could "be the first crack in the tobacco industry's fifty-year-old dam." The announcement of the settlement contributed to a steep drop in tobacco companies' stock prices. However, the impact of the settlement was more symbolic than economic.

The Industry Response

The other tobacco companies named in the lawsuit—American Tobacco, R. J. Reynolds, Brown and Williamson, Philip Morris, and Lorillard Tobacco—said they planned to continue with the lawsuit and would fight further federal regulation of tobacco. Philip Morris, the leading tobacco company in the world, contended that the settlement addressed "only a limited piece of the tobacco landscape" and that it would not have a great impact on other pending issues facing tobacco. Tobacco companies noted that Liggett, whose only name brands are Chesterfield and Eve, had only a 2 percent share of the market. In addition, the settlement represented only a small part of Liggett's profits and only about three hours profit for the entire industry.

Although Liggett is a small company, its parent company, the Brooke Group, was seeking control of RJR Nabisco Holdings, owner of R. J. Reynolds, the second largest company in the industry. If the Brooke Group succeeded in taking over RJR Nabisco, the terms of the agreement would apply both companies and the settlement would be worth about $1 billion and cover about 30 percent of the industry. Members of the Tobacco industry, however, criticized the settlement as a takeover tactic on the part of Liggett, designed to influence RJR Nabisco shareholders.

Consequences

David M. Burns, a professor at the University of California, San Diego, medical school who aided the negotiations, maintained that the settlement's public health benefits offer more to the public than the money paid to individual smokers. Burns said that the yearly cost to society

AP/Wide World Photos

Ernie Perry, who smoked for thirty-five years, was part of the class-action suit against Liggett.

for smoking-related problems is $50 billion, more than the profits of the industry. Therefore, he argued, rather than simply financially penalizing the tobacco companies, it is to society's benefit to gain the industry's aid in fighting smoking.

FDA commissioner David A. Kessler found the settlement encouraging because it demonstrated that Liggett recognized the impact of tobacco advertising on children. The agency hoped that the settlement would aid its efforts to regulate the tobacco industry. President Bill Clinton supported the FDA's 1996 assertion that it had jurisdiction to protect the public health from tobacco products, which it termed "nicotine-delivery devices." Each year, these devices, the FDA asserted, are addicting millions of Americans to nicotine. The FDA's efforts encountered a serious setback in 1998 when the Fourth U.S. Circuit Court of Appeals ruled that the FDA could not regulate tobacco. In 2000, the U.S. Supreme Court upheld the lower court's decision, 5-4, stating that tobacco regulation was the responsibility of Congress, not the FDA.

Two days after announcing its settlement in the class-action lawsuit, Liggett reached a settlement with five states—Florida, Louisiana, Massachusetts, Mississippi, and West Virginia—that had filed suit to recover more than $10 million spent on the treatment of Medicaid patients with smoking-related diseases. Liggett agreed to pay the four states 2.5 percent of its yearly pretax profits for the next twenty-five years. In another settlement in 1997, Liggett agreed to pay twenty-two additional states 25 percent of its pretax profits for the next twenty-five years.

The company also agreed to cooperate with investigators and turn over internal industry documents that are likely to help prosecutors in other lawsuits. The following year, fourteen additional states settled lawsuits against Liggett, which also agreed to numerous restrictions on marketing and advertising. By 2000, all fifty states had reached similar settlements with tobacco companies. The industry faced numerous other lawsuits, including one filed in 1999 by the U.S. Justice Department, which sought repayment of funds spent on federal health insurance.

Frank A. Salamone

Britain Announces Mad Cow Disease Has Infected Humans

The British government revealed that a fatal brain disease that affects cattle, popularly known as mad cow disease, had spread to humans through contaminated beef and meat products.

What: Medicine; Food science; Health
When: March 20, 1996
Where: London, England
Who:
John Wilesmith, head of epidemiology in the Great Britain's Central Veterinary Laboratory
Richard Southwood (1931-   ), zoologist at the University of Oxford
Lord Douglas Phillips of Sudbury, chairperson of the 1998-2000 official BSE inquiry

Prion Diseases

In 1986, the first case of mad cow disease, or bovine spongiform encephalopathy (BSE), was diagnosed in Great Britain. Irritability, fearful behavior, and a staggering gait preceded death in affected cows, and autopsies showed holes in their brain tissue. Food scientists, including John Wilesmith, head of epidemiology in Britain's Central Veterinary Laboratory, believed that the cows got BSE from their feed, which contained meat and bone meal from sheep that had died of scrapie, a disease similar to BSE.

BSE and scrapie are infectious diseases, but microbes do not cause them. A protein molecule called a prion transmits the disease from one animal to another. The prions that cause disease are identical to the prions found in normal cells in their chemical makeup, but they fold into a different shape. In sheep, the normal prion has a spiral backbone, and the scrapie type forms a flat sheet.

Scrapie and BSE take a long time to produce symptoms. Years pass while normal prions change shape. Still more years go by before the misshapen prions accumulate in numbers large enough to cause brain damage. Exactly how the damage occurs is unknown. In cell cultures, the shape change happens inside nerve cells, where misshapen prions accumulate in pockets called lysosomes. The bursting of lysosomes could kill brain cells. The clearance of dead cells could leave the spongy spaces in the brain characteristic of the prion diseases.

The human disease with symptoms similar to those of BSE is Creutzfeldt-Jakob Disease (CJD). CJD begins with numbness and unexplained mood swings. Next come hallucinations, staggering, and pain. Before death in one to two years, people with CJD lose their sight, memory, personality, and control of their bodies. There is no treatment or cure. Before the 1990's, CJD was extremely rare. Worldwide, it occurred only once in every one million people, nearly always between the ages of fifty and seventy-five. Autopsies show spongy tissue in the brain of CJD victims similar to that found in BSE-infected cattle and scrapie-infected sheep.

Crisis

In 1988, the British government took steps to curb the spread of BSE. It banned animal feeds made from the brains and spinal cords of cattle and sheep. Slaughterhouses often ignored the ban, which was poorly enforced. Nervous-system remains from cows that might have been infected continued to be processed into human and animal food. The number of BSE cows continued to rise, reaching a peak of 36,680 in 1992.

As BSE numbers grew, some people worried that eating meat from BSE cows might cause CJD in people. In 1989, a panel led by Richard Southwood, a zoologist at the University of Oxford, assured the public that BSE could not cross the species barrier—that is, it could not infect humans. The government's chief medical officers pointed out that scrapie does not affect people who eat diseased sheep, and therefore, they insisted, British beef was safe to eat. Early in the 1990's, a disease resembling BSE was found among cats. This warning sign that prion diseases could cross the species barrier was ignored. The incidence of CJD among humans in Britain nearly doubled between 1990 and 1994.

By 1996, laboratory investigations had revealed a tragedy. A new strain of CJD called variant CJD (vCJD) had turned up in the British population. It arises much earlier in life than the previously known CJD and is caused by misshapen prions in beef. On March 20, 1996, the British government admitted that vCJD had been found in ten people under the age of forty-two, half of them associated with the meat or livestock industry. Some of the people were teenagers.

Consequences

By the year 2000, nearly 200,000 British cattle were known to be infected with BSE. More than four million cows had been slaughtered, and eighty-one people had died from vCJD. The British beef industry lay in ruins. Because the disease might take fifteen years or more to produce symptoms in people, experts predict a rising human death rate from vCJD in the twenty-first century.

In November, 2000, a committee headed by Lord Douglas Phillips of Sudbury, a member of the House of Lords, published the findings of its two-year, $40 million investigation into BSE. The Phillips report said that the first case came not from sheep but from a mutation, or change in genetic material, of a single cow—probably some time in the 1970's. Some scientists dispute that conclusion, but they agree that BSE spread because brains and spinal cords of dead cows were fed to other cattle. Transmission to humans occurred through brain and organ meats and "mechanically recovered beef" scraped from the spines of cattle.

"BSE spread like a chain letter," Lord Phillips said, while the public was "sedated by the official presentation of risk." The Phillips report said that the government did not purposely lie, but that its campaign of public reassurance was misguided. Lord Phillips blamed the tragedy on the uncritical acceptance of the species barrier and the reluctance of officials to incite panic. In an effort to preserve the British beef industry, the government lulled the public into eating hazardous beef products and blinded scientists to crucial data. The report also alleged that complacency delayed research that would have disclosed the risk sooner.

Editorials in the *British Medical Journal* and the *New Scientist* agreed, saying that the BSE crisis taught an important lesson: Information on health risks should be made freely and openly available to all and that government decision making should be transparent. Everyone—including scientists, elected officials, civil servants, and ordinary citizens—should understand what the risk is, how it can be assessed, and what steps can be taken to reduce it.

Faith Hickman Brynie

AP/Wide World Photos

Cattle from a herd in which evidence of mad cow disease has been found are taken to a slaughterhouse to be destroyed.

FBI Arrests Suspected Unabomber

The man suspected of sixteen bombings over a seventeen-year period was finally arrested in the wilderness of Montana because of a tip from his brother. This ended the longest, most expensive manhunt for a serial killer in U.S. history.

What: Crime; Terrorism
When: April 3, 1996
Where: Scapegoat Wilderness, near Lincoln, Montana
Who:
THEODORE J. KACZYNSKI (1942-), former mathematics professor
DAVID KACZYNSKI (1949-), Theodore Kaczynski's brother
GARLAND BURRELL, JR. (1947-), U.S. district court judge

Early Years

Theodore Kaczynski was born in Chicago, Illinois, and attended high school in Evergreen Park where he was a member of several clubs and played the trombone in the band. He graduated and entered Harvard University when he was only sixteen years old. After receiving his undergraduate degree, Kaczynski attended the University of Michigan and earned both a master's degree and a doctorate. He then joined the faculty at the University of California, Berkeley, where he taught mathematics for two years. After resigning his teaching position, he went to Montana in 1971 to live in a small cabin without electricity or running water.

Reign of Terror

The first bomb attributed to Kaczynski exploded at Northwestern University, injuring one person, on May 25, 1978. This was followed by a series of bombs, either planted or mailed to people and places all over the country including Illinois, Utah, Tennessee, California, Washington, Michigan, Connecticut, and New Jersey. His last bomb, which killed Gilbert Murray, president of the California Forestry Association, exploded in Sacramento on April 24, 1995. In all, he was responsible for sixteen bombs that killed three people and injured at least twenty-two others. The bombings were rather sporadic with periods of great activity followed by periods of inactivity. One or more bombs were sent or placed each year between 1978 and 1982. Then following a two-year hiatus, there were four bombings in 1985, the Unabomber's most active year. One bombing occurred between 1986 and 1993 followed by one or more bombings each year from 1993 through 1995.

Most of Kaczynski's bombs were directed against university professors. However, airline executives and computer industry employees were also targeted on more than one occasion. A task force was formed, composed of agents from the Federal Bureau of Investigation (FBI), the Bureau of Alcohol, Tobacco, and Firearms, and the U.S. Postal Service. Because his favorite targets were universities and airlines, Kaczynski was given the nickname "Unabom" (university/airline bomber). The profile developed by the FBI described the "Unabomber" as probably being a middle-aged, note-keeping, white, male loner. This profile proved to be extremely accurate.

Agents noted that the bombs were rather crudely constructed using bits and pieces of discarded materials. The explosives themselves were homemade from commonly available chemicals. One characteristic trait of the Unabomber was to include wood in the construction of his bombs. This could take the form of homemade wooden containers or used matchsticks built into the mechanisms. At times, pieces of metal such as nails or razor blades appear to have been added to make the bomb more deadly.

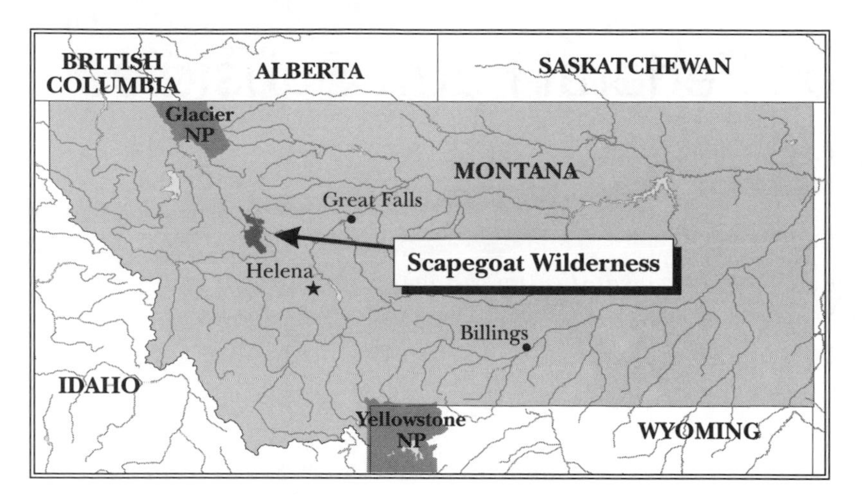

Agents were sent who searched the property and found bomb-making materials in a shed behind the house. David informed them of his brother's cabin in Montana. A surveillance of Theodore began, with agents posing as lumberjacks, mountain men, and postal workers. Sensors and microphones were also used. Kaczynski was found to be a loner who kept to himself but occasionally worked at odd jobs, frequented the public library, and was known locally as the "hermit on the hill."

The surveillance was cut short when it was learned that there had been a news leak to a Columbia Broadcasting System (CBS) reporter. Kaczynski was arrested and everything, including the cabin itself, was removed as evidence. Among the evidence seized were a pipe bomb and bomb-making materials, a typewriter believed to be the one used to write the manifesto, a copy of the manifesto, and a journal in which Kaczynski admitted to being the Unabomber.

Theodore Kaczynski was charged with four bombings in which two people were killed and two injured. After several delays and unsuccessful attempts to reach an acceptable plea bargain, a trial date was set. In addition, Judge Garland Burrell, Jr., refused to allow Kaczynski to fire his two defense lawyers and defend himself. The lawyers wanted to present a defense of mental illness, which Theodore adamantly denied. In fact, he wrote a sixty-nine-page document objecting to being portrayed as insane, even though a federal psychiatrist diagnosed him as a paranoid schizophrenic.

Just before the trial was to begin, a plea bargain was agreed upon. In the agreement, Kaczynski pleaded guilty to the charges in exchange for no further charges and no execution. Judge Burrell gave him four consecutive life sentences without parole plus thirty years. This agreement was criticized by many who believed Kaczynski should have received the death penalty.

Consequences

After the bombing of the Federal Building in Oklahoma City by Timothy McVeigh, the Unabomber began to seek more publicity. He wrote letters to some of his victims as well as to newspapers making threats and demands. Among these was one to the *San Francisco Chronicle* threatening to blow up an airplane leaving Los Angeles. This resulted in increased security at the airport for a period of time. The Unabomber also wrote to the *New York Times* and *The Washington Post* telling them that if they would publish his 35,000-word manifesto titled "Industrial Society and Its Future," a protest against technology and modern civilization, he would stop his reign of terror. The publishers of the *Times* and the *Post* consulted with U.S. attorney general Janet Reno and FBI director Louis Freeh to decide if the document should be published. The decision was made to publish the manifesto in the hope that even if the Unabomber did not honor his promise to cease the bombings, someone might recognize the style or content of the writing. Such was the case.

David Kaczynski, Theodore's brother, lived in Schenectady, New York, where he worked in a shelter for runaway children. David, who had read the manifesto in the newspaper, was helping his mother, Wanda, move from the house in Chicago where the family had lived. There, David found letters his brother had written years earlier to newspapers. Because these sounded like the manifesto, he contacted the FBI through a lawyer in Washington, D.C.

Philip E. Lampe

Congress Reduces Federal Farm Subsidies and Price Supports

Congress enacted the Freedom to Farm Act, marking a substantial change in agricultural policy. This act created a time line for the reduction of government support for crop prices and gave farmers more flexibility in deciding what crops to grow.

What: Agriculture; Government
When: April 4, 1996
Where: Washington, D.C.
Who:
Bill Clinton (1946-   ), president of the United States from 1993 to 2001
Pat Roberts (1936-   ), chairman of the House Agriculture Committee
Richard Lugar (1932-   ), chairman of the Senate Agriculture Committee

Agricultural Policy

On April 4, 1996, President Bill Clinton signed the Federal Agriculture Improvement and Reform Act of 1996, popularly known as the Freedom to Farm Act. This bill represented a significant change in agricultural policy for the United States in that it set forth a seven-year transition period that would end most government payments to farmers.

The first two decades of the twentieth century were prosperous times for American farmers. After World War I ended, European farmers started producing more crops, which cut prices for American farmers. During the war years, many farmers borrowed money to increase production. As prices fell, farmers could not repay their loans, and thousands of families lost their land. The government responded by passing the Agricultural Marketing Act of 1929, which attempted to stabilize prices by encouraging farmers to voluntarily restrict output. This attempt to help failed, and the situation worsened as a severe drought in the central part of the country destroyed crops in the 1930's. During the Great Depression, farm families faced an economic crisis. Thousands of farmers were giving up and moving to cities looking for work. With cities already suffering from high unemployment, the migration of farmers made local conditions worse. The federal government responded in 1933 by beginning to intervene in agricultural markets using three primary strategies—setting legal price floors, buying surplus crops, and limiting production. The government set price floors for individual crops, which prohibited prices from falling below a legal price. If farmers produced more than was sold at the price floor, the government bought the surplus. The government also paid farmers to decrease crops by holding land out of production.

Over time, this government intervention in agriculture became increasingly expensive. New technology, better seeds and fertilizers, and a host of new chemicals to fight weeds and pests increased agricultural output. Production increased faster than demand, so prices for crops kept falling over time. Each year, the government found itself spending more and more money to help farmers. By the early 1990's, the federal government was giving nearly $13 billion a year to farmers—an average of nearly $10,000 for each farm family. Because payments were based on production, much of the money was going to the largest producers. The government was paying farmers to hold 65 million acres out of active crop production.

A Change in Political Climate

With the November, 1994, elections, the Republicans gained control of both the Senate and the House of Representatives. Richard Lugar of Indiana became chairman of the Senate Agriculture Committee and Pat Roberts of Kansas became chairman of the House Agriculture Committee. Both men were interested in reducing the role of the federal government in the lives and production decisions of farmers. Their view was also influenced by the Republican Party's stated goal of creating a balanced budget for the federal government, which would necessarily mean that some spending reductions had to take place.

Debate on a new farm bill began after the 1994 elections. Many senators and congressmen from rural areas expressed concern that the proposed reduction in farm assistance was too great and would hurt farmers if crop prices fell. During the time of discussion, the prices for important crops were stable or even rising. The price of corn was low in November, 1994, with farmers getting only $2.00 per bushel. Corn prices rose steadily throughout 1995 and early 1996, reaching more than $4.00 per bushel by April, 1996. This gave supporters of the new approach confidence that the reduction in government involvement was the right thing to do. Rising prices and the proposed freedom to grow more crops also increased support for the bill from farmers.

Consequences

The Senate and House debated and approved the 1996 farm bill in March. President Clinton expressed concern that the new bill was not flexible enough if prices fell but nevertheless signed the legislation on April 4, 1996. Individual farmers signed a seven-year contract with the Department of Agriculture that contained a schedule for payments over this period. After seven years, most payments would end. The price-floor program would still provide a safety net, but the price floors were set very low so the market price would normally be above the price floor. When the market price is above the price floor, no government payment goes to the farmer.

AP/Wide World Photos

Most of the farmland in southern Minnesota, including that of David Sholtz of rural Mankato (shown here), was enrolled in the federal program ending subsidies.

The 1996 farm bill has had mixed results. Farmers enjoyed the benefits of changing the crops they grow, but government payments were higher than predicted. It was estimated that the amount of money paid by the government would fall to $5.5 billion in 1996 and to $4.0 billion by 2002. These estimates proved to be unrealistic because crop prices started to fall in the late summer of 1996, just after passage of the farm bill. Corn prices, which were $4.50 per bushel in July, 1996, fell below $3.00 in October, 1996, and below $2.00 in August, 1998. To prevent financial crisis in 1999 and 2000, the government increased emergency payments to farmers.

The 1996 farm bill was designed to expire in 2002. Without the enactment of a new bill, government farm programs revert to those described in the 1949 farm bill. A reversion to those programs, based on production capability in the 1940's, which is considerably lower than modern rates, would cost the government billions of dollars in payments.

Allan Jenkins

Clinton Signs Line-Item Veto Bill

A new law allows the U.S. president to veto unwanted parts of bills rather than having to reject the entire bill. This law, the Line-Item Veto Act, is subsequently declared unconstitutional by the U.S. Supreme Court.

What: Government; National politics
When: April 9, 1996
Where: Washington, D.C.
Who:
Bill Clinton (1946-), president of the United States from 1993 to 2001
Bob Dole (1923-), Republican U.S. senator from Kansas, majority leader of the U.S. Senate
Robert Byrd (1917-) Democratic U.S. senator from West Virginia

From Proposal to Reality

On April 9, 1996, President Bill Clinton signed into law a proposal that granted the chief executive of the United States authority to excise individual items from appropriations bills, any new entitlement spending, and a limited tax benefit that met certain criteria. The law, a campaign promise in the 1994 election, was backed by the Republican majority in Congress.

Congressional practices such as House riders—nongermane amendments—and attaching parochial, pork-barrel requests to large spending bills have perpetually frustrated presidents. Starting with Ulysses Grant following the Civil War, U.S. presidents began openly requesting a line-item veto as an additional power to combat wasteful spending. In the middle part of the twentieth century, several constitutional amendments were proposed dealing with the line-item veto. During the administration of Ronald Reagan in the 1980's, the request for a line-item veto made its way into most of the president's State of the Union messages. As annual budget deficits continued and the national debt skyrocketed, the line-item veto was viewed as a device to rein in undisciplined congressional appropriations.

The Contract with America, a compendium of economic, political, and social policy positions promoted by Republican congressional candidates in 1994, included a promise to enact a line-item veto bill. After that election, Republicans gained control of both chambers of Congress for the first time in forty years. Republican Senate leader Bob Dole led the fight to pass the Line-Item Veto Act. The Clinton administration favored the legislation for several reasons. First, the bill could held ferret out wasteful spending, permitting the Clinton team to keep reducing annual budget deficits. Second, the proposal had political benefits in that it could be used to counter an opposition-controlled Congress and could be cited as an example of bipartisan cooperation by the Clinton administration. Finally, the bill would, through the delegation of additional authority to the chief executive, strengthen the institution of the presidency.

Signed on April 9, 1996, the Line-Item Veto Act would go into effect on January 1, 1997. To avoid the constitutional amendment process, congressional authors of the legislation created what was referred to as enhanced-recission authority. The four steps of the procedure were as follows: First, the president would receive a bill and cancel objectionable spending items in appropriations, entitlements, or certain types of tax benefit legislation. Second, Congress could block the president's action with a disapproval bill, which had to be passed in identical form in both chambers, Third, the president would presumably veto the latter bill. Fourth, Congress could attempt to override the veto with a two-thirds vote in both the House and Senate.

Challenging the New Law

Because they believed the new law significantly altered the meaning of the presidential veto clause in Article I, section 7 of the Constitution, several members of Congress—led by Senator Robert Byrd, Democrat of West Virginia—immediately challenged it in the courts, contending that line-item veto authority could be granted to the president only through a constitutional amendment. Just three months after the Line-Item Veto Act became operational, an April, 1997, ruling by U.S. District Court judge Thomas Penfield Jackson found the latter law unconstitutional. On appeal, the U.S. Supreme Court vacated and remanded that decision because of lack of standing by plaintiffs. Subsequently, in a February, 1998, decision, U.S. District Court judge Thomas Hogan agreed with Judge Jackson that the Line-Item Veto Act unconstitutionally transferred spending power from Congress to the executive and violated the separation of powers principle.

AP/Wide World Photos

President Bill Clinton signs the line-item veto bill.

In July, 1998, the Supreme Court affirmed the earlier district court decision, ruling that presidential line-item veto power could be accomplished only through a constitutional amendment. Justice John Paul Stevens wrote the majority opinion, and Justices Stephen Breyer, Sandra Day O'Connor, and Antonin Scalia dissented in a 6-3 decision. The case on which the ruling was made, *Clinton v. City of New York*, emanated from one of President Clinton's line-item vetoes in August, 1997.

Consequences

After the February, 1998, ruling against the Line-Item Veto Act, Clinton agreed to suspend its usage while the courts dealt with the matter. In a little more than a year, he had exercised line-item authority eighty-three times on eleven spending bills. As a result of the Supreme Court's action, those vetoes were rendered moot, as was an estimated $1.9 billion in savings that Clinton's line-item vetoes generated.

Credit for bringing the line-item veto proposal to fruition must be given to President Clinton and Congress, although the law's passage cannot be separated from the political situation of the time. It was promoted by Clinton and Dole, the 1996 Democratic and Republican presidential nominees, respectively, partly because each wanted to be first to employ the device. However, because warnings about the Line-Item Veto Act's constitutionality were ignored by both the chief executive and legislature, the law was overturned just two years after its passage.

Although some lament the law's demise, several existing devices in the president's arsenal take the form of pseudo line-item vetoes. For example, the president possesses regular recission authority as part of his budget powers. Also, when signing a bill, the president can make statements in which he effectively expresses his opposition to or interpretation of selected provisions. Finally, the president may issue executive orders that countermand the law being implemented. Although the legal controversy over the Line-Item Veto Act has ended, the debate continues over whether the tool is needed and how to grant it to the chief executive.

Samuel B. Hoff

Clinton Blocks Ban on Late-Term Abortions

U.S. president Bill Clinton vetoed a bill that would have prohibited all abortions after twenty-four weeks, except for those necessary to save the life of the mother. He wanted the bill to contain an exemption that would have considered the overall health of the mother.

What: Gender issues; Medicine; Religion; Health
When: April 10, 1996
Where: Washington, D.C.
Who:
Bill Clinton (1946-), president of the United States from 1993 to 2001
Bob Dole (1923-), U.S. Senate majority leader
Henry Hyde (1924-), U.S. senator

The Debate

On November 1, 1995, the U.S. House of Representatives passed a bill that proposed a ban on a specific late-term abortion procedure called "intact dilation and extraction" by the medical field but called "partial birth abortion" by its opponents. Under the provisions of the bill, it was illegal to perform this type of abortion procedure except when necessary to save the life of the pregnant woman. Some Senate members, including Senator Bob Dole, wanted to pass the bill immediately, but Senator Arlen Specter recommended that it be sent to the Judiciary Committee for closer review. On December 7, 1995, the Senate passed the bill 54-44 but with some amendments.

The bill then was sent back to the House of Representatives, and on March 27, 1996, the House agreed to three key provisions added by the Senate. The vote fell mostly along party lines, 286-129, a vote that would be nine more than what would be required to override a presidential veto. The 54-44 vote in the Senate, however, would not be enough to override such a veto.

Meanwhile, on February 28, President Bill Clinton wrote to lawmakers, namely Henry Hyde, chair of the House Judiciary Committee, saying that he disapproved of the late-term abortion procedure, but felt that he could not sign a bill unless it also allowed exceptions when the procedure was needed to protect the woman's health or future fertility.

The Veto

In the late afternoon of April 10, 1996, President Clinton called a press conference to announce that he had vetoed House Resolution 1833. He appeared with five women and their families, all of whom had experienced the late-term procedure. The president said that these women represented a small but extremely vulnerable group who desperately wanted to have children. When it became clear to them that their babies would not live and that if they carried them to term, their ability to have children in the future would be jeopardized, they decided to abort. One of the women, Mary Dorothy Line, told the story of how she was overjoyed to find out that she was pregnant with her first child and then was very disappointed to discover that the baby had no chance to live outside the womb. She told of how there was a good chance that the baby might die in utero and that if this happened, toxins might spread, necessitating a hysterectomy. If the baby did not die in utero, a natural labor risked a rupture in her cervix and uterus.

Four other women, Coreen Costello, Claudia Ades, Vikki Stella, and Tammy Watts told basically the same stories. All were thankful to the president and to the doctors who were able and willing to perform the late-term procedure. The bill the president was asked to sign would have addressed the life but not the health of the mother. He said that he would have signed the

AP/Wide World Photos

President Bill Clinton announces his veto, accompanied by families who made the decision to terminate pregnancies with the procedure banned by the legislation.

bill if it had made an exception for cases in which the mother would suffer serious adverse health consequences. These mothers, he said, might have lost all possibility of future child-bearing or might be seriously and permanently injured. Costello said that she and her husband were always extremely opposed to abortion, but in their case, they saw no other choice because the baby was badly malformed and would not live outside the womb.

The president admitted that abortion is a serious problem and reminded his listeners that while governor of Arkansas he had signed a bill to restrict late-term abortions, but that the Arkansas bill exempted women whose life or health was at risk.

Consequences

As expected, Congress attempted to override the president's veto but did not have enough votes. Consequently, various states have proceeded to take up the matter. New Jersey passed a ban against the procedure in June, 1997. However, Governor Christie Whitman vetoed the ban because she did not think it would hold up under constitutional scrutiny. Lawmakers overrode the veto in December, 1997. The law threatens doctors with a $25,000 fine and loss of license. U.S. district court judge Anne Thompson declared the law unconstitutional because its wording was too vague and it placed an undue burden on a woman's constitutional right to obtain an abortion. Nineteen states have won court or state action to partially or totally block the bans because of vague wording or lack of health exceptions.

To date, only three of the country's nineteen contested bans have been heard in a federal appeals court. Justices in Ohio and Wisconsin upheld the lower court's decision to block the laws. In Virginia, an appeals court allowed the ban to go into effect until a district court decided on its constitutionality.

The American Nurses Association supported the president's veto, saying that the legislation was an inappropriate intrusion of the federal government into a therapeutic decision and that when something goes wrong with a pregnancy, the government should not intrude. The American Public Health Association also supported the veto.

Opponents of the procedure contended that it is tantamount to infanticide. The Roman Catholic Church, through its National Conference of Catholic Bishops, fought to uphold the ban. Opponents' main contention is that the health of the mother may be interpreted too broadly, allowing the procedure simply for the sake of the mother's emotional well-being. Protestant and Jewish leaders, however, supported the veto. They believed that this type of abortion was almost always a last resort and that women who are faced with difficult moral decisions must be free to decide how to respond.

Winifred Whelan

South Africa Adopts New Constitution

South Africa began a new era as a democratic nation by adopting a constitution that guaranteed broad civil rights to all South Africans for the first time.

What: Government; National politics; Social reform
When: May 8, 1996
Where: Cape Town, South Africa
Who:
Nelson Mandela (1918-), the first South African president to be elected by a multiracial electorate and a member of the black minority
Frederik W. de Klerk (1936-), deputy president of South Africa and the leader of the largely white National Party
Mangosuthu Gatsha Buthelezi (1928-), a Zulu chief and the leader of the Inkatha Freedom Party

Apartheid in South Africa

In 1652, Europeans began to settle in what later became South Africa. Eventually, South Africa became a colony of Great Britain, and whites, who made up approximately 12 percent of the population, controlled the government and the economic systems. In 1948, the South African government created a form of racial segregation called apartheid, designed to keep people of different races separated. Blacks, Asians, and mixed-race people were forced to use separate doors to public buildings, and they could not eat at the same restaurants as whites. It became illegal for whites and nonwhites to intermarry.

During the 1950's, the restrictions became tighter. Blacks were no longer allowed to live in white areas and could work only in particular areas. A law called the Bantu Education Act created separate, inferior schools for blacks. In some areas where the soil was fertile, farms were taken away from black South Africans and given to whites. In addition, blacks were not allowed to vote, so they had very few ways of bringing about change in their country.

Some black South Africans formed political parties and became activists in an effort to end apartheid. Leaders of these parties, including Nelson Mandela of the African National Congress (ANC), were jailed. Many white citizens also believed apartheid was unfair and worked to end it. They, like the black activists, committed acts of civil disobedience and were arrested for their beliefs. Black South Africans became increasingly frustrated by the injustice of apartheid, and resistance became increasingly violent.

Other countries tried to force South Africa to change its system. In the 1970's and 1980's, many countries refused to do business with South African companies until apartheid was lifted. Stockholders urged large multinational companies to sell their holdings in South Africa because they did not wish to make a profit from an unfair system. Because of the ever-increasing pressure from within and without, South Africa had no choice but to change.

A True Democracy

In 1990, South African leader F. W. de Klerk began moving his country toward democracy. He ordered Mandela released from prison, where he had been for twenty-seven years, and made it legal for blacks to form political parties. He promised that soon the country would hold new elections, and that every citizen—white, black, and "colored"—would be able to vote. Many whites feared that if blacks were given power they would seek revenge for the decades of poor treatment, but under Mandela's charismatic leadership, blacks agreed to help the country transform itself and to avoid retaliatory bloodshed. A national charter was drawn up in 1993, making blacks full citizens.

In April, 1994, South Africa held its first all-race national democratic elections, and Mandela, head of the ANC, was elected president. For the next two years, the Constitutional Assembly negotiated in Cape Town, the nation's capitol, over the terms of the country's new constitution. The strongest voices came from members of Mandela's ANC, de Klerk's white-led National Party, and the Zulu-based Inkatha Freedom Party led by Chief Mangosuthu Gatsha Buthelezi. Finally, on May 8, 1996, a new constitution was approved overwhelmingly, by a vote of 421-2.

With the adoption of the new constitution, South Africa's transition to democracy was officially completed. The constitution called for a strong central government, with a national legislature divided into two houses, much like the division of the U.S. Congress into the Senate and the House of Representatives. It gave broad civil rights to all citizens and prohibited discrimination on the basis of race, gender, sexual orientation, pregnancy, and marital status. It also guaranteed the right to abortion, prohibited capital punishment, and guaranteed free speech, the right to strike, and the right to celebrate one's own culture, religion, and language. South Africa went from being one of the most repressive governments in the world to being one of the most progressive in terms of guaranteeing the civil rights of its citizens.

Consequences

The day after the constitution was approved, de Klerk announced that his National Party was withdrawing from the coalition government. The National Party felt that its influence was too weak to be effective from within, and the new government was clearly strong enough to withstand strong opposition from outside. The Inkatha Freedom Party, led by Buthelezi, had not participated in the vote over the constitution because it believed the central government should

AP/Wide World Photos

President Nelson Mandela (left) and Deputy President F. W. de Klerk after the new constitution was approved.

be weaker and provincial governments should have more power. However, with a strong constitution widely supported by the citizens, South Africa was able to weather these disputes and work toward settling them.

The 140-page constitution was intended to guide the country through its first years as a democracy. Rather than make sweeping changes all at once, the plan called for the constitution to take effect gradually over three years. During those years, the government would be ruled by a coalition of different political parties rather than solely by the ANC, Mandela's party. On June 2, 1999, Mandela retired as president, replaced by Thabo Mbeki of the ANC, who had won a landslide election.

After Mbeki's election, much work remained to be done. The 1996 constitution had celebrated the right to adequate housing, food, water, education, and health care, but the country did not have the resources or the infrastructure to make the needed reforms.

As the twenty-first century began, many poorer South Africans were still without electricity and access to fresh water, especially in the rural areas. The crime rate and the rate of deaths because of acquired immunodeficiency syndrome (AIDS) were rising, and the economy was in a slump. Internal tensions were still felt in the fledgling democracy.

Cynthia A. Bily

Arsonists Burn Thirtieth Black Church in Eighteen Months

An arsonist burned the sanctuary of the Matthews-Murkland Presbyterian Church, prompting the federal government to intensify efforts to stop church arson in the 1990's.

What: Crime; Civil rights and liberties; Religion
When: June 6, 1996
Where: Charlotte, North Carolina
Who:
Larry Hill, pastor of the Matthews-Murkland Presbyterian Church
Gary Fields, *USA Today* reporter
Morris Dees (1936-), civil rights activist and cofounder of the Southern Poverty Law Center

Burning Churches

Church arson occurred frequently in the twentieth century. Many African American churches in the South were targets of racial violence during the Civil Rights Movement in the 1950's and 1960's. Some supporters of segregation expressed their anger at legal decisions that initiated integration of public facilities by vandalizing black churches. Arsonists often burned churches in which civil rights activists met in an attempt to intimidate them. Perhaps the best-known incident was the 1963 firebombing of the Sixteenth Street Baptist Church in Birmingham, Alabama, in which four girls died.

By the 1990's, many arsonists were not motivated by racial intolerance but rather were thrill seekers or copycats. According to Morris Dees, civil rights activist and cofounder of the Southern Poverty Law Center, some arsonists burned vulnerable churches as acts of meanness rather than to achieve any political agenda. Many arsonists set fires to feel superior and powerful, to collect insurance money, or because of peer pressure. The National Fire Protection Association estimated that two thousand church fires occurred annually at churches with congregations consisting of various races and sects. One-fourth of these fires were arson.

Because so many African American churches were burned in the mid-1990's, black pastors testified at congressional hearings. They alleged that federal investigators did not consider the matter seriously enough. The pastors emphasized that African American communities valued churches as peaceful refuges from external conflicts.

Churches heightened security, especially at night when many of the fires occurred. Nationally, businesses offered rewards for information about church arson, and newspapers reported on each fire, speculating that there was a pattern or conspiracy. The National Church Arson Task Force, consisting of federal agents affiliated with the Bureau of Alcohol, Tobacco and Firearms (ATF) and the Federal Bureau of Investigation (FBI), and state and local personnel, coordinated efforts to solve the crimes. Federal grants were created to rebuild burned churches.

Matthews-Murkland Presbyterian Church

An hour before midnight on Thursday, June 6, 1996, flames engulfed the sanctuary of the Matthews-Murkland Presbyterian Church. Built around 1903, the sanctuary had been replaced by a newer building and was no longer used for worship when it burned. The church was historically significant because it had been built by former slaves and belonged to the first all-black Presbytery in the United States. Located in an affluent southeast Charlotte neighborhood, the church had a congregation of mostly African Americans.

Witnesses saw the fire and alerted emergency personnel to extinguish the blaze. Local broad-

AP/Wide World Photos

The Reverend Larry Hill examines the burned remains of the Matthews-Murkland Presbyterian Church in Charlotte, North Carolina.

cast journalists arrived on the scene in time to film the fire. This was the first church arson recorded on videotape. Footage of the church fire was broadcast internationally on the Cable News Network (CNN).

No people were injured in the blaze; however, the sanctuary was destroyed with damages totaling approximately $200,000. The ATF and FBI surveyed the scene with city and state law enforcement and fire investigators. They declared the fire an arson. North Carolina governor Jim Hunt promised a $10,000 reward for information leading to the arrest of the arsonist.

Two days after the fire, a thirteen-year-old white girl confessed that she had set the blaze as part of a Satanic ritual. She also expressed antiblack opinions but admitted that she did not know that the Matthews-Murkland Presbyterian Church was an African American church.

Law enforcement officers arrested and charged the girl, whose name was not released, with arson. They found lighter fluid and other incriminating evidence in her room. During a medical evaluation, health professionals decided that she was emotionally disturbed. Based on this examination, investigators determined that the fire she set was not connected to any others. She pleaded guilty in juvenile court on July 1, 1996, and her family sent her to a mental health facility. Three months later, she was sentenced to twelve months of probation and two hundred hours of community service.

Consequences

On the Saturday after the fire, President Bill Clinton devoted his weekly radio address to outlining how the federal government would expand its efforts to end church burnings. He identified the Matthews-Murkland Presbyterian Church as the thirtieth southern black church razed within an eighteen-month period. Clinton accorded church arson investigation a higher priority and pledged his support for legislation that eased efforts to prosecute federally people who assaulted religious structures.

At the Sunday service at Matthews-Murkland Presbyterian Church following the fire, the Reverend Larry Hill preached a sermon about forgiveness. About two hundred people, including members of Congress and prominent civil rights leaders, gathered near the sanctuary's remains. Hill later expressed concern for the young arsonist and hope for her rehabilitation.

After the girl's confession revealed the fire was not racially motivated, President Clinton canceled a speaking engagement at the North Carolina church. Church leaders chose to use donations to build a community center for at-risk children instead of rebuilding the sanctuary.

Most people still believed that the church fires were set by white supremacists to express their racist beliefs. Reporter Gary Fields of *USA Today* investigated the church fires and determined that many journalists had incorrectly concluded that each fire fit into a larger pattern. Although some arsonists did specifically target black churches, many other arsonists acted individually for racially unrelated reasons. Both whites and African Americans were arsonists, and white churches and synagogues had also been damaged. The Charlotte arson reinforced his findings.

Arsonists set fire to three North Carolina churches within one month of the Charlotte fire. The National Church Arson Task Force increased its efforts to educate people about arson and to implement preventive measures. More Americans became aware of church arson as a result of the media's coverage of the Matthews-Murkland fire. That fire caused politicians to focus on arson. Both federal and state officials announced plans to implement stricter penalties for church arsonists.

Elizabeth D. Schafer

Last Members of Freemen Surrender After Long Standoff

A group of militant, antigovernment extremists resisted arrest by barricading themselves on a 960-acre ranch near Jordan, Montana, for eighty-one days before surrendering to federal authorities who were attempting to apprehend the group for a host of crimes including check fraud and threatening to kidnap and kill federal officials.

What: Civil strife; Crime; Law; Social reform; Political reform
When: June 13, 1996
Where: Jordan, Montana
Who:
LeRoy Schweitzer (1939-), a tax resister who became leader of the Freemen
Daniel Petersen, Jr. (1944-), the second-in-command of the Freemen
Rodney Skurdal (1953-), a highly racist member of the Freemen
Edwin Clark (1949-), a member of the Freemen and the original co-owner of the ranch that became the Freemen compound

The Radical Right Movement

Before the Freemen standoff in Montana, several high-profile events signaled the growing presence of radical right extremists in the United States. These included the 1992 shootout between the family of Randy Weaver and the Federal Bureau of Investigation (FBI) at Ruby Ridge, Idaho, and the 1995 Oklahoma City bombing. Also noteworthy was the tragic 1993 confrontation between the Branch Davidians and the FBI in Waco, Texas.

The Freemen of Montana were one of a host of extremist, antigovernment groups associated with the radical right. Heavily influenced by the ideology of the Posse Comitatus, a racist and anti-Semitic organization that was active in the 1970's and 1980's, the Freemen advocated "sovereign citizenship" and "common law." They believed that the federal government was a highly corrupt, illegitimate tyranny. Therefore, they resisted the federal government's authority and claimed the right to establish their own legal and financial systems. To the Freemen, the only legitimate type of government was a local government headed by the county sheriff. To demonstrate their beliefs, they resisted paying taxes and registering their automobiles. They also created common law courts through which they leveled bogus criminal charges and sometimes death threats against government agencies and individuals who they believed had wronged them. To create havoc in the federal banking system, they placed fraudulent liens against the property of others and cashed phony checks.

The Standoff

The 960-acre farm belonging to Edwin Clark and his family was foreclosed on in 1994 for almost $2 million in unpaid government loans. Devastated by this loss, the Clark family was vulnerable to the persuasion of antigovernment advocates, including LeRoy Schweitzer. Although their farm was sold at auction, several members of the Clark family refused to leave the land. They also joined Daniel Petersen, Jr., Rodney Skurdal, and other extremists in forming a common law court and creating their own local sovereign government. Other Freemen also plotted to kidnap and hang a local judge.

Because of these activities, arrest warrants were issued for several Freemen. To escape arrest, Schweitzer, Skurdal, and others took up residence on the Clark farm, which they renamed Justus Township. Well armed, they challenged local officials to remove them. Local officials, who possessed few resources and feared injury or the loss of life, were powerless to respond. For this reason, the FBI was asked to intervene.

Federal officials, however, were hesitant to take immediate action given the tragedies that had recently unfolded at Ruby Ridge and Waco. Therefore, they initially only kept the Freemen compound under surveillance. This allowed the Freemen to conduct business as usual. They were frequently joined by sympathizers, other fugitives, and people attending the Freemen's workshops on how to engage in check fraud and other financial schemes.

On March 25, 1996, the FBI arrested Schweitzer and Petersen when the two men were inspecting a radio antenna located on the ranch. The Freemen did not realize that the men whom they hired to install the antenna were undercover FBI agents. After this peaceful capture, approximately 150 FBI agents, state officials, and staff constructed a loose perimeter around the compound, cut telephone lines to the farm, and denied access to the media. As the pressure on the Freemen increased, several members voluntarily left. The remaining two dozen members were reluctant to end the standoff, a stance that was supported by several vocal radical right leaders across the country.

Over the next two months, the Freemen met several times with state legislators and even Bo Gritz, a well-known member of the patriot movement, to negotiate an end to the siege. These negotiations were not successful. Finally, in early June, 1996, the group engaged in whirlwind negotiations with authorities. During these negotiations, several Freemen children and other members were escorted from the ranch by the FBI. On June 11, 1996, Clark convinced the FBI to allow him to meet with Schweitzer, the Freemen's jailed leader, to discuss terms of surrender. Finally, on June 13, 1996, the last sixteen members of the Freemen peacefully surrendered to federal and state authorities.

Consequences

The eighty-one-day standoff between the Freemen and the FBI was the longest such siege in U.S. history. During the standoff, the lives of the

AP/Wide World Photos

Freeman group members surrender to FBI agents on the compound outside Jordan, Montana.

five hundred people who lived in this small, tightly knit community were greatly disrupted. Not only was the town inundated by federal agents, the media, and Freemen sympathizers but also the media portrayed the beleaguered townspeople as gun-crazed vigilantes. After the standoff, the residents of Jordan breathed a large sigh of relief. As one resident stated, "I'm glad it's over. I'm glad it ended peacefully, and not like Waco." As this comment suggests, the peaceful conclusion to the standoff also helped repair the FBI's reputation, which was badly damaged after the disaster in Waco.

Although the standoff helped turn the Freemen into radical right martyrs, the court's response to their crimes was a sobering blow to the sovereign citizens movement. Members of the Freemen faced numerous charges, including participation in bank fraud, threatening to kidnap public officials, and aiding federal fugitives. Determined to send "a loud and clear message to those who pass this hatred and ugliness around," U.S. district judge John Coughenour advised members of the radical right, "Be forewarned, your personal liberty is at stake." The judge then levied heavy sentences on the Freemen leadership. Schweitzer was sentenced to 22.5 years in prison. Petersen and Skurdal each received fifteen-year sentences. Most of the remaining Freemen received lighter sentences.

Many state governments have sent similar messages by attempting to quash common law activities. For example, new laws have been established to discourage the filing of fraudulent liens and the intimidating of public officials.

Beth A. Messner

Truck Bomb Explodes at U.S. Base in Saudi Arabia, Killing Nineteen

Islamic militants attempting to drive the United States out of Saudi Arabia bombed the Khobar Towers located on the King Abdul Aziz Air Base near Dhahran.

What: Terrorism; War
When: June 25, 1996
Where: King Abdul Aziz Air Base, Dhahran, Saudi Arabia
Who:
BILL CLINTON (1946-), president of the United States from 1993 to 2001
WILLIAM J. PERRY (1927-), U.S. secretary of defense
FAHD BIN ABDUL AZIZ AL-SAUD (1922-), king of Saudi Arabia from 1982

Background

After the Persian Gulf War ended in 1991, U.S. forces remained in Saudi Arabia for the express purpose of patrolling the no-fly zone between Iraq and Kuwait. In 1995, Islamic militants who opposed the continued U.S. presence in the region attacked a U.S. base outside of Riyadh, killing five Americans and two Indians. Four men were charged with the murders of the servicemen and were convicted in the Saudi courts. Terrorist groups warned that they would carry out an attack against the United States if the Saudi government executed the prisoners. However, in 1996, the men were beheaded as required by the strict Islamic *sharia* code.

The Bomb and the Aftermath

On June 25, 1996, less than a month after the executions, two men parked a fuel truck at the northeast corner of the perimeter of the King Abdul Aziz Air Base just outside of Dhahran. As a guard watched, the driver and his companion jumped out of the vehicle and fled the area in a getaway car. The sentry notified base security, and an attempt was made to evacuate two apartment buildings located just 80 feet (24 meters) from the abandoned truck, but the explosion occurred too quickly. The attack resulted in the deaths of 19 servicemen and injuries to 160 military personnel, most of whom were American. The air base housed American, British, French, and Saudi troops. The U.S. Air Force personnel included the 4404th Air Wing. The total number of Americans stationed on the base exceeded two thousand.

U.S. secretary of defense William J. Perry arrived in the Persian Gulf to investigate the incident. In a press conference aboard the USS *George Washington*, Perry announced that the incident could have been prevented if a U.S. request had been granted that would have extended the perimeter of the base from 80 (24 meters) to 400 feet (122 meters) away from the apartment buildings. Responding to Arlen Specter, a Republican senator from Pennsylvania, who criticized his handling of the Defense Department, Perry assured reporters that the welfare and safety of the troops remained paramount and that he resented Specter's remarks, which blamed him personally for the attack before an investigation could be conducted. Perry, who met with Saudi king Fahd, informed the press that U.S. investigators would be allowed to interrogate any suspects arrested for the bombing. In contrast, in the 1995 incident, the Saudis had refused access to the four terrorists who bombed the base outside of Riyadh. Perry also received assurances that the Saudi government would implement stronger measures to combat attacks against foreign troops within its borders.

Air Force secretary Sheila Widnall released the news of the attack at a House National Security Committee hearing. President Bill Clinton

AP/Wide World Photos

A truck bomb ripped a hole in apartment buildings in the Air Force complex in Dhahran, Saudi Arabia, killing nineteen Americans.

called the bombing an act of cowardice and vowed to punish the individuals responsible for the deaths of the U.S. servicemen. Presidential candidate Bob Dole also expressed the need for the capture of the terrorists involved in the attack and expressed his sorrow at the loss of American lives.

Consequences

Continued threats against U.S. military personnel in Saudi Arabia forced the United States to implement tougher security arrangements. In February, 1997, U.S. forces from Dhahran and Riyadh initiated a $150 million emergency move, evacuating more than 4,000 personnel, 78 aircraft, and 250,000 tons of equipment to the remote Prince Sultan Air Base on the edge of the Empty Quarter Desert, 80 miles (129 kilometers) south of the Saudi capital. The United States occupied 25 square miles (65 square meters) within the 250-square-mile (648-square-kilometer) base that the Saudis had just started developing.

Security remained high even though the base was located behind an 8-foot-high (2.4-meter-high) fence and tons of barbed wire with the inner perimeter located more than 3 miles (4.8 kilometers) from the outer boundary of the facility. All deliveries into the base were transferred to military trucks miles away from the facility to prevent the possibility of truck bombs killing or injuring additional Americans. Even the latrines on the base are cleaned out by the military and then transported to sanitation vehicles away from the compound.

The Air Force initiated a three-month rotation for soldiers at this location because of the increased difficulties and inconveniences endured by the soldiers who live in seven hundred dun-colored, air-conditioned tents on the edge of the desert, where temperatures reach as high as 118 degrees Fahrenheit (48 degrees Celsius) in the summer. The commander of the forces, Brigadier General Daniel M. Dick, assigned 10 percent of his personnel to security detail on a full-time basis, and additional troops were used to create the appearance of dummy security patrols. According to Dick, his primary concern was the protection of U.S. forces regardless of the inconveniences.

Although the terrorists from the June 25, 1996, truck bombing were never apprehended, increased security measures protected U.S. lives in the region and allowed the U.S. Air Force to fulfill its mission. After the move to the Prince Sultan Air Base, no American lives have been lost in Saudi Arabia because of terrorist attacks. The Air Force personnel continued to fly their F-15's and F16's on two hundred missions a day patrolling the no-fly zone between Iraq and the border between Saudi Arabia and Kuwait.

Cynthia Clark Northrup

Yeltsin Wins Runoff Election for Russian Presidency

Boris Yeltsin won reelection to the presidency of the Russian Federation, drawing 54 percent of the vote versus 40 percent for his opponent, Gennady Zyuganov, leader of the Russian Communist Party.

What: Government; International relations; Political independence
When: July 3, 1996
Where: Russia
Who:
BORIS YELTSIN (1931-), incumbent Russian president from 1991 to 1999
GENNADY ZYUGANOV (1944-), Russian Communist Party leader

The Campaign

The 1996 presidential election was the first in Russia after the collapse of the Soviet Union. Boris Yeltsin was completing his first presidential term under the constitution that operated when Russia was part of the Soviet Union. He was first elected president in 1991, the year that the Soviet Union disintegrated. When he announced his candidacy for a second term on February 15, 1996, his prospects were very poor. The economic reforms, commonly known as "shock therapy," had led to considerable economic hardship. During his first term, the country's gross domestic product fell 50 percent, unemployment grew, and hyperinflation wiped out the savings of millions of Russians. Yeltsin's program of privatization, which was designed to transfer industry from state to private ownership, led to the accumulation of great wealth in the hands of a few oligarchs. In 1995, Yeltsin suffered two heart attacks. On top of those problems was a bitter war in the rebellious province of Chechnya that the Russians seemed unable to win. In January, only 6 percent of the public planned to support Yeltsin.

Yeltsin's unpopularity and vulnerability attracted seventy-eight potential opponents. Of that number only ten qualified as presidential candidates. They were Gennady Zyuganov, leader of Russia's Communist party; Vladimir Zhirinovsky, leader of the Liberal Democratic Party, a right-wing nationalist organization; Gregory Yavlinsky, leader of Yobloko, a reform party; Alexander Lebed, a retired general with nationalist leanings; Mikhail Gorbachev, former president of the Soviet Union; Svyatoslav Fyodorov, noted eye surgeon; Martin Shakkum, director of a reform fund; Vladimir Bryntsalov, legislator and millionaire; and Yuri Vlasov, Olympic weight lifter.

From the beginning, Zyuganov was the leading challenger. The Communist Party, the sole legal party throughout the Soviet period, was the only party with a national organization. Yeltsin acknowledged Zyuganov's challenge and portrayed himself as the only viable alternative to communist repression.

Yeltsin's transformation from a probable loser to a winner was one of the most dramatic political comebacks of modern democratic politics. How it happened is the subject of considerable analysis and speculation. One factor was Yeltsin's success in convincing the emerging middle class that the election of a communist president meant the end of economic reform and the probable demise of democracy in Russia. Yeltsin had an advantage in the overwhelming support of the print and television media, which feared a Zyuganov victory would lead to press censorship as it had existed in the Soviet period. He also had the benefit of generous financial support from the wealthy oligarchs as well as encouragement from the West.

Yeltsin shamelessly tailored his policies to maximize their popular appeal. He promised to end the war in Chechnya, signed decrees eliminating conscription by the year 2000, and made military service in dangerous places such as Chechnya optional for conscripts. Studies show that the sharpest rise in his popular support (in April and May) came as a result of massive increases in social spending by the government. Among the benefits he conferred were doubling the minimum pension, compensating those whose savings were devalued by hyperinflation, and most important paying the arrears in wages of millions of workers. The state largesse promised (and delivered) by Yeltsin benefitted students, teachers, medical workers, pensioners, war invalids, veterans, children, single mothers, scientists, Cossacks, and the unemployed, among others.

Finally, Yeltsin proved to be a more effective campaigner than his rivals. In the spring, he made an energetic series of trips to many regions of Russia. In several of these regions, Yeltsin signed bilateral agreements with regional leaders conferring additional powers on the regions.

The Election Results

With ten candidates in the race, it was unlikely that any single candidate would garner a majority vote as required by the constitution, and none did. After the first round of voting on June 16, 1996, Yeltsin led the group with 35.28 percent of the vote, and Zyuganov followed closely with 32.03 percent. Taking the middle ground were Lebed with 14.52 percent of the vote, Yavlinsky with 7.34 percent, and Zhirinovsky with 5.7 percent. Those bringing in less than 1 percent of the vote were Fyodorov, 0.92 percent; Gorbachev, 0.51 percent; Shakkum, 0.37 percent; Bryntsalov, 0.16 percent; and Vlasov, 0.2 percent. A total of 75.7 million people voted, or 69.8 percent of the eligible voters.

A runoff election between the top two men was required. Between the first and second round of voting, a great deal of political maneuvering took place, much of it on the part of Yeltsin. He particularly wanted the support of

AP/Wide World Photos

Boris Yeltsin.

those who had voted for Yavlinsky and Lebed in the first round. Earlier Yeltsin had tried to obtain Yavlinsky's backing by offering to bring him into his administration, but Yavlinsky refused to back out of the race. Within days after the first round, Yeltsin achieved success with Lebed. The general endorsed Yeltsin and in turn was given the high-ranking posts of security adviser and security council secretary. That more than ensured Yeltsin's victory. After the second round of voting on July 3,1996, Yeltsin had received 54 percent of the vote and Zyuganov 40 percent; 5 percent of the voters registered their opposition to both men (figures rounded).

Consequences

At age sixty-five, Yeltsin became independent Russia's first popularly elected president. The term was for four years, but Yeltsin chose not to complete his second term. On New Year's Eve, 1999, he resigned the office to permit his prime minister, Vladimir Putin, to become acting president. Yeltsin's second administration was undermined by a succession of serious health problems that severely weakened his capacity to govern. Unknown to the public at the time, Yeltsin had suffered a heart attack just before the runoff vote. Later that year, he had heart bypass surgery. Though he recovered from the surgery, he remained in bad health throughout much of his term in office. His later years in office were turbulent. In 1996, he was forced to fire Lebed. In 1998, he was confronted with a financial crisis that forced the devaluation of the ruble. In the last few years of his administration, Yeltsin appointed and dismissed five different prime ministers. When he retired, the war in Chechnya was still raging.

At the beginning of the twenty-first century, it remained unclear whether democracy in Russia would survive or what forms it would take. However, overall, Yeltsin's accomplishments were substantial. No leader in a thousand years of Russian history had done more to promote democracy in that country.

Joseph L. Nogee

House of Representatives Passes Legislation Barring Same-Sex Marriages

The U.S. House of Representatives defined marriage as the union of a man and woman and permitted states to decide whether to recognize same-sex marriages performed in other states.

What: Human rights; Social reform
When: July 12, 1996
Where: Washington, D.C.
Who:
Bob Barr (1948-), a Republican member of the House of Representatives from Georgia
Barney Frank (1940-), an openly gay Democratic member of the House of Representatives from Massachusetts

Defining Marriage

On July 12, 1996, the U.S. House of Representatives voted to prohibit federal recognition of same-sex marriages. The legislation, known as the Defense of Marriage Act, also permitted each state to ignore same-sex marriages performed in other states. In approving the Defense of Marriage Act, the House provided the first definition of marriage written into federal law.

Defining marriage had been a power reserved to state governments. The Defense of Marriage Act clearly defined marriage as a union between a man and a woman. It also defined "spouse" as a person of the opposite sex who is a husband or wife. This definition would make it impossible for homosexual couples to qualify for spousal benefits from the federal government. In addition to defining marriage, the bill also told states that they may refuse to recognize same-sex marriages performed in other states.

Representative Bob Barr, a Republican from Georgia, introduced the legislation in part because of concern that a court in the state of Hawaii would require that state to recognize marriages between two people of the same sex. Hawaii would have been the first state to grant this recognition, creating a constitutional issue for other states. Article IV of the U.S. Constitution requires that "full faith and credit shall be given in each state to the public acts, records, and judicial proceedings of every other state."

The legal action in Hawaii began in December, 1990, when three homosexual couples filed applications for marriage with the Hawaiian Department of Health, the agency responsible for administering the state's marriage laws. The department denied the application because it claimed that marriage laws did not allow for same-sex marriages. The couples sued, taking the case to the Hawaii supreme court, which decided to review the case. Before the Hawaii supreme court reviewed the case, the House of Representatives passed the Defense of Marriage Act.

The Progress of the Legislation

Representative Barr introduced his bill on the floor of the House on May 7, 1996. The bill had 117 cosponsors, other representatives who agreed to provide their names to the bill as a show of support. Cosponsors included 105 Republicans and 12 Democrats. A companion bill was introduced in the U.S. Senate on the same day, a fairly common practice.

One week after the bill's introduction, the House Subcommittee on the Constitution held hearings. The witnesses presented information on the social and constitutional issues raised by the legislation. The discussion largely centered on morality and family values. On May 30, 1996, the subcommittee approved the bill for review by the House Committee on the Judiciary. The Judiciary Committee approved it on June 12, 1996, and it was sent to the floor of the House for consideration and debate.

The debate on the House floor started on July 9, 1996. Supporters of the bill argued that the Defense of Marriage Act was necessary to protect the American family. Representative Barr said the bill was needed because "the flames of self-centered morality are licking at the very foundation of our society, the family unit." Opponents included the three openly gay members of the House, Representatives Barney Frank and Gerry E. Studds, both Democrats from Massachusetts, and Steve Gunderson, a Wisconsin Republican. These opponents contended that the bill was not necessary and was most likely unconstitutional.

On July 12, 1996, the House voted on the measure. The bill won passage with the backing of 224 Republicans and 118 Democrats. Sixty-five Democrats, one Independent, and one Republican (Gunderson) voted against the bill. Representatives Sheila Jackson Lee of Texas and Major R. Owens of New York, both Democrats, voted present. The next step for the legislation was the U.S. Senate.

Consequences

The Senate Judiciary Committee held hearings on the Defense of Marriage Act on July 11, 1996. Like the House hearings, the hearings in the Senate were charged with moral overtones. The bill was reported to the floor of the Senate for debate on September 10, 1996. The Senate approved the bill by a vote of 85 to 14, with all 53 Republicans joined by 32 Democrats voting "yes." One senator, Arkansas Democrat David Pryor, was absent. President Bill Clinton, a Democrat, signed the bill into law on September 21, 1996.

The consequences of the Defense of Marriage Act emerged slowly. Because he signed the bill, President Clinton was able to keep the gay rights issue out of his 1996 reelection campaign. As of five years after its adoption, the legislation had not been challenged. In 1999, the Hawaii supreme court ruled that the original lawsuit was moot because of a constitutional amendment ratified by Hawaiian voters in 1998, allowing the state to limit marriages to couples consisting of a man and woman.

Most of the reaction to the Defense of Marriage Act occurred at the state level. A majority of state legislatures passed similar state acts, as allowed by the federal law. The Vermont legislature took a different approach. In 2000, under pressure from the Vermont supreme court to grant gays and lesbians the same rights as heterosexuals, the legislature passed a bill providing for civil unions, which was signed into law by Democratic governor Howard Dean. A civil union offers state-given rights of marriage such as inheritance and next-of-kin status, but it avoids the controversial word "marriage." Legal battles could ensue when gay couples travel to Vermont for civil unions and demand that their states recognize the unions. Rather than settling the question, the passage of the Defense of Marriage Act marked the beginning of additional debate on the right of same-sex couples to marry in the United States.

John David Rausch, Jr.

Prince Charles and Princess Diana Agree to Divorce

The marriage of Britain's Charles and Diana, the prince and princess of Wales, ended when they agreed to divorce on July 12, 1996, after nearly fifteen years of marriage.

What: Government
When: July 12, 1996
Where: London, England
Who:
CHARLES (1948-   ), prince of Wales, heir to Great Britain's throne
DIANA (1961-1997), princess of Wales, the former Lady Diana Spencer

A Fairy Tale Marriage

What began as a fairy-tale romance and marriage for Prince Charles and Lady Diana Spencer, Great Britain's prince and princess of Wales, ended in a much-publicized divorce almost exactly fifteen years after it began. On July 29,1981, millions of people all over the world watched on television as England's future king married Lady Diana Spencer, thirteen years his junior, in an elaborate ceremony at Saint Paul's Cathedral in London. The royal couple quickly won the approval of both British subjects and people around the world. Within a year of their marriage, Charles and Diana produced an heir to the throne, Prince William. Two years later, the couple added a second son, Prince Harry.

The combination of Diana's beauty, youth, and demeanor caused the public and the press alike to scrutinize her every move. Although Charles had grown accustomed to the press's presence in every aspect of his daily life, the experience was new and troubling to Diana. In addition to adjusting to a new husband, new official duties heading numerous charitable organizations, and two young sons, Diana also had to adjust to the celebrity she had achieved in such a short period of time.

The stress resulting from all these adjustments soon took its toll on the pair's marriage. Not long after Harry's birth, reports of marital problems began to surface in the media. These reports and others that Charles had reestablished his relationship with a former companion, Camilla Parker-Bowles, served only to further diminish the storybook status of their marriage.

By the late 1980's, it was apparent that the royal marriage was seriously damaged. Amid speculation from the media and from royal-watchers, Charles and Diana began to lead separate lives if only unofficially. However, in 1992, rumor became reality when British prime minister John Major announced that the couple had officially separated. After both Charles and Diana admitted to having been unfaithful in their marriage, Queen Elizabeth recommended in December of 1995 that the couple divorce.

A Fairy Tale No Longer

In December of 1995, following the queen's recommendation, Prince Charles and Princess Diana, the fairy-tale couple no longer, began the process of ending their marriage. It would take nearly two years for the details and settlement to be worked out. Difficult decisions had to be made concerning all aspects of the couple's lives. Especially difficult were the decisions concerning Diana, who would still be the mother of England's future king, and her standing in the royal family as well as a financial settlement. Between December of 1995, when the process began, and July of 1996, when Charles and Diana agreed to the terms of their divorce, progressively more details of their troubled marriage emerged, and it became apparent that their union was never what the world had believed it to be.

Ultimately, Diana was allowed to keep the title "princess of Wales" but was stripped of the title "her royal highness," a move that was much criticized by Diana's many supporters. By denying Diana the latter title, Queen Elizabeth, who could have allowed the princess to retain it, effectively moved Diana one step further from official status in the royal family. In addition to reporting these facts, the press also immediately released unofficial reports that Diana would receive a lump-sum settlement of between $22 million and $27 million along with approximately $600,000 a year to maintain her official office. Charles and Diana were granted joint custody of their sons, Prince William and Prince Harry, and Diana retained the right to live at her home, Kensington Palace.

On July 12, 1996, Charles and Diana agreed to all the conditions of the divorce decree. The preliminary divorce decree was then issued on July 15, 1996, and on August 28, 1996, the divorce officially became final. The most celebrated royal marriage of the century had ended after just more than fifteen years.

AP/Wide World Photos

Princess Diana.

Consequences

The impact that Prince Charles and Princess Diana's marriage and divorce had on Great Britain's monarchy is significant. The monarchy had, for several decades, been considered obsolete and in need of modernization. Although it was never an option to eliminate the monarchy completely, many believed changes needed to be made. The marriage of Charles and Diana in 1981 breathed new life into the ailing monarchy, and the presence of Diana, a young and modern commoner, added the element of glamour that had long been missing from the royal family. Thus, the country and the world viewed the union as a new beginning for England's monarchy.

However, as with many hopes and dreams, the marriage was unable to maintain the necessary strength to sustain the ideal for long. As their marriage failed, the public's confidence in the monarchy began to disintegrate as well. By July 12, 1996, when the couple agreed to the terms of the divorce, the country once again faced questions about the future of the monarchy. Diana, who had so enlivened the royal family, wanted to continue to play a part in the future of the monarchy. In a press conference, the princess who would never share England's throne expressed her wish to become, if not England's queen, then "the queen of people's hearts."

The difficult aftermath of their divorce, in which Charles struggled to regain the confidence of his subjects and Diana worked to define a new identify, was short-lived. Little more than a year after the divorce became final, Diana was killed in an automobile crash in a Paris tunnel. For Great Britain, the question of how to come to terms with Charles and Diana's divorce suddenly became one of how to deal with the death of the woman who had become the "People's Princess." Since that time, Charles has worked to repair his image as the country's future king, and Diana's legacy as a compassionate crusader for the less fortunate continues through the charities she worked for so faithfully during her lifetime.

Kimberley H. Kidd

Pipe Bomb Explodes in Park During Summer Olympics in Atlanta, Georgia

Two people died and more than one hundred were injured when a bomb exploded in Atlanta's Centennial Olympic Park during the 1996 summer Olympic Games. Officials considered closing the park and other Olympic venues to prevent further bombings but instead increased security to keep music and other cultural activities open.

What: Terrorism; Crime; Sports
When: July 27, 1996
Where: Atlanta, Georgia
Who:
Eric Robert Rudolph (1967-), a suspected domestic terrorist
Richard Jewell (1958-), an Olympics security guard

Olympic Security Concerns

The Olympic Games symbolize international peace and goodwill as they bring together athletes from all over the world to compete. Sponsoring the Games is an honor for the host city and host nation, which earn international recognition for their efforts. The Games require tremendous financial expenditures for athletic and cultural facilities, the organization of athletic competitions, and the effective mobilization of thousands of volunteers and paid workers. The 1996 summer Games represented a tremendous investment by the city of Atlanta, the state of Georgia, and the United States.

Threats of violence to athletes, officials, visitors, and local residents during an Olympics can cause economic loss, political conflicts, and the loss of national honor. The international media covers the athletic competitions and any events associated with the Games. The potential to have such a large impact can attract domestic and international terrorists as well as others interested in disrupting the games. In 1972, Palestinian terrorists attacked athletes during the Olympic Games in Munich, Germany. Israeli athletes were taken hostage and killed in the shootout between the terrorists and German police. The tragic events were watched on television by millions of people all over the world. The images from Munich in 1972 have encouraged all Olympic hosts to invest heavily in security to protect athletes, officials, and visitors.

Threats to the 1996 Olympic Games

Preparations for the 1996 Olympic Games in Atlanta included large investments in security precautions by the Atlanta Committee for the Olympic Games, the state of Georgia, the city of Atlanta, the federal government, and other agencies. Security was provided by more than thirty thousand federal, state, and local law enforcement officials, military personnel, private security guards, and volunteer law enforcement officers from many nations as well as thousands of volunteers who manned gates and watched fences. Military and police counterterrorist forces were posted around the city, and officials monitored the whereabouts of known international and domestic terrorists.

On July 27, 1996, a bomb exploded in Centennial Olympic Park. One person was killed by shrapnel, a television cameraman died from a heart attack, and more than one hundred people were injured by shrapnel or in the scramble to get out of the park. A threat had been called in to authorities from a public telephone near the park shortly before the explosion, and law enforcement officers had located the suspected bomb. Many of those hurt in the explosion were law enforcement officers who were close to the bomb, trying to determine its nature or in the

AP/Wide World Photos

Centennial Park in Atlanta, after an explosion hit a tower near the stage at a concert during the Olympics.

process of moving park visitors to safety. Dozens of bomb threats had been made before the bombing, and authorities were uncertain whether the telephoned threat was a hoax. Problems with the city's emergency communication center also caused the message to be delivered late to law enforcement authorities and for the location of the bomb to be uncertain because the center did not have an address for the park.

After the explosion, emergency medical personnel attended to the injured, and law enforcement officers cleared the park in case there were more bombs. Officials initially feared that there would be more bombings in the park or at other venues and closed the park. Within a few days, Olympic authorities announced that the "Games will go on," and the park was reopened with heightened security. Security was also heightened at Atlanta's forty athletic venues and at the cultural events. City officials declared that terrorists would not ruin the Games.

Consequences

The Olympic Games in Atlanta were expected to attract groups hostile to the international athletes and visitors and prepared to use violence to achieve their political agendas. Authorities felt that domestic terrorists might see the Games as an opportunity to embarrass government officials and to scare visitors to the games. Similarly, international terrorists might see the Games as an opportunity to carry out attacks against foreign enemies or to embarrass the U.S. government. Domestic extremists carrying firearms and explosives had been arrested on their way to Atlanta, and Georgia-based groups had also been active before the Games. Before the bombing, several threats were made, but none was viewed as presenting a serious risk to the Games' participants or spectators.

Initially, law enforcement officials reviewed videotapes taken by visitors to the park, trying to identify the bomber. The Federal Bureau of Investigation (FBI) targeted Richard Jewell, a park security guard who had helped point out the bomb, but after months of investigation and derogatory news coverage, he was found to have had no involvement in the bombing. The attention of law enforcement officials shifted when similarities were noted between the park bombing and other bombings. A similar bomb had been used in an attack on a women's clinic in Birmingham, Alabama, just before the Olympics. In that explosion, a nurse was very seriously injured and an off-duty police officer working as a security guard was killed. On January 16, 1997, a bombing took place at a women's clinic in Sandy Springs, north of Atlanta. In this case, a man was seen acting suspiciously in the vicinity of the clinic. The bombing involved two devices, a smaller one that damaged the clinic and a larger one placed where emergency responders and law enforcement officers were likely to be. Fortunately, cars parked near the second bomb protected most of the people who were close by. Two bombs were also used in a February, 1997, attack on a gay and lesbian club in Atlanta. Ultimately, authorities tied the bombings to Eric Robert Rudolph and a manhunt began in the mountains of North Carolina. As the search passed the four-year mark, Rudolph had not been found, and law enforcement authorities began to believe that he died while in hiding.

William L. Waugh, Jr.

Clinton Signs Bill to Raise Minimum Wage

Nearly 10 million American workers learned they would get a raise when a new federal minimum-wage rate schedule was announced on August 2, 1996. Minimum wages would increase from $4.25 to $4.75 in October, 1996, and to $5.15 in September, 1997.

What: Business; Labor; Economics
When: August 2, 1996
Where: Washington, D.C.
Who:
BILL CLINTON (1946-   ), president of the United States from 1993 to 2001
FRANKLIN D. ROOSEVELT (1882-1945), president of the United States from 1933 to 1945

Minimum Wage Legislation

On Saturday, June 25, 1938, U.S. president Franklin D. Roosevelt signed 121 bills. Among these was a landmark law in the United States' social and economic development, the Fair Labor Standards Act of 1938 (FLSA). Enacted after a fierce political battle with Congress, the act banned most child labor, set the minimum wage at twenty-five cents per hour, and fixed the maximum workweek at forty-four hours. Frances Perkins, Roosevelt's secretary of labor, was the chief architect of the new policy. At first, the new wage legislation applied to only about one-fifth of all workers. Over time, amendments extended coverage to approximately 80 percent of all workers. The FLSA does contain exemptions from the minimum wage that apply to some workers. For example, the law includes a youth subminimum wage that employers can pay employees under the age of twenty during their first ninety consecutive calendar days of work for that employer.

No long-term formula has been developed for increases in the minimum wage. However, since 1938, the minimum wage has been increased nineteen times, reaching $1.00 an hour in 1956, $2.00 an hour in 1974, $3.10 in 1980, and $4.25 in 1991. Both Republican and Democratic administrations have been active on this issue. Presidents Harry S. Truman, Dwight D. Eisenhower, John F. Kennedy, Jr., Lyndon B. Johnson, Richard M. Nixon, Jimmy Carter, George Bush, and Bill Clinton signed minimum-wage increases into law. Each time the minimum wage increased, questions arose about its impact on unemployment, particularly its effect on teenage unemployment. The major concern was that the increase in the minimum wage would benefit workers who kept their jobs but would also force businesses to reduce the number of employees, making it harder for teenagers to find work.

Congress Adjusts the Minimum Wage

In 1989, Congress passed legislation raising the minimum wage from $3.35 to $3.80 in 1990 and $4.25 in 1991. After reaching $4.25 an hour, the minimum wage remained at that level for more than five years. In any discussion regarding wages, the impact of inflation must be taken into account. If prices rise, but people's wages stay the same, their ability to purchase goods declines, or as economists would say, their real wages decline. In terms of purchasing power, the minimum wage hit a peak in 1969 but then began a sharp decline. By 1995, the minimum wage adjusted for inflation was approaching its lowest level in forty years. Inflation had largely wiped out the last increase in the minimum wage approved by Congress in 1989. In February, 1995, President Clinton proposed raising the minimum wage by ninety cents in two steps in 1996 and 1997. This would not be an insignificant action, for by 1996, approximately 10 million U.S. workers were earning between $4.25 and $5.14 per hour.

Because a conservative group of Republicans controlled Congress in 1995, many observers felt that an increase in the minimum wage was not likely. As debate on the issue began, it became ev-

AP/Wide World Photos

President Bill Clinton signs a bill raising the minimum wage as Vice President Al Gore and children of minimum-wage earners watch.

ident that overall public support for an increase was quite strong. Even after the proposed increase to $5.15, the minimum wage remained low. A person who worked full-time for the entire year (roughly two thousand hours) at this wage would earn only $10,300. This was not enough money to keep a family above the poverty level. After a lengthy debate, Congress included the proposed increases as part of the Small Business Job Protection Act of 1996. Both the Senate and the House of Representatives passed this legislation on August 2, 1996. President Clinton signed it into law on August 20, 1996. With the new legislation, the minimum wage increased from $4.25 to $4.75 in October, 1996, and to $5.15 in September, 1997.

Consequences

Economists debate the impact of increasing the minimum wage. Traditional economic analysis holds that the increase in minimum wage will increase unemployment. The argument is that many workers produce just enough to cover their wages and to generate a very slight return to the owner of the business. If wages rise, revenue will not rise enough to cover all the increase in wages. Workers then cost the business more than the value of their output. Therefore, some workers will be laid off, or the business will fail. This impact is seen as particularly troublesome for teenagers, who, because they are inexperienced, tend to be less productive than older workers.

However, studies that examine the actual impact of increases in minimum wage often find results opposite of those predicted by traditional theory. Researchers often find that each increase in minimum wage is followed by an increase in employment. There are at least two explanations for this outcome. First, if inflation exists, what matters is the real wage, not the unadjusted wage. If prices rise by 3 percent and wages rise by 1 percent, real wages go down, so each worker actually becomes cheaper for the business. Second, if forced to pay higher wages, business owners and managers may provide better equipment or more training to employees, which increases their productivity. If workers are more productive, a higher wage is justified. In turn, a higher wage may encourage workers to work harder or to care more about their jobs.

What is the evidence following the latest minimum-wage increase? In August, 1997, just before the latest increase, the national unemployment rate was 4.8 percent. Over the next three years, it fell to 4.0 percent. Although there are many factors that influence the unemployment rate, this decrease in unemployment after an increase in the minimum wage is an outcome that cannot be easily explained by traditional economic theory.

Allan Jenkins

NASA Announces Possibility of Life on Mars

NASA scientists found traces of life processes and possible microscopic fossils in a meteorite believed to have come from Mars.

What: Astronomy; Biology
When: August 7, 1996
Where: National Aeronautics and Space Administration (NASA) headquarters, Washington, D.C.
Who:
David S. McKay (1936-), a NASA geologist
Richard N. Zare (1939-), a Stanford University chemistry professor
Daniel S. Goldin (1940-), the administrator of NASA

Life on Other Planets

Is Earth the only planet that supports life? Scientists have pondered that question ever since they found that the other planets, like Earth, are bodies orbiting the Sun. Only in the last century, however, have scientists been able to gather the evidence needed to answer the question.

Astronomers found that of all the planets, Mars is most like Earth and therefore offers the best environment for possible extraterrestrial life. Although Mars now has no surface water, is extremely cold, and has thin air, between 4 billion and 3.5 billion years ago it had a much thicker, warmer atmosphere and large rivers. Similar conditions on Earth spawned single-cell organisms at about the same time.

NASA scientists thus hoped to find signs of life when they landed two space probes, Viking 1 and Viking 2, on Mars in 1976. The landers scooped up samples of Martian soil and tested it for traces of life-related chemicals. None was found. Space probes sent to other planets—particularly the Galileo probe to Jupiter and its moons—also searched for conditions suitable for life, detecting promising locations but no definite life signs. Rocks brought back from the Moon by Apollo astronauts showed no life signs at all.

Rocks from Mars

On August 7, 1996, at NASA headquarters in Washington, D.C., a panel of scientists, headed by NASA administrator Daniel S. Goldin, held an unexpected news conference. At the news conference, David S. McKay told reporters that while inspecting a meteorite (a rock that came to Earth from outer space), he and his NASA research team had found strong evidence of bacteria-like life; moreover, because of its composition, the meteorite was believed to have come from Mars. Both McKay and Goldin, however, cautioned that there was not yet enough evidence to prove beyond doubt that they had found proof of the existence of Martian microorganisms.

The meteorite is called Allan Hills 84001 (ALH84001), after the place in Antarctica where it was found in 1984. About the size of a potato, it weighs 1.9 kilograms and was formed on Mars 3.6 billion years ago, at just the time when Mars had an Earth-like environment and could have supported life. About sixteen million years ago, an asteroid hit Mars and knocked ALH84001 into space; the meteorite landed on Earth about thirteen thousand years ago.

After two years analyzing ALH84001, McKay's team found four signs of life deep inside it. First, surrounding small orangish-brown beads of carbonate were bands of two minerals, magnetite and iron sulfide, that living organisms sometimes make. Other processes can form the minerals, but not so close together.

Second, chemical compounds called polycyclic aromatic hydrocarbons (PAHs) were also found next to the carbonate beads. PAHs are common by-products of chemical processes on Earth; for example, they are produced when

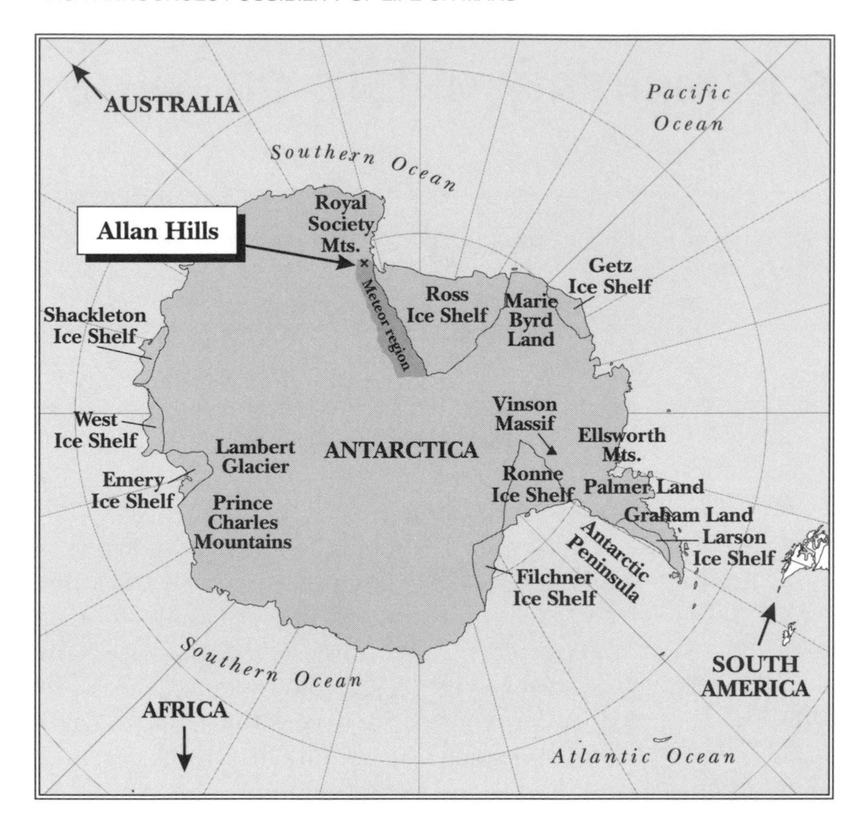

the minerals, PAHs, and fossil-like shapes lie within a few hundred-thousandths of a centimeter of one another—seemingly too close together to be a coincidental accumulation of by-products from nonbiological processes. It is most likely, Zare and McKay argue, that the shapes are fossils of Martian organisms that made the minerals and PAHs.

Goldin promised that more work would be done to test the conclusions of McKay's team. He also described spacecraft that NASA planned to send to Mars; at least one would be designed to dig up soil samples for return to Earth. The samples could perhaps finally settle the question of whether life ever existed on Mars.

some microorganisms turn into fossils. Yet according to Richard N. Zare, the leader of the NASA team's chemists, the PAHs in ALH84001 are an unusual mixture of lightweight types and are not likely to have come from Earth organisms.

Third, when McKay's team looked inside the carbonate beads with a high-power electron microscope, they spotted tube-like and egg-like shapes. Some of these shapes are similar to the shapes of fossilized Earth bacteria, although much smaller. These shapes, McKay and Zare concluded, could be fossils of Martian microorganisms, but the scientists were not sure; the shapes could also be bits of dried clay or even contaminants from the Antarctic environment.

These first three types of evidence, if considered one by one, do not make a strong argument for life on Mars. Taken together, they do. That is the fourth type of evidence: Deep inside the meteorite, where contaminants from Earth's environment are unlikely to be found,

AP/Wide World Photos

This meteorite, in a Johnson Space Center lab in Houston, contains organic elements that may indicate that life once existed on Mars.

Consequences

The news conference created a worldwide sensation. The media speculated about the possible existence of life on Mars and elsewhere in the universe. Excited scientists contributed to the speculation; many noted that, if confirmed, the discovery would mark a turning point in human history by showing that life is not unique to Earth.

If it is shown that life did develop on Mars, scientific theories about the origin and evolution of life in general will have to be changed to account for that profound fact. Moreover, if life could have existed on Mars, the chances appear good that it also could arise on planets circling other stars in the universe.

Yet not all scientists were enthusiastic. Like McKay, many biologists, chemists, and meteorite researchers insisted that much more work would have to be done to prove the discovery genuine. Some critics have doubted that the dating of the fossils is correct. Others have devised nonbiological explanations for the presence of the minerals and PAHs in the meteorite, while still others argued that the fossils are really Earth bacteria that seeped into the meteorite after it fell on Antarctica. Nevertheless, the announcement generated enormous interest in and enthusiasm for the study of Mars and for the space sciences generally, as many fascinated observers eagerly awaited the resolution of the debate.

Roger Smith

Taliban Leaders Capture Capital of Afghanistan

Only four years after the establishment of an independent government in Afghanistan, the Taliban surprised the Afghan government and the world when it took the capital city of Kabul.

What: Civil war; National politics; Coups
When: September 27, 1996
Where: Kabul, Afghanistan
Who:
Taliban, the Islamic ruling group in Afghanistan
Mullah Mohammad Omar (1962-   ), leader of the Taliban

Problem of Political Stability in Afghanistan

By the time the Taliban, also known as the Taliban Islamic Movement of Afghanistan (TIMA), took control of Kabul, Afghanistan's people had suffered many years of war and political turbulence. In the late nineteenth century, the British and Russian empires fought over control of Afghanistan. In 1919, Afghanistan gained full independence from Britain, and in 1926, a monarchy was established. Because the new country was an amalgamation of many different groups of people, it experienced difficulty establishing a stable and peaceful government. A coup felled the monarchy in 1973, and its leaders established a republic. Another coup, led by pro-Soviet leftist Afghans, took place in 1978. The following year, the Soviet Union invaded Afghanistan, supposedly at the behest of the Afghan government, to subdue anticommunist revolts. With the help of Pakistan and the United States, a resistance movement, the Mujahideen, fought the Soviets and finally pushed them out by 1989.

At that point, the different groups in the Mujahideen, who had worked together during the resistance, each wanted their own leaders to form the new government. By 1992, some of them established a Mujahideen government, but conditions in the country were very difficult for the people. Millions had died, been disabled, or were refugees from their home areas. Land mines were everywhere. Some areas of Afghanistan were still controlled by local warlords. Corruption and smuggling were common. In this context, a group of Islamic students and academics formed the Taliban with the help and training of the Pakistani government. The group promised to bring peace, stability, and honesty through a strict and fundamentalist application of Islamic law.

The Taliban Take Afghanistan

Starting in 1994, the Taliban took over some of the small towns and provinces southwest of Kabul. Early in 1995, it quickly took control of thirty-one provinces. Then, slowly but surely, it pushed closer and closer to the capital city. The first time the Taliban attacked Kabul, it lost. However, on September 21, 1996, it gained control of Jalalabad, a key city on the road to Pakistan. The leaders in Kabul became so worried that they joined with their old enemies, who also opposed the Taliban, in order to hold on to the capital and maintain a supply link with their northern borders. This was not enough. After just six days, the Taliban took control of the capital city.

The government leaders and the world community were surprised. All along, no one believed that the students, though well organized, could win against the battle-hardened soldiers from the Soviet resistance days. The Taliban set up their own government and renamed the country the Islamic Emirate of Afghanistan. Although the Taliban, under the leadership of Mullah Mohammad Omar, proclaimed itself the new government of Afghanistan, it did not control the entire country, and most other countries did not recognize the Taliban government. One of the older warlords still controlled territory in the north along the border of Uzbekistan, and the remnant of the Mujahideen government retained control of a larger territory in the northeast along the border of Tajikistan. In 1996, the Taliban began its fight for control of these areas. By 2001, the Mujahideen controlled only about 10 percent of Afghanistan; however, only three countries had recognized the Taliban as the legitimate government of Afghanistan: Pakistan, Saudi Arabia, and the United Arab Emirates.

Consequences

The impact of Taliban control in Afghanistan was immense. By 2001, most of Afghanistan was at peace, smuggling and looting were largely gone, and corruption had been reduced dramatically though not eliminated. However, this peace had come at a great price for everyone. Because the students who made up the Taliban were strict fundamentalists, they instituted a penal code based on that of Hammurabi, an eighteenth century B.C.E. Babylonian leader who wrote down an early set of laws that were strict and severely enforced. According to the Taliban's code, the punishment for stealing was the severing of the offender's hand or arm. Anyone who spoke against the Taliban or who broke any of its rules could be executed. These executions were public and could be quite gruesome. All men had to grow beards if they wanted to appear in public.

The rules for women were more restricting. As soon as the Taliban seized the capital, it ordered all women to stay in their homes. Women and girls were not allowed to attend school, and women could not work even if they had to support their families. When walking down the street or going to the market, all women had to be escorted by male members of their families and completely covered from head to toe in *burqas* (a long, thick cloth that covers the body and face). If they broke these rules, women could be stoned to death in public.

After gaining power in 1996, the Taliban regime is known to have aided international terrorists, particularly those of Osama bin Laden's al Qaeda organization, which is believed to have been responsible for bombings of the U.S. embassies in Kenya and Tanzania in August, 1998. The regime also forced international relief workers—many of whom were not Muslims—to leave the country, resulting in a drop in supplies of medical and food resources. Meanwhile, The regime ordered the eradication of the opium poppy, which is used to make heroin and morphine. The poppy was a common crop in Afghanistan, which formerly produced nearly 70 percent of the world's heroin. As late as May, 2001, the U.S. government actively supported the Taliban's anti-opium program.

When unknown terrorists hijacked four American airliners and launched destructive suicide attacks on the Pentagon Building and the World Trade Center in New York City on September 11, 2001, suspicion immediately fell

AP/Wide World Photos

Taliban rebels exchange greetings near the bodies of President Najibullah (right) and his brother Shahpur Ahmedzai.

upon bin Laden's al Qaeda movement, whose headquarters were believed to be in Afghanistan. Backed by an international alliance dedicated to ending international terrorism, the U.S. government issued an ultimatum to the Taliban to surrender bin Laden and his followers or face the full wrath of the U.S. military. The regime refused to comply with the ultimatum and prepared for war. Allied missile and bomber strikes against Taliban targets began in October, while Afghan rebels who had been fighting the regime advanced on Taliban centers from the north. By December, the Taliban regime had collapsed, and international plans were underway to install a new, democratic government in Afghanistan.

Carolyn V. Prorok

Supreme Court Declines to Hear Challenge to Military's Gay Policy

The U.S. Supreme Court refused to review a decision upholding the constitutionality of the "Don't Ask, Don't Tell" policy under which an avowed homosexual was dismissed from the U.S. Navy.

What: Law; National politics
When: October 21, 1996
Where: Washington, D.C.
Who:
Paul G. Thomasson (1965-), a homosexual naval officer
Bill Clinton (1946-), president of the United States from 1993 to 2001
Colin Powell (1937-), chairman of the Joint Chiefs of Staff from 1989 to 1993

Political Background

During the 1992 presidential campaign that brought him to office, U.S. president Bill Clinton announced that if elected, he would alter the military's policy on homosexuality. At that time, homosexuals were officially barred from membership in the armed forces; Clinton wished to allow them in. Nine days after his inauguration, he directed Defense Secretary Les Aspin to reevaluate the policy of excluding homosexuals from service in the armed forces. From March to July, 1993, hearings on the matter were held by the armed services committees in both the House of Representatives and the Senate. Testimony was received from military commanders, gay rights activists, experts in military personnel policy, and others. The president's plan met with immense political resistance, particularly from General Colin Powell, chairman of the Joint Chiefs of Staff. President Clinton was forced to agree to a compromise, the "Don't Ask, Don't Tell" policy.

The new policy was announced by the president on July 19, 1993. Under his directive, the armed services were to cease any inquiry into the sexual orientation of service personnel or new recruits. Homosexual servicepeople were still subject to separation from the service in three circumstances: if they were found to have engaged in or solicited homosexual acts, if they stated that they were homosexual or bisexual, or if they "married or attempted to marry a person of the same biological sex."

The new policy was passed by Congress in the National Defense Authorization Act for Fiscal Year 1994, which President Clinton signed into law on November 30, 1993. The central difference between the new policy and the existing policies was that the Department of Defense was to stop asking questions about sexual orientation as part of enlistment processing. In turn, homosexual recruits were expected to remain silent and not reveal their sexual preferences.

The Thomasson Case

On March 2, 1994, the day after a naval regulation was promulgated implementing the new statute, Lieutenant Paul G. Thomasson delivered a letter to four U.S. Navy admirals with whom he had served in the bureau of naval personnel in Washington, D.C. The letter stated that he was a homosexual. In accordance with military policy, the Navy began proceedings to dismiss Thomasson from the service. An administrative board was convened, which after hearing evidence on Thomasson's service record (which was excellent) and on homosexuality in general, unanimously found that he had made statements that he had "a propensity or intent to engage in homosexual acts" as prohibited by the policy. The board recommended that Thomasson be honorably discharged. A Navy review board upheld the decision, and Thomasson filed suit in U.S. District Court for the Eastern District of Virginia to prevent the Navy from discharging him.

Thomasson raised three issues. He claimed that the "Don't Ask, Don't Tell" policy violated his right to free speech, that the ban on homosexuals in the armed forces violated the equal protection clause of the Constitution, and that his discharge would deprive him of due process of law. The heart of Thomasson's position was the equal protection argument. He believed—and wanted to establish—that the equal protection clause forbids the government from discriminating on the basis of sexual preference or orientation. After the district court decided against him on all three questions, Thomasson appealed his case to the U.S. Court of Appeals for the Fourth Circuit in Richmond, Virginia.

The court of appeals also rejected Thomasson's claims. It dealt first with the central equal protection issue. As long as there was a rational basis for Congress to make its finding that homosexual activity is harmful in the military service, the regulations are constitutional, according to the majority of the court. The court, reciting all the constitutional clauses that confer military powers on Congress and the president, deferred to their judgment as to appropriate policy in this area of law. The court also pointed out that the elected branches of the government have a stronger claim to resolve this kind of questions especially when, as here, a long national debate had resulted in the fashioning of a compromise.

The court also gave short shrift to Thomasson's First Amendment argument. Thomasson's statement gave rise to a presumption of future homosexual conduct. He could have attempted to rebut that presumption but refused to do so; moreover, prior cases had established that expressive conduct can be limited by the necessities of military life. Under the Navy's policy, Thomasson would have been free to criticize the policy or to affiliate with and publicize a group that opposed the policy. What he could not do was state that he was a homosexual and expect to remain in the Navy.

Thomasson petitioned the U.S. Supreme Court for a writ of certiorari, which would have set the case for further review by the Court, but on October 21, 1996, the Court refused to grant the writ.

Consequences

The votes of four of the nine Supreme Court justices are required to grant a petition for certiorari. There are many reasons why the Court might decide not to hear a case besides a decision on the merits of an issue. The justices may feel that the important questions are not presented clearly enough or that the case is not important enough to warrant review at a particular time. However, when a case is rejected, the decision of the next lower court down—in this case, the Fourth Circuit Court of Appeals—remains in effect. Therefore, the Court's decision not to hear Thomasson's case meant that the military's policy of separating homosexuals from the service would remain in effect indefinitely. The Court's reluctance to raise the standard of equal protection review for sexual orientation from the rational basis to one of the higher and more demanding levels, reflects both deference to the political process and the Court's unwillingness to establish homosexuality as a new constitutional right.

Robert Jacobs

Clinton Is Reelected

Incumbent president Bill Clinton easily defeated Senator Bob Dole, the Republican candidate, becoming the first Democratic president since Franklin D. Roosevelt to be elected to two terms.

What: National politics
When: November 5, 1996
Where: United States
Who:
Bill Clinton (1946-   ), president of the United States from 1993 to 2001
Bob Dole (1923-   ), senator from Kansas, majority leader of the U.S. Senate, and Republican presidential candidate
Ross Perot (1930-   ), Reform Party presidential candidate
Newt Gingrich (1943-   ), Republican speaker of the House of Representatives

Four More Years

On November 5, 1996, Bill Clinton was reelected president of the United States by a significant margin over Kansas senator Bob Dole, although for the first time in U.S. history, less than 50 percent of the eligible voters cast a presidential ballot.

President Clinton's victory was perceived by many commentators as a considerable comeback. In his early presidency, Clinton had been unable to establish his own policy priorities and had faltered badly over health care reform. In addition, a significant number of Americans viewed him and his administration as both incompetent and corrupt. In 1994, the Republicans, led by Georgia representative Newt Gingrich, had proposed a bundle of conservative reforms called the Contract with America and had succeeded in taking over both houses of Congress from the Democrats for the first time in decades. Clinton learned from that defeat, and in the next two years, he publicly exhibited a modesty and a willingness to listen. However, Gingrich and many Republicans failed to learn from their victory, becoming overly confident and arrogant.

After the 1994 congressional elections, many Republicans assumed that Clinton could be easily defeated in 1996. However, some leading Republican candidates, notably General Colin Powell, declined to run. Dole had challenged for the Republican nomination in 1988, losing to Vice President George Bush, who was later elected president. Dole, Senate majority leader and a World War II veteran, had been a congressional figure for many years and was well-respected in Washington, D.C., but was little known to most Americans. However, his Republican opponents were even less known, and Dole wrapped up the Republican nomination by early spring. Jack Kemp, a former New York congressman and former professional football player, became Dole's vice presidential candidate.

The Campaign

In spite of his problematic first term, Clinton faced no serious opposition to renomination, and Vice President Al Gore was again chosen as his running mate. A major issue of the campaign was Clinton himself, particularly his character. In 1992, critics attacked him for not serving in the military in Vietnam. During his first term, he also faced allegations that he had been unfaithful to his wife, Hillary Rodham Clinton. An ongoing investigation examined the extent and nature of his involvement in a failed real estate deal called Whitewater in his home state of Arkansas, and during the 1996 campaign, Clinton and his supporters, including Vice President Gore, were accused of obtaining campaign funds illegally.

Dole's campaign never caught fire. In the spring, he was 15 percentage points behind in the polls. After the Republican convention in the summer, Dole crept to within 3 percentage points of Clinton, but after the Democratic convention a few weeks later, the president's lead returned to double figures, and there it remained.

Clinton, for all the questions about his personal character, was one of the most charismatic politicians in U.S. history, easily comparable to the master communicator of the previous decade, Ronald Reagan. Many voters did not trust Clinton entirely, but they liked him. In an era in which personal foibles and failures were the subject of television talk shows, Clinton, with all of his personal faults, had an advantage over the older and more diffident Dole, a figure out of an earlier America. In addition, Clinton capitalized on the widespread suspicion of Gingrich and his Contract with America. As bright as Clinton and equally fascinated with public policy, Gingrich came across in public as abrasive and too radical in his conservatism. In the presidential campaign, Clinton chose to campaign against Gingrich and his policies, largely ignoring candidate Dole.

Clinton came to the presidency with the public image of a moderate Democrat, and after the Democrat's congressional defeat in 1994, he moved even further away from the traditional liberalism of the Democratic Party. He supported welfare reform that reduced benefits and required recipients to actively seek employment. He also promised to put more police on the streets to fight crime and strongly supported a balanced budget. In essence, he successfully coopted many traditional Republican programs and policies, making them his own.

Most important, the country was both prosperous and at peace. The economy had emerged from the doldrums of the early 1990's, a situation reflected in the rapidly rising stock market, a symbol of the times. The Cold War was over, and current foreign policy matters were of little concern to most Americans. As the incumbent, Clinton was the beneficiary of that peace and prosperity, and most voters were unwilling to

AP/Wide World Photos

President Bill Clinton (left) and Vice President Al Gore celebrate their second electoral victory.

depart from what appeared to be a satisfactory status quo.

Consequences

The election outcome was no surprise. The political polls had predicted the final result several weeks before voting day. Americans reelected Clinton, who garnered 49 percent of the vote to Dole's 41 percent. The popularity and support experienced by the Republicans in 1994 did not translate into victory in the 1996 presidential election; however, the Republicans maintained their majority in Congress. Ross Perot, a Texas businessman and the Reform Party candidate, received far fewer votes than he had in 1992.

Men were equally divided between Clinton and Dole, with 44 percent voting for each candidate, but women chose Clinton by a significant margin over Dole, largely because of social issues such as abortion rights, which the congressional Republicans, if not Dole himself, seemed eager to abolish. Whites voted slightly in favor of Dole, 46 percent to 43 percent for Clinton, but African Americans decisively supported the incumbent over the challenger, 83 percent to 12 percent.

The number of voters dropped to an all-time low, with less than 50 percent of those eligible casting their ballots. Interpretations varied on the meaning of such a low turnout. Some claimed that the voters were angry at traditional politicians with their misleading policy promises and their money-centered activities, and others argued that Americans were generally satisfied with things as they were and therefore had no strong preferences among the candidates and their programs. As a result, incumbency was a decided advantage, and Clinton became the first Democratic president to be reelected to a second full term since 1936.

Eugene S. Larson

Texaco Settles Racial Discrimination Lawsuit

Texaco, one of the nation's largest oil companies, agreed to pay the plaintiffs in a class-action racial discrimination suit a record settlement of $176.1 million.

What: Law; Civil rights and liberties; Labor
When: November 15, 1996
Where: White Plains, New York
Who:
Bari-Ellen Roberts (1952-), the lead plaintiff
Sil Chambers, co-plaintiff
Peter Bijur (1942-), chief executive officer of Texaco
Richard Lundwall (1941-), a Texaco executive

Recognizing Discrimination

Bari-Ellen Roberts, an African American professional in the finance office at Texaco Oil, recognized that there was a racial divide within the company's corporate offices and that African Americans like herself were not being promoted. Likewise, company records indicated that many African American Texaco employees were not being paid as much as the minimum salary for their employment grade (job level). Roberts became more alarmed when her own performance evaluation was altered after she had already seen it. Sil Chambers, another African American Texaco professional, shared with Roberts his own experiences of discrimination.

Upon further investigation, Roberts found that other Texaco employees had also been repeatedly denied opportunities to supervise junior staff. She and other African American employees reported being subjected to stereotyping, racially motivated comments, jokes and slurs, and intimidation by white male executives. When Roberts and fellow plaintiff Chambers complained about the treatment that they received, they were labeled as troublemakers. This prompted the African American employees to consult with an attorney and then file a lawsuit charging Texaco with racial discrimination.

The Lawsuit

In 1994, Roberts and Chambers were joined by four other Texaco employees as the named plaintiffs in the racial discrimination lawsuit filed against this corporate giant. Over the next two years, the employees, Texaco executives, and other witnesses were questioned regarding the work environment and the possibility that Texaco might be discriminating against African Americans. The attorneys from both sides attempted to discredit their opponents. By this time, many of the African American employees of Texaco had stepped forward and joined the class-action lawsuit. During the discovery process, the plaintiffs acquired evidence that Texaco had been involved in race discrimination issues, sexual harassment issues, and unfair labor practices in the past and in multiple divisions within the corporation. This evidence suggested that Texaco was guilty of systematic discrimination (discrimination that is planned and approved of by the top executives and is widespread).

The judge in the case ordered the parties to submit to mediation with the hope that a meaningful, reasonable, and just agreement could be reached without the case actually going to trial. However, by January of 1996, the judge had ordered the two sides back into court to explain why an agreement had not been reached.

The case was at a near standstill through the first half of 1996. The Equal Employment Opportunity Commission (EEOC) had begun an investigation of the widespread claims against Texaco related to unfair labor practices. In June of 1996, the EEOC released its report, indicating that Texaco was guilty of failing to promote Afri-

AP/Wide World Photos

In November, 1996, Texaco chairman Peter I. Bijur publicly apologized for racist remarks made by top executives of the company.

can American employees and other types of discrimination. These findings by the EEOC gave the case a much-needed burst of energy.

Another important twist occurred in August, 1996, when one of the white male executives, Rich Lundwall, came forward with audiotapes that he had secretly recorded during meetings among various white Texaco executives. These tapes exposed Texaco executives discussing a plan to destroy evidence that would be damaging to their defense in the discrimination case. The executives were heard using racial slurs and referring to African American employees as "Black Jelly Beans." This evidence was the turning point for the plaintiffs in the lawsuit. Now, in addition to the EEOC report, the African American employees had the voices of Texaco executives on tape making the same types of statements they had been accused of making. These individuals were also planning to obstruct justice by destroying evidence.

Consequences

The case against Texaco came to a quick end just two and a half months later when Texaco agreed to settle out of court with the plaintiffs. By this time, national leaders of the African American community had rallied in support of Roberts and the other plaintiffs and had begun protesting and picketing against Texaco and had also planned a nationwide boycott of Texaco products and service stations. The new Texaco chief executive officer, Peter Bijur, was under serious pressure from stockholders to settle the suit. The lawsuit was settled on November 15, 1996, for $176.1 million. This was the largest settlement in a race discrimination lawsuit in U.S. history to date. Texaco's African American salaried employees who worked for the company between March 23, 1991, and the settlement date were given a lump-sum settlement of $115 million, giving each employee approximately $63,000. In addition, Texaco agreed to raise the pay of African Americans still employed by the company by 11 percent over five years, a total of about $26 million. Texaco was also to pay $35 million for a five-year task force that was to be responsible for revising Texaco's personnel procedures.

The judge in the case approved the settlement amount, and the case was over. However, the lasting effects of the case remained. Several Texaco shareholders filed suit against the board of trustees for failure to protect the company and its profits from such a lawsuit. Lundwall was indicted for obstruction of justice for his alleged role in the destruction of evidence. Roberts was forced to resign from Texaco to preserve her right to speak about this event and subsequently wrote a book about her experience. This case set a precedent for future cases of racial and other forms of discrimination and made the nation aware that not all men and women were being treated equal in corporate America

Darlene E. Hall

Ghana's Annan Is Named U.N. Secretary-General

On December 17, 1996, acting on a Security Council recommendation the General Assembly appointed Kofi Annan of Ghana as the secretary-general of the United Nations.

What: International relations
When: December 13, 1996
Where: New York
Who:
Kofi Annan (1938-   ), U.N. secretary-general
Boutros Boutros-Ghali (1922-   ), former U.N. secretary-general
Francesco Paolo Fulci (1931-   ), envoy, U.N. Security Council
Razali Ismail (1939-   ), envoy, U.N. General Assembly

Background

In December, 1996, Kofi Annan was selected the secretary-general of the United Nations. He arrived at his new post with vast professional experience as well as a distinctive educational and social background. He was born in Kumasi, a provincial capital of the Ashanti region in Ghana and is descended from a family of traditional chiefs and hereditary nobles. As a child, Annan enrolled in Ghana's leading boarding school. Later he attended the University of Science and Technology in Kumasi. At age twenty, he won a Ford Foundation scholarship, which enabled him to complete his undergraduate education at Macalester College in St. Paul, Minnesota. Between 1961 and 1962, Annan pursued graduate studies in economics at the Institut Universitaire des Hautes Études Internationales in Geneva, Switzerland. In 1971-1972, he was a Sloan Fellow at the Massachusetts Institute of Technology, where he earned a master of science degree in management.

He joined the United Nations in 1962 as an administrative and budget officer for the World Health Organization in Geneva and was classified as P1, the lowest professional designation. Later he served with the U.N. Economic Commission for Africa in Ethiopia, the U.N. Emergency Force in Egypt, and the office of the U.N. High Commissioner for refugees in Switzerland. At U.N. headquarters in New York, he held the positions of assistant secretary-general for human resources management, security cocoordinator for the U.N. system, and assistant secretary-general for program planning, budget, finance, and the comptroller.

AP/Wide World Photos

Kofi Annan.

Annan served on numerous boards of the United Nations and other organizations and received several honorary awards from educational and other institutions. He gained professional experience outside the U.N. system between 1974 and 1976, when he returned to his home country of Ghana, where he served as the managing director of the Ghana Tourist Development Company. During the same period, he served on the board of that company and on the Ghana Tourist Control Board.

While in the service of the United Nations, Annan was given a number of sensitive diplomatic assignments relating to conflicts in Iraq and the former Yugoslavia. Before being confirmed as secretary-general, he served as assistant secretary-general for Peacekeeping Operations and then as undersecretary-general with the same task force. When he was selected secretary-general, Annan, fluent in English, French, and several African languages, became the chief administrator of an international parliamentary system and the boss of ten thousand international civil servants.

Appointment as Secretary-General

In November, 1996, U.N. secretary-general Boutros Boutros-Ghali's bid for reappointment was vetoed in the Security Council. Because of an informal tradition that a specific region of the world fills the post of secretary-general for two consecutive terms, the Organization of African Unity (a grouping of African nations) was asked to suggest a candidate to replace the Egyptian Boutros-Ghali. Annan became that candidate.

After weeks of debate among members of the U.N. Security Council, Italy's Francesco Paolo Fulci, president of the council, announced that the world body had agreed on Annan as Boutros-Ghali's replacement. At a private meeting on December 13, 1996, the Security Council unanimously recommended that Annan be appointed the new secretary-general of the United Nations for a term of office from January 1, 1997, to December 31, 2001. Once he was formally approved, his name was sent to the 185-member General Assembly for ratification. He was appointed secretary-general on December 17, 1996. On December 18, 1996, the General Assembly, presided over by Razali Ismail of Malaysia, elected Annan as the secretary-general of the United Nations.

Annan thus became the seventh person to assume the role of secretary-general. Previous secretaries-general are Boutros-Ghali (1991-1996), Javier Pérez de Cuéllar of Peru (1981-1991), Kurt Waldheim of Austria (1971-1981), U Thant of Burma (1961-1971), Dag Hammersjold of Sweden (1953-1961), and Trygve Lie of Norway (1946-1952). Annan was the first secretary-general to be elected from the ranks of the U.N. staff.

Consequences

Secretary-General Annan came to office amid demands for U.N. reform. Having worked within the organization for more than three decades, he was highly familiar with the missions and challenges surrounding its operations.

As secretary-general, his first major initiative included an effort to reform and renew the United Nations to enhance coherence and coordination. He also acted to improve the status of women in the U.N. secretariat. His other top priorities included strengthening the organization's efforts in promoting peace and development and in supporting and advocating human rights and other universal human values as expressed in the organization's charter. Also, in line with his effort to restore public confidence in the organization, he organized the Millennium Summit. In his post, Secretary-General Annan became a highly visible African dignitary. This veteran U.N. civil servant established a reputation for fairness, efficiency, and effective world-class diplomacy.

Formal recognition of Annan's exceptional performance came in May, 2001, when members of the General Assembly voted unanimously to appoint him to a second five-year term, six months before his first term was due to end. In October, Annan, along with the United Nations as a whole, was named recipient of the 2001 Nobel Peace Prize. The timing of the prize was significant, as it represented the one hundredth year of the award.

Kwasi Sarfo

Television Industry Panel Proposes Program Rating System

The American television industry adopted a rating system intended to promote better parental guidance for younger viewers. These ratings were to appear on television screens at the beginning of each program and were designed to work in conjunction with newly created V-chip technology.

What: Communications; Entertainment
When: December 19, 1996
Where: Washington, D.C.
Who:
JACK VALENTI (1921-), president of the Motion Picture Association of America
EDWARD J. MARKEY (1946-), member of the U.S. House of Representatives

Demand for Ratings

In a press conference held in Washington, D.C., on December 19, 1996, Motion Picture Association of America president and television industry spokesperson Jack Valenti announced that American television networks would begin displaying content ratings that would alert viewers to programs that might be considered inappropriate for certain age groups. After five decades of commercial broadcasting, this marked the first time that the television industry committed itself to such content ratings for television programming.

Nearly thirty years earlier, also under Valenti's leadership, the motion picture industry had replaced its Production Code with a rating system administered by an independent board. Television networks had no industrywide production code, but each network policed itself through the work of a so-called standards department. As increasing amounts of sex, violence, and offensive language crept into television programs over the years, demands for a rating system for television became more insistent from parent groups, professional associations, religious organizations, and ultimately from elected officials.

In 1995, U.S. representative Edward J. Markey, a Democrat from Massachusetts, won approval of an amendment to what became the Telecommunications Act of 1996. His amendment required newly manufactured television sets to contain a V-chip that would provide the hardware and software to decode television rating information and allow viewers, particularly parents, to block objectionable programs. On February 29, 1996, a conference was convened by President Bill Clinton at the White House to encourage the television industry to come up with a rating system that would address public concerns about program content and at the same time allow implementation of the V-chip technology. Representatives of the television industry promised to develop such a system, knowing that failure to do so could result in a government-designed system being imposed on them.

The Rating System

The rating system announced on December 19, 1996, was somewhat similar to the rating system used for motion pictures, especially in its emphasis on the suitability of programming for certain age groups. The proposal called for the use of ratings to begin the following month, with six different labels: TV-Y, for programs designed to be appropriate for children of all ages; TV-G, for programs that most parents would find suitable for all ages of viewers; TV-14, for programs that many parents would find unsuitable for children under the age of fourteen because of adult themes or intense violence; TV-Y7, for programs designed for children age seven and older; TV-PG, for programs that many parents would find unsuitable for younger children because of coarse language, limited violence, or sexual content; and TV-M, for programs specifically designed to be viewed by adult viewers and therefore unsuitable for children under the age of seventeen. These warning labels would appear in the upper left-hand corner of the television screen for fifteen seconds at the beginning of each television program.

Each of the approximately five hundred broadcast and cable networks was given responsibility for assigning these labels to its own programs, with thousands of hours of programs to be rated each day. News programs were exempted from the system because of concerns that ratings could prompt networks to manage the news to fit a particular rating and hence cause the public to get less information.

Consequences

Before his announcement, Valenti had warned that the television industry would consider any attempt to challenge its ratings system as an assault on the industry's First Amendment rights. Following his announcement, opponents of the system quickly held their own press conference and declared that its reference only to age and not to the specific content would be of little help to parents. These critics of the plan, including the Center for Media Education, the American Psychological Association, the Children's Defense Fund, and Representative Markey, argued that the system would work better if it also included the letters "V" for violence, "S" for sex, and "L" for language. However, the television industry claimed that these letters might frighten viewers away even when the violence was minor or intended to be funny, as in a cartoon, and that including all sexual situations would be overwhelming and blur any sense of intensity or context.

In the months following the introduction of the ratings system, public opinion surveys found that most parents were not using the system to guide their children's television viewing. To give parents more information that might prove useful and thus satisfy the many organizations and experts who criticized the original system, in July, 1997, most television networks agreed to revise the age-based ratings by adding letters to alert viewers to violence ("V"), sexual situations ("S"), coarse or indecent language ("L"), and suggestive dialogue ("D").

Adoption of the rating system left several questions to be answered. First, to what extent would parents actually take advantage of the system? Parents might find the system too confusing or uninformative or simply not care what programs their children viewed. Second, what would be the result of letting networks rate their own programming, rather than assign that task to an independent board? Networks might assign ratings in a way that would lessen the risk of losing audiences and advertisers. Third, what would the effect be on the quality of television programming? Ratings might prove a way for the television industry to simply maintain its usual fare of sex and violence while defusing public opposition to such content. Given these questions, the impact of the rating system would clearly be dependent on the use made of it by both the television industry and the viewers for whom it was intended.

Lawrence W. Haapanen

Guatemalan Government and Rebels Sign Accord

Following the longest and most deadly civil war in the history of Latin America, the Guatemalan government reached a peace settlement with a coalition of rebel forces.

What: Civil war; Human rights
When: December 29, 1996
Where: National Palace, Guatemala City, Guatemala
Who:
Alvaro Arzu (1946-), president of Guatemala from 1996 to 2000
Otto Perez Molina, inspector general of the Guatemalan army
Rolando Moran, commander of the Guatemalan National Revolutionary Unity (URNG)

A Coup Leads to Civil War

In 1954, the Central Intelligence Agency of the United States assisted in the overthrow of the democratic nationalist government of Jacobo Arbenz Guzmán. He had recently been elected president of Guatemala, and the United States considered his plans for land reform to be a threat to its economic and political interests in the region. These interests included the protection of large plantations owned by U.S. corporations and containment of the Communist nations of Soviet Union and Cuba. The coup brought a military-led regime to power. Attempts to overthrow the new regime began in 1960, starting a civil war that lasted longer than any other in the history of Latin America. The war claimed more lives than any other in the region and led to widespread and well-documented abuses of human rights.

The early years of the conflict were characterized by low-intensity guerrilla warfare between loosely organized opposition forces on one side and a military dictatorship on the other. By the 1970's and 1980's, the conflict escalated as opposition forces became gradually more unified under various organizations, eventually coming together as the Guatemalan National Revolutionary Unity (URNG). The nature of the rebel attacks precluded traditional military engagement. Rather, the war was waged in an escalating series of small-scale actions and counteractions.

The government instituted Civil Defense Patrols, which were small, paramilitary organizations trained by the United States and charged with controlling the rebels. In many cases, civilian men living in the vicinity of rebel activity would be forced by the government to join the Civil Defense Patrols. Because it was often difficult to identify specific participants in the loosely organized rebel movement, these patrols frequently retaliated against entire villages located in areas of rebel activity. The brutal response of the death squads, as these patrols came to be known, actually helped insurgent groups recruit new members, perpetuating a cycle of violence.

At the end of the conflict, Guatemala had a population of approximately 10 million people. Because of the nature of the conflict, exact casualty figures will never be known. At least 100,000 people were killed and an additional 40,000 disappeared and are assumed dead. Although both sides in the conflict committed human rights violations, the majority of casualties resulted from actions of the government-sponsored death squads. Approximately 1 million people were displaced from their homes, including between 45,000 and 100,000 who fled the country, many seeking refuge in the United States. Because of its involvement in the conflict, the U.S. government was reluctant to grant asylum status.

AP/Wide World Photos

Guatemalan president Alvaro Arzu (left) and rebel commander Rolando Moran, after the signing of the peace accord.

The Event

In March, 1990, the Guatemalan government and URNG signed the Oslo Peace Accord in Norway, inaugurating a peace process in which government representative Jorge Serrano held direct negotiations with the URNG. In May, 1993, however, Serrano attempted to reimpose military rule in a coup, known as the Serranazo or *auto-golpe*. His purpose was to seize absolute control of the country and to impose a cease-fire without addressing the demands of the URNG. The political crisis of 1993 was resolved by the ascendance to the presidency of former human rights ombudsman Ramiro de León Carpio, and the peace process continued.

Between 1991 and 1996, the government and the opposition signed at least eleven agreements, each moving Guatemala closer to resolving the conflict. During this time, four international church organizations worked with Guatemalan organizations to strengthen the peace process, though they were not part of the formal negotiations. The tenth such agreement was an Accord on Strengthening Civilian Power and the Role of the Army in a Democratic Society, signed in Mexico City on September 19, 1996. This began a process of demobilizing almost 3,000 remaining guerrilla troops and reducing the armed forces from 46,000 to approximately 30,000. More important, the remaining government forces would focus exclusively on defending Guatemala's borders, with internal security becoming the responsibility of the civilian interior ministry.

Following a definitive cease-fire on December 3, 1996, the Accords for a Firm and Lasting Peace was signed on December 29, 1996. Military leaders from both sides of the conflict, including Otto Perez Molina, inspector general of the Guatemalan army, and Rolando Moran, commander of URNG, joined Guatemalan president Alvaro Arzu in the ceremony. Because the conflict had been the focus of international attention for many years, about 1,200 dignitaries, including Prime Minister José Maria Aznar of Spain, President Ernesto Zedillo of Mexico, and the presidents of many other Latin American countries, were present to witness the occasion.

The final accords established a number of goals for Guatemalan government and civil society, including full respect for human rights, deactivation of insurgent forces, reducing military forces, removal of paramilitary forces, and constitutional reforms. Although the population of Guatemala was greatly relieved to end the conflict, the celebration was muted. The tremendous casualties that had been suffered led to a mood that was more somber than celebratory, and many lacked confidence that the accords would be fully implemented.

Consequences

The demobilization of the rebel troops was achieved, and the cease-fire held, but demobilization of government troops was only partially achieved. In the years following the accords, occasional political assassinations continued. Most Guatemalans did not participate in a 1998 referendum on constitutional reforms, and the reforms failed in the polls. A clarification commission created by the accords presented its Memory of Silence Report on February 25, 1999. This report was more thorough and accurate than similar reports following conflicts in other countries. It presented the history of the conflict and government abuses accurately, but the government was slow to prosecute those found responsible for atrocities during the war. At the time the accords were signed, 2 percent of landowners controlled two-thirds of the land, and three of four Guatemalans lived in poverty. The formal end of the conflict did not resolve these problems.

James Hayes-Bohanan

Researchers Announce Development of Atom Laser

Physicists at the Massachusetts Institute of Technology announced the success of an elementary version of a laser that produced a beam of atoms rather than a beam of light.

What: Physics; Technology
When: January 27, 1997
Where: Massachusetts Institute of Technology (MIT), Cambridge, Massachusetts
Who:
Wolfgang Ketterle (1957-), a research group leader at MIT's Research Laboratory of Electronics
Marc-Oliver Mewes and
Michael Andrews, graduate students in Ketterle's research group

Chilling Gas Atoms

Air is primarily oxygen and nitrogen. At room temperature, molecules of these gases zip about with the astonishing average speed of 1,600 feet (490 meters) per second—one and one-half times the speed of sound. Cooling the air will cause its molecules to travel slower. Scientists once thought that if a gas were cooled to absolute zero on the Kelvin temperature scale (minus 459.67 degrees Fahrenheit), all motion of atoms and molecules would cease. Modern science shows that this is not quite true; instead, motion is reduced to a minimum but not eliminated at absolute zero.

The first step in constructing an atom laser is to cool a cloud of atoms to within a whisker of absolute zero. Wolfgang Ketterle and his group at the Massachusetts Institute of Technology's Research Laboratory of Electronics begin the process by vaporizing sodium in an oven. Sodium atoms then stream from a small hole in the oven into a vacuum chamber where they are bombarded by laser light. The sodium atoms both absorb and emit light, but the laser frequency is carefully chosen so that the atoms emit slightly more energy than they absorb, something that they can do only by using up some of their energy of motion. In this fashion, the cloud of sodium atoms is cooled down to one hundred millionth of a kelvin.

To maintain this low temperature, the vacuum must be ultra-pure, and the sodium atoms must be kept away from the chamber walls. The small cloud of sodium atoms—less than 0.1 inch (0.25 centimeters) across—is pushed into place by lasers shining from six different directions and then trapped there by a carefully shaped magnetic field. The cloud is now bombarded with radio waves to drive off the most energetic atoms and further cool the cloud in a process called evaporative cooling. When the temperature reaches about one millionth of a kelvin, the cloud undergoes a phenomenon called Bose-Einstein condensation.

The modern science that describes the behavior of atoms and molecules is called quantum mechanics. Fundamental principles of quantum mechanics include that particles have a wave nature and that the wavelength of a particle is equal to a small number (Planck's constant) divided by the product of a particle's mass and its velocity. This means that slowly moving gas particles (near absolute zero temperature) should have relatively long wavelengths. This is not to say that if you could see them, a group of particles would necessarily look like a series of water waves, but it does mean that the movements of the particles can be correctly predicted by assuming that the particles behave like waves.

Subatomic particles can be divided into two groups depending on how they behave when their waves overlap. Two fermions—named for Enrico Fermi—cannot be in exactly the same energy state, whereas any number of bosons—

named for Satyendra Nath Bose—can be in exactly the same energy state. Electrons are fermions; photons and nuclei that have an even number of protons plus neutrons are bosons. Generalizing Bose's work, Albert Einstein predicted that as the wavelengths of atoms grow and begin to overlap near absolute zero, a cloud of atoms would behave like a single coherent entity, now called a Bose-Einstein condensate. Ketterle's pioneering work in evaporative cooling led his group to achieve a Bose-Einstein condensate in September of 1995.

The Atom Laser

An ordinary laser has a resonant cavity in which photons replicate themselves and thereby amplify the brightness of a light beam. Whereas any number of photons can be created if sufficient energy is supplied to the resonant cavity of an ordinary laser, new atoms cannot be created in the atom laser. However, the number of atoms in the Bose-Einstein condensate naturally increases—or amplifies—to include all available suitable atoms. Like teenagers adopting a new fad, the more atoms in the condensate, the more likely it becomes that new atoms will join the condensate.

Laser light can be focused into intense, tight beams because the light is coherent; that is, the beam's photons all march in lock step. A microscopic observer watching the beam pass would see all of the light waves in the beam rise and fall in unison as they marched by. Graduate student Michael Andrews led the effort to show that the Bose-Einstein condensate is coherent. After twenty hours of aligning and focusing the lasers and magnetic fields, the group achieved success about 3:00 A.M. on November 16, 1996. They formed a condensate and then shined a thin laser beam up through it, cutting it in two. Turning off the magnetic trap, the two condensates began to fall downward and to expand because of the slight repulsion between atoms. When they had expanded enough to overlap, a laser beam shined from the side revealed a startling zebra-striped pattern, exactly the behavior expected when coherent waves overlap. Regions in which wave crests overlapped were thick with atoms, but regions in which wave crests and troughs overlapped were empty.

In an ordinary laser, the light beam escapes from the resonant cavity because the mirror on one end is only partially silvered. Graduate student Marc-Oliver Mewes led the effort to extract an atom beam from the condensate. Each sodium atom behaves like a tiny bar magnet, and these magnets are all aligned in the condensate. The group hit the condensate with short bursts of radio waves, which knocked a fraction of the atomic magnets out of alignment. No longer properly aligned, these atoms fell out of the trap. In analogy with a pulsed laser, each time the condensate was hit with a burst of radio waves, a pulse of atoms fell from the trap.

Consequences

The experiment had created an atom laser, a device that produced a pulsed beam of coherent atoms, and the team announced its results on January 27, 1997. Later, improved versions that beamed atoms in a controlled direction were made. Eventual applications may include precision measurements of fundamental constants and the manufacture of ultra-small integrated circuits.

Charles W. Rogers

Chinese Leader Deng Xiaoping Dies

The chief architect of the economic reforms launched in China in 1978, Deng left a legacy of economic growth and new institutions to the country.

What: Economic reform; Government; International relations
When: February 19, 1997
Where: Beijing, China
Who:
DENG XIAOPING (1904-1997), Communist leader of China from 1978 to 1997
JIANG ZEMIN (1926-), Deng's successor

Rise to Power

The death of Deng Xiaoping marked the end of an era in China. Deng was born in 1904 in Sichuan Province in southwestern China. As a young man he went to France in 1921 under a work-study program. He joined the Chinese Communist Youth League Branch in France in 1992. Upon his return to China in 1924 he joined the Communist Party. He was sent to the Soviet Union to study in 1925 and returned to China in 1926.

His career as a party leader went through a number of ups and downs. Generally speaking, he was successful rising through the ranks in the Communist Party and the Red Army and reached the position of a member of the standing committee of the politburo and secretary-general of the secretariat of the party's central committee. Before his purge during the Cultural Revolution of 1966-1976, he was one of the top leaders in the Chinese Communist Party and the government. During the Cultural Revolution, Deng was removed from all his leadership positions and was sent to the countryside to do manual work. He was accused of being a "capitalist roader." He was reinstated briefly in 1975 and was purged again 1976 by the party radicals who were then in control of the party. After the death of Chairman Mao Zedong in 1976 and the arrest of the Gang of Four radicals, Deng was finally reinstated to all his leadership positions.

Economic Reform

After consolidating his leadership within the party, Deng launched economic reform programs in 1978 that had a profound impact on national development In essence, he wanted to carry out "four modernizations": agriculture, industry, science and technology, and national defense. His primary goal was to make China strong, modern, and prosperous under communist rule, but not to promote a Western type of democracy in China.

Deng's political reforms were mainly administrative changes designed to make the government more efficient to perform its functions, not to expand people's right to political participation in government decision making. His basic policy in economic reform was to liberalize the economic system by providing economic incentives to the people and to open up China for foreign trade and foreign investment.

Deng began his reform work in the agricultural sector. He dissolved the more collectivized commune system and introduced the agricultural production contract responsibility system. This was a popular system with the farmers who worked harder to increase production. The result was the substantial increase of agricultural production and improvement of the living standard of the farmers. The farmers later invested their surplus capital to develop village industries and services. The rapid expansion of the rural economy benefited China's overall national economic development.

Encouraged by the success of the agricultural reform, Deng implemented the contractual responsibility system in the industrial sector. Due to the more complex conditions in industries, it was not as successful as in the agricultural sector.

However, many factories improved their productivity. Opening China's door to more foreign trade and investment was also successful. By the beginning of the twenty-first century, China ranked ninth among nations in world trade. It is the second-largest recipient country of foreign direct investment in the world, next only to the United States.

Foreign trade and foreign investment were instrumental in China's rapid economic growth. Between 1978 and 1998, China's annual economic growth rate averaged 9 percent, the fastest economic growth rate in the world through that period. By 2001 China ranked seventh in the world in gross domestic product and ranked third in the world in terms of purchasing power parity.

China's science and technology also improved during Deng's tenure. More than 90 percent of the school-age children attended schools. Illiteracy was reduced to less than 20 percent throughout the country, and most of the illiterate people were the elderly. The government established new colleges and universities. Equally important, Deng started a policy of sending college graduates, teachers, and other professionals to industrialized countries for advanced studies to help China's ambitious modernization projects.

Deng also launched an ambitious plan to modernize China's large but poorly equipped and poorly trained military forces. China upgraded its conventional and nuclear weapons and improved the training of its military forces. China began launching its own satellites, and in 2000 became the third country in the world to put manned ships into space.

Deng also negotiated with Great Britain and Portugal for the return of Hong Kong and Macao, European-ruled enclaves that had long been challenges to Chinese pride. The recovery of Hong Kong and Macao restored all the territories that the Chinese empire had held during its peak under the eighteenth century Qing dynasty, except Taiwan and Outer Mongolia.

Deng adopted an independent foreign policy and advocated the doctrine of peaceful coexis-

AP/Wide World Photos

Chinese leaders file by the coffin of Deng Xiaoping.

tence, multipolarism, and multiculturalism. In particular, he emphasized territorial integrity, national sovereignty and noninterference with each other's internal affairs among nations. His policies helped make China a major player in international affairs.

Another important contribution of Deng Xiaoping to the development of contemporary China was his farsightedness in preparing for a peaceful transition of political power in China after his death. His handpicked party leadership, with Jiang Zemin as the core, has carried out Deng's reform policies successfully since Deng's death.

Consequences

The Chinese people are generally grateful to Deng Xiaoping for his courage to revise the Marxist and Maoist communist doctrines and practice Deng's "socialism with Chinese characteristics." Only Deng, during his era, had the wisdom, prestige, influence, power, and skill to change the course of China by adopting a more pragmatic policy for China's national development. Without his reforms, China probably would not have been able to achieve the economic success that has led to its being chosen to host the 2008 Olympic Games.

There are, however, problems that have been created by Deng's reforms. These include inflation, growing economic inequality among individuals and geographical regions, government corruption, and environmental degradation. The Tiananmen Square massacre of 1989, in which the government used excessive force to suppress the student-led prodemocracy demonstration resulted in many deaths. Although the demonstration was suppressed, it left a black mark in Deng's legacy. On balance, however, the Chinese people regard Deng Xiaoping as a great leader who was the chief architect of reforms that have made China a great power.

George P. Jan

Galileo Makes Closest Pass to Jupiter's Moon Europa

After a six-year journey through interplanetary space, the unmanned spacecraft Galileo passed within 370 miles of Jupiter's moon Europa, revealing an ice-enshrouded world whose surface characteristics suggest an underlying planetary ocean that may harbor extraterrestrial life.

What: Astronomy; Space and aviation
When: February 20, 1997
Where: Jupiter's moon Europa
Who:
John Anderson, planetary scientist, Jet Propulsion Laboratory
Galileo Galilei (1564-1642), Italian physicist and astronomer, discoverer of Jupiter's moons and Saturn's rings
James Head (1941-), professor of geological sciences, Brown University, and a group leader of the Galileo research team

Planetary Exploration

Following the successes of the Apollo manned lunar lander missions (1969-1972), the Skylab orbital space station (1973-1979), and the manned space shuttle program (beginning 1981), the National Aeronautics and Space Administration (NASA) turned its attention to robotic exploration of the solar system. Principal among the early unmanned probes to Jupiter were the Pioneer 10 and 11 missions of the early 1970's and the twin Voyager spacecraft (1977-1989) that made close approaches to Jupiter and the outer planets Saturn, Uranus, and Neptune before finally heading outward to interstellar space. Named after the seventeenth century astronomer Galileo Galilei, discoverer of Jupiter's four largest moons, the spacecraft Galileo was finally launched in October, 1989, after many months of delay caused by engineering setbacks and shifting administrative priorities following the space shuttle *Challenger* explosion (January, 1986).

After traveling more than 400 million miles, Galileo entered the orbit of Jupiter, largest of the nine known planets and fifth in distance from the sun, on December 7, 1995. The spacecraft deployed a descent probe to analyze Jupiter's dense outer atmosphere, which is composed mainly of hydrogen and helium, then settled into orbit around the massive gas giant, observing its shifting cloud bands and its neighboring moons, Callisto, Ganymede, and Europa. Of these Jovian moons, Europa has proven the most surprising.

Europa appeared in Galileo's long-range television cameras as a pale ivory globe crisscrossed by curving networks of rust-brown striations. Unlike the heavily cratered surfaces of Callisto and Ganymede, the entire surface of Europa is so remarkably uniform that its overall topography varies only a few kilometers in height, making it one of the smoothest planetary objects in the solar system. Although only slightly smaller than Earth's Moon, Europa is many times brighter, which has led planetary scientists to conclude that Europa is covered by a mantle of ice probably several miles thick. Except for a handful of small, strangely shallow craters that become visible only at close range, virtually no high-relief surface features are evident. Also visible on Europa's frozen surface are a number of smooth irregularly shaped areas (maculas) that resemble frozen puddles, perhaps marking impact sites flooded by darker overflow from beneath the surface ice.

Europa Flyby

On February 20, 1997, the Galileo passed within 364 miles (585 kilometers) of Europa, transmitting high-resolution television images of the planetary surface to the Jet Propulsion Laboratory (JPL) in Pasadena, California. (It would later pass even closer to the surface, 124 miles, or 200 kilometers, on December 8, 1997.) Like the Voyager images a decade earlier, those from the Galileo revealed that Europa was an arctic wasteland, its uniform surface marked by a network of rust-brown streaks and fissures (linea). Gaps between the larger wedges of ice, in which recent new crust appears to have formed, suggest the surface of Europa is undergoing, or has recently undergone, plate tectonic-like activity similar to the forces responsible for the formation of Earth's continents. This hypothesis is supported by the presence of widespread "chaotic terrain," chiefly blocks of crust that appear to have been tilted/rotated out of position, like floating icebergs. According to another theory, a layer of ice 62 miles (100 kilometers) thick would be more than sufficient to cover all of the principal topographic features. Darker mottled areas, apparent on the surface, may be regions where the ice (or subsurface slush) is relatively thin, thus revealing a darker surface beneath.

It has been theorized that tidal flexure, caused by the massive gravitational field of giant Jupiter acting in concert with its larger Jovian moons, may be responsible for continual dynamic changes of Europa's surface. These forces, in combination with radiation effects caused by Jupiter's huge magnetosphere, would likely cause a warming of Europa's interior. Such factors might contribute to the likelihood that Europa's surface may conceal a subsurface water ocean, heated by geothermal flexing of the planetary crust. "These new results from the gravity data are very consistent with the idea of subsurface oceans on Europa," noted John Anderson, a planetary scientist at JPL. "We know that Europa has a very deep layer of water in some form, but we don't know whether that water is liquid or frozen."

Consequences

Volcanic ice flows and apparent melting (rafting) of ice on Europa's surface seem to support the theory of a liquid water ocean beneath the surface, either now or at some point in Europa's history. The mottled terrain features appear to represent regions of most recent geological activity on Europa formed as chaotic areas of icy crust break apart to expose a darker underlying substance. Numerous iceberg-like formations (massifs) visible on Europa's surface also seem to support the theory of a subsurface water ocean.

James Head, a group leader of the Galileo research team noted, "The combination of interior heat, liquid water, and infall of organic material from comets and meteorites means that Europa has the key ingredients for life." Europa's great distance from the Sun (its surface temperature is −260 degrees Fahrenheit, or −162 degrees Celsius) might seem to preclude the possibility of life developing on Europa. However, if a water ocean does exist on Europa, the tidal forces that release geothermal energy on Earth may also do the same on Europa. In the Earth's oceans, tube worms and other deep-sea fauna have been discovered that thrive in isolated hydrothermal environments near subocean vents; this fact has fueled speculation among planetary scientists that Europa's ocean waters, if indeed they do exist, may likewise harbor some form of extraterrestrial life.

On the subsequent Europa orbiter mission, scheduled for launch in 2003, scientists planned to use special radar imaging equipment able to see beneath the surface ice.

Larry Smolucha

Scottish Researchers Announce Cloning Breakthrough

A team of scientists stunned the world with the news that they had succeeded in creating an exact copy of an adult mammal, a sheep named Dolly.

What: Biology; Genetics
When: February 22, 1997
Where: Roslin Institute, near Edinburgh, Scotland
Who:
Ian Wilmut, an embryologist with the Roslin Institute
Keith H. S. Campbell, an experiment supervisor with the Roslin Institute
J. McWhir and
W. A. Ritchie, researchers with the Roslin Institute

Making Copies

On February 22, 1997, officials of the Roslin Institute, a biological research institution near Edinburgh, Scotland, held a press conference to announce startling news: They had succeeded in creating a clone—a biologically identical copy—from cells taken from an adult sheep. Although cloning had been performed previously with simpler organisms, the Roslin Institute experiment marked the first time that a large, complex mammal had been successfully cloned.

Cloning, or the production of genetically identical individuals, has long been a staple of science fiction and other popular literature. Clones do exist naturally, as in the example of identical twins. Scientists have long understood the process by which identical twins are created, and agricultural researchers have often dreamed of a method by which cheap identical copies of superior livestock could be created.

The discovery of the double helix structure of deoxyribonucleic acid (DNA), or the genetic code, by James Watson and Francis Crick in the 1950's led to extensive research into cloning and genetic engineering. Using the discoveries of Watson and Crick, scientists were soon able to develop techniques to clone laboratory mice; however, the cloning of complex, valuable animals such as livestock proved to be hard going.

Early versions of livestock cloning were technical attempts at duplicating the natural process of fertilized egg splitting that leads to the birth of identical twins. Artificially inseminated eggs were removed, split, and then reinserted into surrogate mothers. This method proved to be overly costly for commercial purposes, a situation aggravated by a low success rate.

Nuclear Transfer

Researchers at the Roslin Institute found these earlier attempts to be fundamentally flawed. Even if the success rate could be improved, the number of clones created (of sheep, in this case) would still be limited. The Scots, led by embryologist Ian Wilmut and experiment supervisor Keith Campbell, decided to take an entirely different approach. The result was the first live birth of a mammal produced through a process known as "nuclear transfer."

Nuclear transfer involves the replacement of the nucleus of an immature egg with a nucleus taken from another cell. Previous attempts at nuclear transfer had cells from a single embryo divided up and implanted into an egg. Because a sheep embryo has only about forty usable cells, this method also proved limiting.

The Roslin team therefore decided to grow their own cells in a laboratory culture. They took more mature embryonic cells than those previously used, and they experimented with the use of a nutrient mixture. One of their break-

AP/Wide World Photos

Dolly, the genetically cloned sheep.

throughs occurred when they discovered that these "cell lines" grew much more quickly when certain nutrients were absent.

Using this technique, the Scots were able to produce a theoretically unlimited number of genetically identical cell lines. The next step was to transfer the cell lines of the sheep into the nucleus of unfertilized sheep eggs.

First, 277 nuclei with a full set of chromosomes were transferred to the unfertilized eggs. An electric shock was then used to cause the eggs to begin development, the shock performing the duty of fertilization. Of these eggs, twenty-nine developed enough to be inserted into surrogate mothers.

All the embryos died before birth except one: a ewe the scientists named Dolly. Her birth on July 5, 1996, was witnessed by only a veterinarian and a few researchers. Not until the clone had survived the critical earliest stages of life was the success of the experiment disclosed; Dolly was more than seven months old by the time her birth was announced to a startled world.

Consequences

The news that the cloning of sophisticated organisms had left the realm of science fiction and become a matter of accomplished scientific fact set off an immediate uproar. Ethicists and media commentators quickly began to debate the moral consequences of the use—and potential misuse—of the technology. Politicians in numerous countries responded to the news by calling for legal restrictions on cloning research. Scientists, meanwhile, speculated about the possible benefits and practical limitations of the process.

The issue that stirred the imagination of the broader public and sparked the most spirited debate was the possibility that similar experiments might soon be performed using human embryos. Although most commentators seemed to agree that such efforts would be profoundly immoral, many experts observed that they would be virtually impossible to prevent. "Could someone do this tomorrow morning on a human embryo?" Arthur L. Caplan, the director of the University of Pennsylvania's bioethics center, asked reporters. "Yes. It would not even take too much science. The embryos are out there."

Such observations conjured visions of a future that seemed marvelous to some, nightmarish to others. Optimists suggested that the best and brightest of humanity could be forever perpetuated, creating an endless supply of Einsteins and Mozarts. Pessimists warned of a world overrun by clones of self-serving narcissists and petty despots, or of the creation of a secondary class of humans to serve as organ donors for their progenitors.

The Roslin Institute's researchers steadfastly proclaimed their own opposition to human experimentation. Moreover, most scientists were quick to point out that such scenarios were far from realization, noting the extremely high failure rate involved in the creation of even a single sheep. In addition, most experts emphasized more practical possible uses of the technology: improving agricultural stock by cloning productive and disease-resistant animals, for example, or regenerating endangered or even extinct species. Even such apparently benign schemes had their detractors, however, as other observers remarked on the potential dangers of thus narrowing a species' genetic pool.

Even prior to the Roslin Institute's announcement, most European nations had adopted a bioethics code that flatly prohibited genetic experiments on human subjects. Ten days after the announcement, U.S. president Bill Clinton issued an executive order that banned the use of federal money for human cloning research, and he called on researchers in the private sector to refrain from such experiments voluntarily. Nevertheless, few observers doubted that Dolly's birth marked only the beginning of an intriguing—and possibly frightening—new chapter in the history of science.

Jeff Cupp

CIA Official Admits Espionage Activity

A high-ranking Central Intelligence Agency officer pleaded guilty in federal court to having been a Russian spy since June, 1994.

What: International relations; Crime
When: March 3, 1997
Where: Washington, D.C.
Who:
Harold J. Nicholson (1950-), officer with the Central Intelligence Agency (CIA)
Aldrich H. Ames (1941-), former CIA agent convicted in 1994 of having spied for the Soviet Union

Hero Gone Bad

On March 3, 1997, Harold J. Nicholson, an officer in the Central Intelligence Agency (CIA), admitted to selling secret U.S. documents to Russian intelligence agents. Nicholson spent two and a half years working for the Russians, even going as far as to hack into the CIA's computer system in search of new information. Nicholson's admission was one in a series of spy scandals that rocked the CIA during the 1990's, causing serious damage to the agency's reputation. On June 5, 1997, a federal judge sentenced Nicholson, the highest-ranking CIA officer ever convicted of espionage, to twenty-three years in prison for his crime.

Nicholson had enjoyed a rapid rise through the agency. He joined the CIA in 1980 after serving as a captain in the U.S. Army with the elite 101st Airborne division. He spent most of the first half of his CIA career overseas, including posts in Thailand, Japan, and the Philippines. Nicholson landed an important promotion in 1990 when he was made station chief, the CIA equivalent of an ambassador, in Romania. After Romania, Nicholson served as assistant station chief in Kuala Lumpur, Malaysia, before returning to the United States to become an instructor of new agents at the CIA's training center.

The Crime

In 1994, while working as an instructor, Nicholson began his espionage activity. His marriage had disintegrated in a bitter divorce earlier that year, separating him from his three children and straining his finances. Over the course of the next eighteen months, he passed along a continuous stream of information to Russian intelligence sources in exchange for more than $300,000. He came under suspicion in 1995 after a routine lie-detector test raised questions about his loyalty. On June 27, 1996, Federal Bureau of Investigation (FBI) agents who had been following him observed Nicholson putting a briefcase in the trunk of a car bearing Russian diplomatic plates in Singapore. After continued investigation, the FBI arrested Nicholson on November 16, 1996, at Dulles International Airport near Washington, D.C. He was on his way to Zurich, Switzerland, where he was to meet his Russian contacts and pass along fresh information, including rolls of film containing images of top-secret documents and a computer disk bearing classified CIA information.

Nicholson's arrest came at the same time that the full story regarding another CIA agent turned spy, Aldrich H. Ames, became public knowledge. Ames passed secrets to the Soviets and the successor Russian intelligence agencies for nine years, between 1985 and 1994. The information Ames gave the Soviets revealed the name of every important agent working for the United States within the Soviet Union. At least ten of those men were quickly arrested and executed. The CIA was working to reestablish its intelligence-gathering network in Russia when Nicholson began working for the Russians. Nicholson's treachery was not as destructive as that of Ames, but Nicholson did compromise existing operations, and he gave the Russians information about the training classes he had supervised, forcing the

CIA to reassign many officers. He also passed along summaries of Ames's jailhouse interviews, allowing the Russians to repair any damage caused by Ames's arrest.

Consequences

During his interrogation, Nicholson revealed that he had met with his Russian contacts on four occasions to pass along secrets in India, Indonesia, Switzerland, and Singapore. The FBI had witnessed the last rendezvous but had no idea about the other three meetings. In another surprise, the government discovered that Nicholson had made much more than the $180,000 he was thought to have collected. The final total was more than $300,000. The FBI and CIA also learned that Nicholson had given the Russians information endangering American officers working under secret identities. The officers that Nicholson trained and betrayed found their careers seriously damaged. Many could not hope to work undercover anywhere in the world because of the information Nicholson gave to the Russians. Other operations had to be called off because Nicholson had revealed too much information.

In court, Nicholson claimed he had intended to use the money he made spying to benefit his children. He felt guilty for having spent so much time away from home and allowing his marriage to fail. The government agreed to ask for a reduced sentence in exchange for Nicholson's cooperation in the investigation. Nicholson's lawyer asked the judge to further reduce the sentence, based on Nicholson's service in the army and his fourteen years of loyal service to the CIA before turning traitor. On June 5, 1997, Nicholson received a sentence of twenty-three years, seven months in prison for his treason, the term requested by the government.

Nicholson's activities did not damage the CIA nearly as badly as Ames's actions had. The Ames case represented the biggest single blow to the CIA's field operations in the agency's history. The information Nicholson gave the Russians did not directly result in the arrest of any American agents, but the CIA did have to restructure both operations and postings to compensate for Nicholson's exposure of U.S. plans. The acting director of the CIA in 1997 concluded that the government might never learn how much Nicholson's treachery had damaged the agency's operations.

Matthew G. McCoy

Hale-Bopp Comet's Appearance Prompts Scientific and Spiritual Responses

An unusually bright comet approaching Earth resulted in both exciting new astronomical information and age-old fears of the extinction of human life.

What: Astronomy
When: March 22, 1997
Where: Northern Hemisphere
Who:
ALAN HALE, director of the Southwest Institute, Space Research, Cloudcroft, New Mexico
THOMAS BOPP (1941-), an Arizona amateur astronomer

Brilliant Celestial Snowball

Comets, with their eye-catching luminous globes and ghostly tails, have stirred a sense of wonder in sky watchers throughout human history. A particularly awe-inspiring comet was the one discovered on July 23, 1995, by Alan Hale and Thomas Bopp, two amateur astronomers. Comet Hale-Bopp had last been visible from Earth about 4,200 years ago and would not again be observable from the planet for about 2,500 years. Although its closest approach to Earth—about 122 million miles (197 kilometers) on March 22, 1997—was not startling, the record-breaking intrinsic brightness of this comet (peaking at -0.7 or -0.8 magnitude in early April) gave an illusion of its unique nearness to the planet.

Like other comets, Hale-Bopp is a big, dirty snowball composed of a fragile aggregate of rocks, frozen water, and volatile substances such as carbon monoxide and carbon dioxide. Hale-Bopp is thought to derive from the Oort cloud—a hypothetical cometary region surrounding the planets of the solar system—and it typically orbits around the Sun in a large elliptical pattern. Its orbit is inclined nearly 90 degrees from the ecliptic (the plane of the solar system), and at its fastest, the comet travels more than 98,000 miles an hour (158,000 kilometers per hour). As the comet is traveling toward the Sun, its frozen carbon monoxide converts from a solid into a gas, and as it nears the Sun, its frozen water likewise vaporizes. The result is a remarkably brilliant coma (a head of jetting dust and gas) with extensive tails. One tail is composed of freed reflective dust particles driven by sunlight. Another, more spectacular tail is made up of light-releasing ionized gas molecules blown by the solar wind.

Frightening Celestial Omen

Because of its unprecedented brightness, Hale-Bopp aroused public as well as scientific interest. Throughout history so-called blazing stars have often been imagined to be omens of ruinous upheaval threatening human existence. Hale-Bopp's extraordinary luminosity, its seeming large appearance, and its false impression of nearness to Earth's orbit all contributed to a sense of degeneration and decay some people tend to sense at the end of a millennium.

In a manner similar to the reaction to Halley's comet in 1910, some people feared that Earth and the tail of Hale-Bopp would intersect with catastrophic consequences. However, even at its closest approach, the comet was not especially near Earth. If such an intersection had occurred, moreover, the comet's tail would have behaved like a harmless vacuum, and its poisonous traces would have had no effect.

Several Christian fundamentalists, in spite of the disappointment that resulted from similar expectations about Comet Kohoutek in 1973, interpreted Hale-Bopp as a sign of the end of time as prophesied in the Book of Revelation in the Bible. They specifically identified the comet with the "star called Wormwood" (Rev. 8:11), the fall-

AP/Wide World Photos

The Hale-Bopp comet and the Grand Teton, as seen from Moose, Wyoming.

ing celestial body that is supposed to strike and destroy Earth, the Moon, and even the Sun.

Heaven's Gate, a group combining a belief in Christianity and Unidentified Flying Objects (UFOs), interpreted Hale-Bopp as the return of the Star of Bethlehem that signaled the birth of the Christ child two thousand years ago. Hoping to be taken aboard a spaceship that they believed would follow Hale-Bopp, this sect committed mass suicide in March as the comet made its closest approach.

Other UFO believers whose claims received attention from the mainstream press spoke of Hale-Bopp as either an alien mother-ship or as an object under alien control. They pointed to alleged "course corrections" in the comet's path as an indication that this celestial object was behaving abnormally. In fact, however, they were misinterpreting astronomers' corrections of early estimates of the comet's trajectory.

News stories also pointed to the electronic images made through a telescope that revealed a saucer-shaped (Saturn-like) object behind Hale-Bopp. In fact, this object was simply a high-magnitude nearby star, its apparent rings the result of a diffraction distortion common in overexposed stellar photography. When comet discoverer Hale and the popular astronomer Carl Sagan, among others, explained this phenomenon, they were accused by various UFO believers of conspiring with national governments to cover up the truth.

Consequences

The passage of Hale-Bopp neither ended nor renewed life on Earth. It was, however, particularly rewarding for astronomers, who appreciate comets as message-bearing time travelers from the very distant past. One such message was the discovery that comets contain nitrogen sulfide, which is also found in dense interstellar clouds. A related finding based on Hale-Bopp's spectra (chemical fingerprints) provides fresh evidence of the actual existence of the Oort cloud.

Concerning the comet itself, astronomers were fascinated by its unique straight tail of sodium atoms. From its water-carbon ratio, astronomers have concluded that Hale-Bopp first formed in the region between the planets Jupiter and Neptune about four billion years ago. They were surprised to learn that the comet possesses a double nucleus and rotates on more than one axis. This nucleus of primordial solar material accounts for the unusual size of Hale-Bopp's brilliant coma.

Amateur astronomers and other sky watchers benefited from the spectacle of this coma. Because most comets are faint when viewed from Earth, Hale-Bopp presented an unusual opportunity. Numerous photographs of Hale-Bopp were taken as countless people found themselves watching the sky in wonder, much as their ancestors did aeons ago.

William J. Scheick

Seven Climber Deaths on Mount Everest Echo Previous Year's Tragedy

About a year after eight climbers lost their lives in a single day, seven mountaineers from several countries died while trying to ascend the world's tallest mountain.

What: Disasters; Environment
When: April-May, 1997
Where: Mount Everest, Nepal
Who:
Aleksandr Toroshchin, a Russian member of the Kazak military expedition
Nikolai Shevchenko, a Russian member of the Kazak military expedition
Ivan Plotnikov, a Russian member of the Kazak military expedition
Peter Kuwalzik, a twenty-nine-year-old German mountaineer
Ang Nima, a climbing Sherpa-porter from Nepal
Malcom Duff, a British climber

The Lure of Everest

The seven mountain climbers who died during the spring of 1997 brought Mount Everest's known death toll to 154. Since 1852, the year a surveyor discovered Mount Everest to be the world's tallest mountain at 29,035 feet (8,850 meters), adventurers have naturally been drawn to it. In 1953, Edmund Hillary, a New Zealand beekeeper, and Tenzing Norgay, a Sherpa, reached the summit of Everest after the members of fifteen expeditions over a one-hundred-year period had failed to scale the peak. Since then, more than 700 people have made it to the summit, but thousands more have tried without successfully reaching their goal.

Climbing Everest is always risky and difficult. The weather can be harsh, often turning fierce and dangerous within minutes. Blizzards, with blinding snow, can prevent climbers from seeing even a few feet in front of themselves. Strong winds, often as high as 125 miles per hour (201 kilometers per hour), can propel climbers to their deaths. Temperatures are usually below freezing. The weather is so harsh that ascents of Everest are possible only in May and October, between the winter snowstorms and summer monsoons. The terrain on Everest is also treacherous. Cliffs, exposed rock, icy slopes, and hidden cracks (crevices) can cause climbers to fall to their deaths. Besides contending with difficult weather and demanding terrain, climbers must be able to endure a lengthy ordeal, hiking and climbing in demanding conditions for several weeks, just to reach base camp, which puts them into a position from which they can attempt to scale the peak. From base camp, the going becomes even more difficult, in part because of the high elevation, which results in oxygen deprivation and can cause pulmonary and cerebral edema, life-threatening forms of altitude sickness. Physical exertion becomes more difficult, and judgment is often impaired.

The 1997 Tragedy

Five of the seven people involved in the 1997 tragedy headed for the summit of Mount Everest on May 8 and never returned to base camp, dying on May 13. Two of the Russian climbing victims, Nikolai Shevchenko and Ivan Plotnikov, reached the summit at approximately 6:00 P.M., a time much later than that recommended by many experts, who advise turning around and starting down the mountain sometime in the early afternoon. The third Russian victim, Aleksandr Toroshchin, became exhausted, so he turned around and started back down the mountain one thousand feet below the summit. Although

Toroshchin died, another Russian climber who accompanied him survived. Peter Kuwalzik, the twenty-nine-year-old German victim, and Ang Nima, the Sherpa victim, also appear to have died while climbing that day. All five were climbing the north face of the mountain.

For the most part, sketchy and conflicting information is all that is known about the tragedy. Cyclone-force winds at 125 miles per hour (201 kilometers per hour) were reported around the time of the tragedy and are believed to have caused the deaths. The climbers appear to have fallen off the mountain while trying to descend in the dark. A French team found Toroshchin's body. At the time of these deaths, twelve teams on the Tibetan side of the mountain and ten teams on the Nepalese side were waiting for a break in the weather so that they could attempt their ascents.

The other two casualties making up the 1997 tragedy are Malcolm Duff, a British climber who died of a heart attack at base camp in late April, and a Sherpa with a Malaysian team, who fell to his death a week before the five-person tragedy.

Consequences

In many ways, the tragedy in 1997 is a larger but similar version of the fatalities that take place on Everest every year. However, the tragedy in 1997 happened almost exactly a year after eight people died in a single day on Everest. For a variety of reasons—the presence of numerous Americans, including a magazine reporter, on the expeditions; the deaths of well-known mountaineering guides Scott Fisher and Rob Hall; the survival of a man given up for dead; and cell phone and Internet communications with the participants—this 1996 tragedy became the subject of numerous books and a film. The second tragedy, just a year after the 1996 deaths, added to the debate about who should be able to climb Everest and how they should do it.

This debate started in large part because the climbing of Everest had become an industry. By the end of the twentieth century, Everest attracted not only the experienced mountaineer but also those wealthy professionals who have done little or no mountaineering but are able to pay up to $65,000 to hire a guide to take them up Everest. These climbers expect to ascend to the top of the peak, and the guides who take them up are under enormous pressure to help them achieve their goal because the reputations of their guiding businesses are based, in part, on the number of clients who reach the summit. Critics of the adventure travel industry are worried that decisions about whether to attempt a summit or other aspects of an expedition may be unduly influenced by business rather than safety concerns.

In spite of the 1996 and 1997 tragedies on Everest, the numbers of people who want to climb the world's highest peak continues to grow. The tragedies have, however, caused many members of the mountaineering community as well as the governments of Nepal and Tibet to consider the wisdom of strengthening existing rules regarding climbers and imposing new regulations on those attempting the ascent.

Cassandra Kircher

Attempted Ascents of Mount Everest, 1922-2000

Period	*Ascents*	*Deaths*
1922-1924	0	11
1930's	0	1
1940's	0	0
1950's	6	1
1960's	18	4
1970's	78	22
1980's	183	59
1990's	887	55
2000	142	2
Totals	1,314	155

Source: EverestNews.com

Presbyterian Church Groups Adopt Law Against Ordaining Homosexuals

A majority of regional bodies of the Presbyterian Church voted to amend the church's constitution to ban from ordination anyone who is sexually active outside marriage, notably homosexuals.

What: Religion; Gender issues; Social reform
When: April 1, 1997
Where: United States
Who:
Roberta Hestenes (1939-), senior pastor, Solana Beach Presbyterian Church, Solana Beach, California
John Shelby Spong (1931-), Protestant Episcopal bishop of the Newark, New Jersey, diocese
Bruce Bawer (1956-), religious author, commentator on the Church's action

The Church and Ordination

Until 1861, the Presbyterian Church in the United States was a single entity. With the onset of the Civil War, the church split into northern and southern branches over the question of slavery. In 1983, these two bodies, one largely northern and headquartered in New York, the other largely southern and headquartered in Atlanta, merged into a single entity.

In matters of doctrine, the congregations of this newly merged church, consisting of 2.7 million members, were still divided into two definite groups, the religiously fundamental and the religiously liberal. The former groups were found largely in the American southeast, Texas, southern California, and Pennsylvania. The liberal groups were essentially from New England, New York, New Jersey, and Northern California. The Rocky Mountain states and the Midwest were about equally divided between fundamentalists and liberals.

Until the merger of the two strands of the Presbyterian Church in 1983, the more liberal branch ordained qualified people with little regard to their sexual orientation. Jack Haberer, coordinator of the Presbyterian Coalition and senior pastor at the Clear Lake Presbyterian Church in Houston, Texas, voiced his support for an amendment banning the ordination of unmarried candidates who are not celibate. He pointed out that in most cases brought before Presbyterian courts, clergy who had been defrocked because of sexual misconduct were heterosexual. He acknowledged that complete and permanent enforcement of the proposed amendment was unlikely. However, he said that if avowed, practicing homosexuals were ordained, charges would be filed against those who ordained them, thereby rendering these ordinations null and void.

The Amendment

Amendment B to the Presbyterian Church's Book of Order, which essentially articulates the rules of the church, is a ninety-word statement devised by a committee chaired by Roberta Hestenes, senior pastor of the Solana Beach Presbyterian Church in Solana Beach, California. It states that anyone ordained as a member of the clergy, an elder, or a deacon shall live in obedience to Scripture and to church standards, which require either marital fidelity between a man and a woman or celibacy among the unmarried.

The Religious Right usually supports a literal interpretation of Scripture, whereas those in the middle or with liberal leanings often interpret at least some of it metaphorically. The same sort of split that occurred in the Presbyterian Church over the matter of slavery resurfaced 122 years later over the matter of who was fit to be admitted to holy offices.

This sort of controversy was not limited to the Presbyterian Church. The social changes of the 1960's and 1970's that affected every aspect of American life had caused many people to reconsider their religious and philosophical stands regarding questions of morality and of individual worth. In 1996, two San Francisco congregations were expelled from the Evangelical Lutheran Church in America for choosing homosexuals as their pastors. In the same year, Episcopal bishop John Shelby Spong was brought before a church tribunal on charges of heresy for having ordained an openly homosexual man as deacon. He was ultimately acquitted by this tribunal.

Presbyterian congregations throughout the United States were asked to vote on Amendment B, which would limit ordination to those who were married to a member of the opposite sex and monogamous or those who were single and celibate. The results of the voting, which took place over several weeks, were announced by a spokesperson at church headquarters in Louisville, Kentucky, on April 1, 1997.

A total of 172 presbyteries were polled. Of these, 87 voted for the amendment and 64 against it; 21 presbyteries either had not voted or failed to send their results to church headquarters. This vote reflected regional attitudes similar to those that had split the church in 1861. The conservatives, who were in the majority, established Amendment B as a rule of the church.

Consequences

Commenting on the establishment of Amendment B as a rule in the church's Book of Order and the effective banning of homosexuals from ordination, Bruce Bawer, author of *A Place at the Table: The Gay Individual in American Society* (1993), differentiated between the Church of Law, which demands a strict, literal interpretation of Scripture, and the Church of Love, which allows a broader interpretation. He pointed out that the distinction is largely a regional one, with the South following the Church of Law and the North the Church of Love. Bawer noted that the split, both regional and intellectual, was evident during not only the Civil War but also such events as the 1925 Scopes trial over the teaching of the theory of evolution in public schools. Although this distinction remains, Bawer identified some differences between the two factions that have faded through the years. For example, conservative Presbyterians once believed that the "elect" were predestined for salvation while all others were consigned to hell. This doctrine collapsed well before the social revolution of the 1960's and 1970's.

In his final analysis, Bawer suggested that the major question posed by the divisions found in the Presbyterian Church in the 1990's boiled down to the fundamental question of what Christianity really is and whether the purpose of religion is to control people or to lead them toward living better, more fulfilling lives.

R. Baird Shuman

U.S. Appeals Court Upholds California's Ban on Affirmative Action Programs

The U.S. Circuit Court of Appeals for the Ninth Circuit ruled that Proposition 209, which banned affirmative action as the law of California, was constitutional.

What: Human rights; Law
When: April 8, 1997
Where: California
Who:
WARD CONNERLY (1939-), regent of University of California, chair of American Civil Rights Institute
PETE WILSON (1933-), former governor of California
MARK DALE ROSENBAUM (1948-), lawyer, U.S. Supreme Court staff counsel for the American Civil Liberties Union
THELTON EUGENE HENDERSON (1933-), federal judge, U.S. District Court for Northern District of California

Ban on Affirmative Action in California

In November, 1996, California voters passed the California Civil Rights Initiative, more commonly known as Proposition 209. The initiative, sponsored by Ward Connerly, a regent of the University of California, and supported by Pete Wilson, former governor of California, amended the California constitution to ban preferences based on race or gender in public sector education, employment, and contracting. The initiative provides that "the state shall not discriminate against, or grant preferential treatment to, any individual or group on the basis of race, sex, color, ethnicity, or national origin in the operation of public employment, public education, or public contracting." Initially enjoined from implementation by a federal district court, the initiative did not go into effect until April 8, 1997, when the Ninth Circuit Court of Appeals declared Proposition 209 to be constitutional. On August 26, 1997, the Ninth Circuit denied emergency motions that would delay implementation of Proposition 209 until the U.S. Supreme Court could rule on its constitutionality. The motions were denied although the Court left open the option of further motions. On November 3, 1997, the Court declined to review the constitutionality of Proposition 209.

Before the Supreme Court's action, the U.S. Department of Justice had formally joined the legal debate over the controversial initiative by filing a brief in federal court that opposed the measure. The brief stated that even before Proposition 209, both race- and gender-conscious state affirmative action programs were required to satisfy rigorous scrutiny as to their constitutionality. The Department of Justice also said that states are free to decide the fate of state affirmative action programs; however, Proposition 209 both repealed California's existing affirmative action programs and violated the equal protection clause of the Fourteenth Amendment by creating obstacles for minorities and women who want to overcome discrimination.

Analyses of the District and Circuit Courts

On December 6, 1996, Federal Judge Thelton Eugene Henderson of the U.S. District Court for the Northern District of California granted a temporary restraining order, halting the implementation of Proposition 209. Lawyer Mark Dale Rosenbaum, member of the U.S. Supreme Court staff counsel for the American Civil Liberties Union, who filed the request for a restraining order, challenged the portion of Proposition 209 that prohibited governmental bodies at every level from taking voluntary action to remedy past and present discrimination through the use of constitutionally permissible race- and gender-based affirmative action programs. The plaintiffs argued that because of the supremacy clause of the U.S. Constitution, Title VII of the Civil Rights Act of 1964 preempts or overrides state law and Proposition 209. The District Court reasoned that Title VII preserved the right of public employers voluntarily to use race and gender preferences. It enjoined Proposition 209 only to the extent that it eliminated programs that granted preferential treatment to individuals on the basis of their race or gender.

The U.S. Circuit Court of Appeals for the Ninth Circuit examined how affirmative action is applied. At one point, the Court stated that a state may eradicate racial discrimination in many ways that do not involve racial preferences. When, for example, a state gives jobs or contracts to the identified victims of state discrimination, the beneficiaries are not granted a preference on the basis of their race but on the basis that they have been individually wronged. The Ninth Circuit went on to say that as a matter of conventional equal protection analysis, there was no doubt that Proposition 209 was constitutional.

The court pointed out that whenever the government treats any person unequally because of his or her race, that person has suffered an injury that falls squarely within the language and spirit of the Constitution's guarantee of equal protection. The court believed, however, that the state's affirmative action programs granted preference to individuals who had not demonstrated that they had suffered any injury. The Ninth Circuit criticized the reasoning of the district court, which reasoned that women and minorities had the right under the equal protection clause to petition local government for preferential treatment. The Ninth Circuit concluded that the district court relied on an erroneous legal premise, and it dissolved the temporary restraining order, allowing the proposition to take effect.

Consequences

In late November, 2000, the California Supreme Court handed down a decision that prohibited public agencies in California from gearing recruitment and outreach programs toward minorities and women. Attorneys said that the decision was a very broad one, and as a result, California cities and counties were likely to avoid from all affirmative action programs so as to avoid possible legal entanglement. Justice Janice Rogers Brown of the California Supreme Court, joined by three other justices, attacked the entire history of affirmative action. The justices wrote that courts over the years had moved from protecting individuals against racial discrimination to setting up entitlement programs for certain groups. In a separate opinion written by Justice Joyce Kinnard, Chief Justice Ronald M. George strongly objected. The justices said that dismissal of all affirmative action as "proportional group representation" is a serious distortion of history and does a grave disservice to the sincerely held views of a significant segment of the populace.

After the implementation of Proposition 209, the percentage of underrepresented minority freshmen at the University of California, Berkeley, began to decline. In 1997, some 257 African American freshmen enrolled; however, for the 2000-2001 school year, the number fell to 148. Similarly, the enrollment figures for Latino students over the same period dropped from 472 to 320. The decision was expected to have a continuing impact on the state's educational system, including the highly regarded University of California.

Attempts have been made to counteract the effects of Proposition 209 and similar popular initiatives in other states. At the University of California, Los Angeles (UCLA), a law fellows program was founded to improve minority recruitment for its law school. For the 2000-2001 school year, minority enrollment at the law school increased for the first time since the passing of Proposition 209. In Washington state, Initiative 200, which resembled Proposition 209, was passed in 1998. Students and faculty at the University of Washington School of Law in Seattle afterward reviewed the UCLA program as a means of increasing the ethnic diversity in the law school's student population.

Dana P. McDermott

Peruvian Government Commandos Free Japanese Embassy Hostages

A Peruvian commando force rescued seventy-two people who had been held hostage in the Japanese embassy in Lima, Peru, for more than four months by the Tupac Amaru Revolutionary Movement.

What: Terrorism; Civil strife; International relations
When: April 22, 1997
Where: Japanese Embassy, Lima, Peru
Who:
Néstor Cerpa Cartolini, leader of the Tupac Amaru Revolutionary Movement
Victor Polay Campos, founder of the Tupac Amaru Revolutionary Movement
Alberto Fujimori (1938-   ), president of Peru from 1990 to 2000
Nancy Gilvonio Conde, common-law wife of Néstor Cerpa Cartolini
Ryutaro Hashimoto (1937-   ), prime minister of Japan from 1996 to 1998
Juan Luis Cipriani, Roman Catholic archbishop of Ayacucho, Peru
Anthony Vincent, Canadian ambassador to Peru
Michel Minning, representative of the International Red Cross
Terusuke Terada, representative of the Japanese government
Domingo Palermo, Peruvian minister of education

Background

On April 22, 1997, an elite Peruvian commando force freed seventy-two hostages who had been held in the Japanese embassy in Lima, Peru, by fourteen members of the Marxist revolutionary group known as the Tupac Amaru Revolutionary Movement. The attack on the embassy by the Peruvian armed forces ended a 127-day ordeal that began on December 17, 1996.

The Tupac Amaru Revolutionary Movement, modeled after the Cuban revolutionary movement, was founded in 1983 by Victor Polay Campos. Following Polay Campos's capture by the Peruvian army in 1992, Néstor Cerpa Cartolini took over leadership of the movement, which is usually referred to as MRTA, the initials of its Spanish name. The name itself was derived from two Peruvian Indian leaders, Tupac Amaru I, who resisted the Spanish conquest in the sixteenth century, and Tupac Amaru II, who led a revolt against the Spanish government of Peru in the 1780's.

In the 1980's and the 1990's, the MRTA was the second largest revolutionary group operating in Peru. The Shining Path (Sendero Luminoso), the larger and more violent group, often resorted to wholesale killings to intimidate peasant communities into joining the movement. Although the MRTA was less violent, the rebel group also resorted to murder and violence. Like the Shining Path, it demanded a tax from the peasants who grew coca leaves, the source of cocaine, and used this money to finance revolutionary activities. The ultimate goal of both groups was the overthrow of the Peruvian government and the establishment of a socialist state.

Alberto Fujimori, the son of Japanese immigrants, had been elected president of Peru in 1990. Throughout his administration, he had implemented a neoliberal economic policy that encouraged foreign investment, the sale of government-owned businesses to private concerns, and the end of subsidies on food and other commodities that were important to the Peruvians. The goal of his policies was to modernize and improve the Peruvian economy. However, the immediate affect was to increase unemployment and worsen the already existing poverty. Both revolutionary groups, who presented themselves

AP/Wide World Photos

Peruvian soldiers help hostages escape from the Japanese ambassador's residence in Lima.

as champions of the poor, terrorized the Peruvian countryside and ultimately took their attacks to the capital city of Lima.

In 1992, President Fujimori dismissed congress, suspended the constitution, and with the support of the army, governed in an authoritarian manner. He launched an effective counterattack against both the Shining Path and the Tupac Amaru Revolutionary Movement. The revolutionaries' terrorist activities alienated many Peruvians and led them to support the president's policies. The leadership of the Shining Path was captured and imprisoned in 1992. By 1995, the strength of the MRTA also was diminished; more than four hundred of its members had been captured and were living in prisons under extremely harsh conditions. Many, including MRTA founder Polay Campos and Nancy Gilvonio Conde, Cerpa Cartolini's common-law wife, were serving life sentences.

The MRTA's plan to seize the Peruvian congress and hold the congressmen for ransom was discovered and crushed in 1995. Probably Cerpa Cartolini and his comrades began planning the attack on the Japanese embassy shortly thereafter. Peruvian intelligence reports later revealed that the MRTA had been bringing heavy weapons into Lima for several months before the attack on December 17. The Peruvian army and Intelligence Service (SIN), perhaps lulled into a false sense of confidence based on past successes, failed to detect the MRTA's activities.

When the Japanese embassy was seized, about 300 guests were attending a reception in honor of the Japanese emperor's birthday. Cerpa Cartolini stated that the attack was meant to force the government to free the more than 400 MRTA prisoners and to end its neoliberal economic policies. If these demands were not met, the MRTA would kill the hostages. Within a few days, 225 hostages were released, and by the end of January, all but 72 hostages—52 Peruvians, 19 Japanese, and 1 Bolivian—had been set free.

Negotiations

President Fujimori's long-standing policy of refusing to negotiate with terrorists gave way in the face of Japanese interests. Meeting with Japanese prime minister Ryutaro Hashimoto in Toronto, Canada, on February 1, 1997, Fujimori agreed to seek a peaceful solution to the crisis. The first of ten meetings between the representative of the Peruvian government, Education Minister Domingo Palermo, and the MRTA took place on February 11. Also attending were three individuals known as the Commission of Guarantees—Archbishop Juan Luis Cipriani of Ayacucho, Peru, Ambassador Anthony Vincent of Canada, and Michel Minning, the representative of the International Red Cross—whose responsibility was to see that the provisions of any agreement were carried out. Terusuke Terada represented the Japanese government at the meetings.

On March 1, President Fujimori made a surprise visit to Cuba and the Dominican Republic and received agreement from Fidel Castro and the president of the Dominican Republic to grant asylum to the MRTA captors of the Japanese embassy.

The direct negotiations stopped on March 6, when Cerpa Cartolini ended the meetings because of the Peruvian government's effort to dig tunnels from neighboring houses into the embassy. After direct negotiations failed, the members of the Commission of Guarantees acted as mediators, trying to find points on which a peaceful solution could be arranged. However, Fujimori's insistence that the MRTA hostages had to be freed and Cerpa Cartolini's refusal to free any additional hostages after reducing their number to twenty were insurmountable barriers to a peaceful solution.

Consequences

At 3:17 P.M. on Tuesday, April 22, President Fujimori gave the order to execute a long-planned and well-orchestrated attack on the embassy. He later explained that he based his decision on the deteriorating conditions under which the hostages were being held, citing the MRTA's new policy of limiting medical visits to the embassy to once a week. The attack consisted of a three-pronged assault: one attack, preceded by a bomb explosion, from the tunnel under the embassy's living room, the second through a second-floor balcony, and the third through the front door. The presence of electronic listening devices that had been smuggled into the embassy allowed the Peruvian army to know the whereabouts of the terrorists and to communicate with the hostages. All fourteen members of the MRTA were killed. One hostage was wounded and died of a heart attack. Three Peruvian soldiers were killed in the assault. The retaking of the embassy by the Peruvian military ended the 127-day hostage taking, the longest ever in Latin America.

Glenn J. Kist

Labour Party's Blair Becomes British Prime Minister

After losing four straight elections while the opposition Conservative Party enjoyed eighteen years in power, the Labour Party of Great Britain achieved a landslide victory in the election of May 1, 1997, and party leader Tony Blair became the youngest prime minister elected since 1812.

What: Government; National politics
When: May 1, 1997
Where: Great Britain
Who:
Tony Blair (1953-   ), leader of the Labour Party and prime minister of Great Britain from 1997
John Major (1943-   ), leader of the Conservative Party and prime minister from 1990 to 1997

A Significant Change

The general election of May 1, 1997, produced a dramatic change in British politics. For almost two decades, the Conservative, or Tory, Party had maintained power, winning the largest number of members of Parliament in the elections of 1979, 1983, 1987, and 1992. It was the oldest modern British political party and tended toward the political right. The Labour Party, which had always been identified with the political left, won a landslide victory in 1997, propelling its leader, Tony Blair, into the position of prime minister. Blair had been elected head of the party in 1994 when its leader, John Smith, died suddenly. Blair led the party in the election campaign, which lasted from March 31 to May 1, 1997.

In the election of 1997, the number of members elected and percentage of the vote received by the top three parties were as follows: Labour, 419 and 44 percent; Conservative, 165 and 31 percent; and Liberal Democrat, 46 and 18 percent. Other parties won 29 members and 7 percent of the vote. This represented a dramatic departure from the 1992 elections, in which the Conservative Party won 336 seats and 42 percent of the vote to the Labour Party's 271 seats and 34 percent of the vote. In the election of 1997, Labour gained 145 seats, and the Conservatives dropped from 323 to 165 seats, their lowest number since 1906. The Liberal Democrats jumped from 26 to 46 seats, the largest number for a third party since 1929. Turnout for the election of 1997 was 71 percent, down from 77 percent in 1992.

British Politics

Great Britain is ruled by Parliament, more specifically, by the House of Commons, in which executive and legislative power lie. The monarch—queen Elizabeth II since 1952—wields no political power. Political parties compete, and the party that holds the most seats in the 659-member House of Commons earns the right to govern. The party leader also becomes the prime minister. The two most prominent parties have been the Conservative and Labour parties. The presence of a strong third party in the elections in the 1990's meant that neither major party achieved 50 percent of the vote.

The Labour Party was historically the party of workers and has advocated socialist principles since its founding late in the nineteenth century. It held power in the 1970's but became plagued by divisiveness and obsolete policies and lost four elections during the 1980's and 1990's. After gaining leadership of the party, Blair changed its policies and made the election of party leader more democratic, establishing a one-member, one-vote process. He abandoned some socialist doctrines such as "common ownership of the means of production" and moved the Labour Party more toward the center, stressing how it would be "tough on crime and tough on the causes of crime." He also conducted American-style political campaigns and formulated the slogan New Labour, New Britain.

As a result, in the 1997 elections, Labour gained the largest majority of any political party since 1935. The losses of the Conservative Party were so great that questions were raised about its future. Six of its former cabinet members, including the foreign secretary and defense minister, lost their seats. Blair succeeded John Major, the leader of the Conservative Party, who resigned in some disgrace. The 1997 election also differed from past elections in that more women were elected, 120, up from 62, and a record nine minority members, including the first Muslim member of Parliament ever. Two members of Sinn Féin, the political wing of the Irish Republican Army, were elected but refused to take their seats.

AP/Wide World Photos

Tony Blair, with wife, Cherie.

Consequences

The election of 1997 was a mandate favoring several Labour political issues and a repudiation of some Conservative positions. The National Health Service was criticized as ineffective. The economy was deteriorating. The crisis in Northern Ireland continued with increasing violence. Privatization, for example, of the railroads, had not led to improved services. Scandals involving the prime minister and problems concerning the handling of the mad-cow disease outbreak plagued the Conservatives. The European Union banned British beef, and millions of British cattle had to be destroyed. Most contentious was the role of Great Britain in the expanding European Union and, especially, in whether to adopt a single currency, presumably leading to the abandonment of the British pound sterling. Leading Conservatives advocated opting out of the single currency; Labour was more favorable and promised a referendum before final action was taken.

Blair's handling of the 1997 campaign is linked with American-style politics. In the United States, President Bill Clinton had restored and returned the Democratic Party to power, partly by moving to the political center. Blair purposely identified himself with Clinton, making a high-profile visit to Washington, D.C., and visiting the Clintons before the election.

Eugene L. Rasor

Zaire President Sese Seko Flees Capital as Rebels Take Over

Long-term tyrant Mobutu Sese Seko, after battling corruption charges, a powerful insurrection, and terminal illness, was forced to abandon Kinshasa, the capital, and surrender the country to rebel leader Laurent Kabila.

What: Civil strife; National politics; Coups
When: May 16, 1997
Where: Kinshasa, Zaire (now Democratic Republic of Congo)
Who:
Mobutu Sese Seko (1930-1997), leader of Zaire from 1965 to 1997
Laurent Kabila (1939-2001), former Marxist and leader of the revolutionary forces; successor to Mobutu Sese Seko from 1997 to 2001

Mobutu Dictatorship

The future president-for-life was born in Lisala, a small village in what was then the Belgian Congo. Educated in Catholic missionary schools, Joseph Désiré Mobutu (his original name), joined the colonial Belgian Congolese army, known as the Force Publique, while still a teenager.

Mobutu excelled at his studies and was eventually promoted to the highest possible rank for a native African, sergeant major. At the same time, he was becoming a talented journalist, first gaining a reputation in that area by writing a series of letters, most of which dealt with what he felt was the Congo's bright future, to various Leopoldville (later Kinshasa) newspapers. After leaving the military, he became a roving reporter for a daily newspaper and simultaneously edited a weekly periodical.

Frustrated by Belgium's oppressive political system, racial bigotry, and mercantile economics, Mobutu came to support independence for the Congo. He allied himself with nationalist leader Patrice Lumumba's Congolese National Movement and, in 1960, stood in for the imprisoned Lumumba at a Brussels conference dealing with the Congo's impending freedom from colonial rule. Lumumba and Mobutu favored a unitary form of government with strong central authority, feeling such a system was necessary to maintain authority.

When independence came June 30, 1960, Lumumba was named premier, and a rival tribesman, Joseph Kasabubu, became president. Mobutu was quickly promoted from Lumumba's private secretary to chief of staff to secretary of state. Nevertheless, Mobutu betrayed his mentor and supported Kasabubu in a mortal power struggle, which led to Lumumba's murder in 1961, an act for which many felt Mobutu was responsible. Mobutu seized power in September, 1960, claiming that the nation was disintegrating and, therefore, his action was regrettably necessary. After imposing a degree of stability, he returned nominal power to the now subservient Kasabubu in 1961.

Mobutu then assumed the position of commander in chief and thereby built a formidable power base in both the bureaucracy and the military. He appeared to be waiting for the right moment to seize power on a permanent basis, and it arrived in 1965. Taking advantage of another power struggle, this time between Kasabubu and the premier, Moise Tshombe, Mobutu usurped both and led a successful coup on November 25. He would never again voluntarily surrender power.

Mobutu's Rise and Fall

At first, the new leader insinuated that his rule would be brief and that his only purpose was to be the harbinger of peace and prosperity, but before the year was over, there were disturbing signs that Mobutu intended to be a long-term presence. He quickly renamed himself president for a five-year term, without recourse to an election, and then unilaterally cancelled the scheduled 1966 elections. In 1966, he dissolved the position of premier in a move to consolidate his power, and in 1967, he crushed an uprising led by white mercenaries. In an attempt to revitalize the economy (and enrich himself) Mobutu nationalized the incredibly rich copper mines of the rebellious Katanga province. He also benefited greatly by huge amounts of foreign aid pumped into the Congo by the West, which saw Mobutu's regime as an anticommunist bulwark.

In 1970, Mobutu established the Popular Movement of the Revolution as the only allowed political party and declared himself reelected to the presidency as a result of a single-candidate poll. To create unity and support African pride (and to divert the nation's attention from real problems), Mobutu led the movement to mandate that both place and human names be changed to reflect the nation's African roots. Accordingly, the Congo was renamed Zaire and Joseph Mobutu became Mobutu Sese Seko. Although not realized at the time, the seeds of his ultimate overthrow were already being sown.

In 1973, Mobutu nationalized two thousand foreign-owned companies, a move that not surprisingly infuriated his Western benefactors. These companies were then distributed among his supporters, who ruined most of the enterprises through mismanagement. At the same time, Mobutu was amassing an incredible personal fortune by stealing the wealth from the sales of the nation's natural resources.

In the 1970's and 1980's, Mobutu had to fight an ongoing war to suppress rebellion in Katanga and deal with border skirmishes against the neighboring states of Angola and Rwanda. Meanwhile, his massive foreign aid (hundreds of millions from the United States alone) dried up abruptly as the Cold War ended.

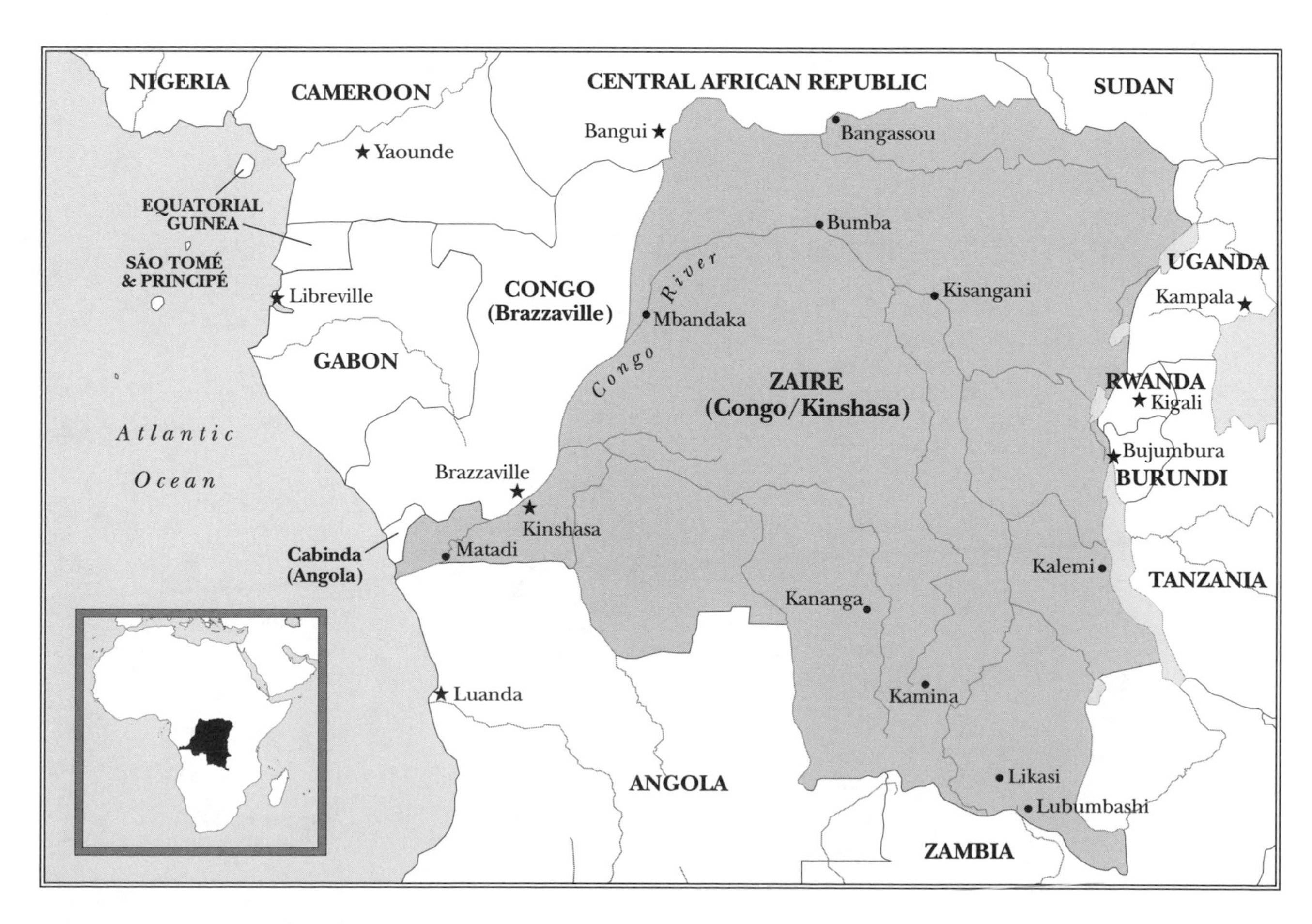

In 1994, over a million tribal Hutus left Rwanda after leading a wave of genocidal attacks against rival ethnic groups. Led by Laurent Kabila, the Hutus settled in Zaire but soon wore out their tenuous welcome. Meanwhile, a Zairian group opposed to Mobutu, the Union for Democracy and Social Progress (led by Etienne Tshisekedi), succeeded in forcing the regime to agree to a coalition government with Tshisekedi as premier.

By 1996, Mobutu's situation was becoming increasingly desperate. Diagnosed with prostate cancer, he was also having to deal with the disappearance of his foreign aid while trying to suppress both domestic insurrection and, now, foreign invasion. When Mobutu tried to push out the Hutus, Kabila retaliated by launching a full assault on the government. Assisted by legions of Mobutu's foreign enemies, Kabila put together a force that the terminally ill and disheartened Mobutu could not overcome. Decades of mismanagement and corruption had alienated the people from the government to the degree that few felt they had a vested reason for fighting for the regime's survival. With Mobutu frequently out of the country to receive medical treatment, organized resistance quickly crumbled. In desperation, the dictator used his own plundered wealth to hire mercenaries, but they quickly proved to be worthless.

Consequences

As Kabila's forces focused on the outskirts of Kinshasa, Mobutu resigned from power May 16, 1997, and fled the country the next day. Exiled in Togo, he died of cancer later that year. Zairians were ecstatic concerning Mobutu's overthrow, but the new Kabila regime would be afflicted with many of the same problems. During his regime, the country reverted to its old name, the Democratic Republic of Congo. Kabila was killed in an attempted coup in January, 2001, and his son, Joseph Kabila, took charge of the government.

Thomas W. Buchanan

Last Common Ancestor of Humans and Neanderthals Found in Spain

Anthropologists discovered the fossil skull of a boy who lived in Spain nearly 800,000 years ago. His skull combined features of both modern humans and earlier human species.

What: Anthropology; Archaeology
When: May 30, 1997
Where: Burgos, Spain
Who:
JOSÉ MARIA BERMÚDEZ DE CASTRO (1952-),
JUAN LUIS ARSUAGA FERRERAS, and
ANTONIO ROSAS, Spanish paleoanthropologists

Discovery of Fossils

In 1976, paleontologists working in the Atapuercan hills near Burgos in Northern Spain discovered the largest known collection of human bones from the Middle Pleistocene period, dating from about 300,000 years ago. They found the bones of at least thirty-two people in a large pit known as Sima de los Huesos (cave of bones). Scientists have guessed that early humans used the pit as a mass burial ground. Between 1994 and 1996, new fossils were discovered at a site called Gran Dolina, in a railway cut near Sima de los Huesos. The layers of earth containing bones were dated (using periodic shifts in Earth's magnetic field as a guide) at about 800,000 years old. On May 30, 1997, a team of Spanish paleontologists, including José Bermúdez de Castro and Juan Luis Arsuaga Ferreras, paleoanthropologists at the National Museum of Natural Sciences in Madrid, Spain, announced in the journal *Science* that they had concluded an examination of stone tools and braincase fragments from six individuals found at the Gran Dolina site. One of the fossils found at Gran Dolina was the nearly complete skull of a young boy about ten or eleven years old. The fossil was unique because it was nearly complete and displayed features that had not been seen in any human fossils found anywhere before.

A Mixture of Traits

The boy's skull showed an unusual mixture of ancient and modern human features. This mixture of traits led the Spanish team that found the fossil to speculate that it represented a new species in the line of human history, which eventually gave rise to both modern humans and the Neanderthals. Modern humans emerged as a species about 40,000 years ago. The Neanderthals were a species of humans who lived from about 200,000 to 30,000 years ago and then disappeared because they were unable to compete with or perhaps interbreed with modern humans. The Spanish scientists named the new species *Homo antecessor*, or "the man who came first."

Antonio Rosas, a paleoanthropologist at the National Museum of Natural Sciences in Madrid, said that the boy's face had some features typical of more ancient human species. His brow ridge stuck out, and he had primitive teeth with multiple roots and a heavy jaw. Other aspects of the boy's face were like those of modern humans or Neanderthals—his cheekbones were sunken and had a horizontal rather than a vertical ridge where the upper teeth attach. In addition, the boy's face was flatter than would be typical of a Neanderthal, but it was not as flat as a modern human face is. The projecting nose and jutting mid-face gave the skull a shape that combines the features of both Neanderthals and modern humans. Like modern humans, the boy had a bony protrusion on the back of his skull. As a result, the boy's skull was an unusual blend of primitive and modern features that was not typical of any known human species. The Spanish paleoanthropologists therefore believed that the fossil proved the existence of a new species—one that came before both modern humans and Neanderthals and was the common ancestor for both of these later species.

Other scientists did not agree that the Burgos fossil represented a new human species. They stated that it was impossible to determine on the basis of a single skull whether other people living at the same time shared the same features as this boy did. More comparison with other adults and juveniles from the same time and place was needed to prove that the boy was a typical member of his species and that these people were indeed a unique human species. Also, these scientist cited the difficulty of making accurate projections about what the adults of Atapuerca looked like based on the appearance of a child. In response, Rosas stated that facial bones from the other five individuals found at Gran Dolina have the same modern features as the boy's face. However, the skulls from the other individuals were not complete, and extrapolating from partial fragments of bone is difficult.

Consequences

The Atapuercan bones are the oldest human fossils ever found in Europe. No fossils had ever been found in Europe and very few in Africa for the time period from about 1.8 million years ago (when the first humans left Africa) until about 500,000 years ago. The Gran Dolina fossils are therefore valuable for the information that they give scientists about what types of humans were living in Europe 800,000 years ago, when the climate was very harsh and survival must have been hard. The existence of these fossils has caused scientists to believe that humans began to move into Europe from Africa about 1 million years ago; previous estimates of that arrival were 500,000 years ago.

Regardless of whether the Atapuerca fossils are those of a new species, they do show that a great deal of variation existed amongst different groups and species of early humans. In this way, humans are no different from other groups of animals, in which diversity of physical traits and behavioral patterns is the norm. Scientists had once thought that humans evolved in a steady line through time. The Gran Dolina fossils show that it is more likely that humans evolved in different ways at different times and in different places, with some groups surviving and others dying out or changing further to develop new variations on the species. Where the Gran Dolina fossils fit in the overall picture of human history is not yet clear, but they do add another piece to what is known about ancient human ancestors. The future discovery of other early hominid sites in southern Europe will no doubt help to clarify the status of the Gran Dolina fossils as a separate species.

Helen Salmon

Caribbean Volcano Erupts, Destroying Much of Island of Montserrat

Montserrat's Soufrière Hills volcano began a series of explosive eruptions that destroyed crops, vegetation, wildlife, homes, and entire towns, forcing many to flee the island.

What: Disasters; Earth science
When: June 25, 1997
Where: Montserrat Island, Soufrière Hills volcano
Who:
FRANK SAVAGE, British governor of Montserrat from 1993

Eruption

For more than two years before its eruption on June 25, 1997, the Soufrière Hills volcano had been spewing lava and ash and casting a shadow over the inhabitants of Montserrat, a tiny island and British colony in the Caribbean Sea. Most of the people living in the southern two-thirds of the island, including Plymouth, the island's capital, had already been evacuated. About four thousand Montserratians—more than one-third of the eleven thousand original inhabitants—had left the island, some emigrating to other Caribbean islands, others moving to Britain. Most of the remaining population was crowded in a safe zone on the north end of the island, living in churches, schools, community centers, and homes of relatives. A few people defied warnings to evacuate so that they could tend crops and livestock in the danger zone. No one had been killed, and some evacuees talked about returning to their abandoned homes and businesses.

In the early afternoon on June 25, a dark ash column suddenly shot up from the northern flank of the volcano, reaching 45,000 feet (13,700 meters) in a matter of minutes. In rapid succession, three major pyroclastic flows—avalanches of scalding gas, rock, and ash—pulsed down the north and west flanks of the volcano, devastating seven villages. Nineteen people were killed. Then, during August and September, the volcano experienced a series of even more violent explosions, sending pyroclastic flows into Plymouth, 20 miles (32 kilometers) away.

Firefighters were unable to stop the flames from burning many homes and businesses, and more than 80 percent of the town was destroyed. The airport was also consumed. Ash deposits were 4 feet (1.2 meters) deep in and around the abandoned city. The weight of the ash was causing roofs to collapse. Pyroclastic flows also threatened other abandoned towns in the southern part of the island. All Montserratians, even those

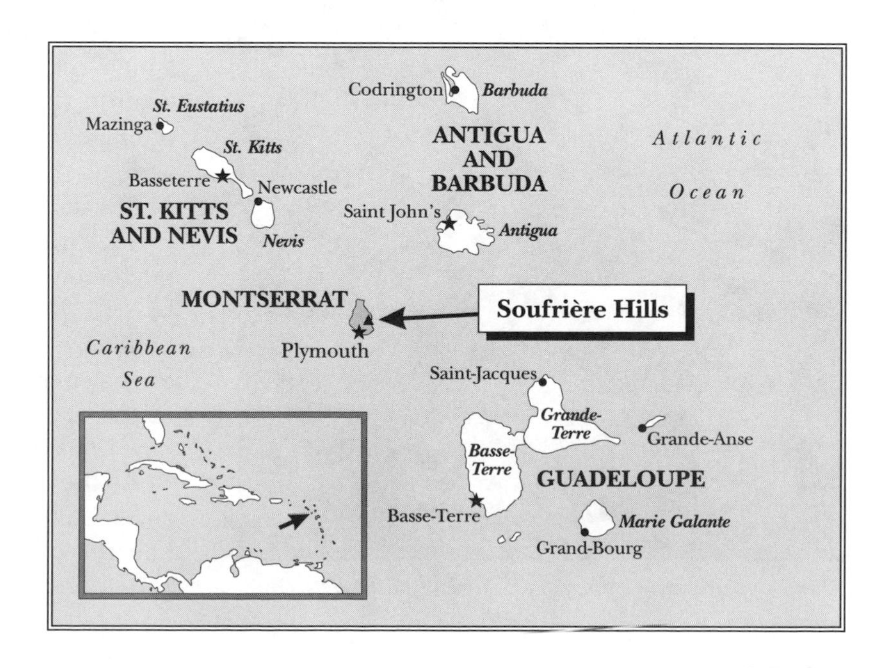

in the safer northern end of the island, were advised to seek shelter under strong roofs, to wear helmets for protection from falling rocks, and to wear masks to avoid inhaling ash. Fortunately, few people were in the most dangerous part of the island, and no one died.

Force of Nature

Montserrat Island is part of the Lesser Antilles, a chain of small volcanic islands that stretches between Puerto Rico and South America. These islands are a direct result of the process of plate tectonics. Earth's lithosphere—an outer layer of rigid rocks—is divided into large sections called plates. Heat energy rising from Earth's hot core causes the plates to move. The motion is very slow, measured in inches per year. Montserrat and neighboring volcanic islands are on the eastern edge of a lithospheric plate that is moving east—the Caribbean plate.

AP/Wide World Photos

Plumes of ash escape from the Soufrière Hills volcano on July 1, 1997.

A second plate, the westward-moving Atlantic plate, is crushing against and descending beneath the eastern edge of the Caribbean plate. Rock in the descending Atlantic plate is melting and forming fluid magma, which is rising through the Caribbean plate to the surface to form volcanoes in the Lesser Antilles. The magma cools and hardens at the surface to form lava called andesite. The volcano from which the magma pours is called a stratovolcano. Stratovolcanoes are the most dangerous volcanoes, because their eruptions can be huge. Each island in the Lesser Antilles group has at least one such volcano; the most famous is Mount Pelée on the island of Martinique. In 1902, flows of hot gas, ash, and lava from this volcano destroyed St. Pierre, Martinique's capital, and killed more than twenty thousand people on the island. Mount Pelée illustrates the potential explosiveness of Montserrat's Soufrière Hills volcano.

Consequences

In the light of the destruction caused by the Soufrière Hills volcano, the British government announced a five-year plan to rebuild on the northern end of the island. In a *National Geographic* article, Frank Savage, Britain's governor of Montserrat, explained, "The north of the island has not been affected by a volcano in two million years, so we've based our contingency plans on that." The British government also offered $4,000 to anyone who wished to relocate off the island. The devastation of the island, which included destruction of a cloud forest around the summit of the volcano and serious damage to coral reefs at its base, attracted worldwide media attention.

Musicians Elton John, Paul McCartney, Sting, Eric Clapton, Mark Knopfler, and Phil Collins all performed at a concert in London on September 15, 1997, to benefit the victims of the Soufrière Hills eruption. The concert was shown on pay-per-view television a few days later and raised around $1.6 million for relocation and new housing.

At the start of the twenty-first century, only about 5,000 people lived on the island, less than half the original population, mostly in the safe zone on the northern end of the island.

Richard A. Crooker

Supreme Court Overturns Communications Decency Act

In Reno v. American Civil Liberties Union, *the U.S. Supreme Court decided that the Communications Decency Act, an attempt by Congress to regulate Internet content, was unconstitutional. The case set the legal standard regarding First Amendment rights on the Internet.*

What: Law; Computer science
When: June 26, 1997
Where: Washington, D.C.
Who:
John Paul Stevens (1920-), associate justice of the U.S. Supreme Court
Janet Reno (1938-), U.S. attorney general

First Amendment and the Internet

The First Amendment protection of freedom of speech is not unlimited. Courts have set boundaries for freedom of expression. Libel and obscenity, for example, are not sheltered by the First Amendment. Courts usually require that any law limiting speech must be as narrow as possible in its application and must serve a compelling governmental interest.

The Internet has had a phenomenal growth rate among both those who provide content and those who use that content. No single entity controls the Internet, and anyone can create a site and post whatever he or she wishes, including sexually explicit material and other material thought to be harmful to children. Such material can easily be accessed by even very young children. This ready access, combined with a lack of monitoring, has caused great concern for parents and, in turn, for the U.S. Congress.

Congress responded to this anxiety by passing the Communications Decency Act (CDA), part of the overarching Telecommunications Act of 1996. In part, the law provided that people sending "obscene or indecent" or "patently offensive" communications over the Internet to anyone under the communications age of eighteen years could be imprisoned for two years as well as fined. Consortiums of nearly fifty organizations involved in the computer or communications industries or in publishing or posting materials on the Internet as well as citizen groups filed suit to challenge the CDA on the very day that President Bill Clinton signed the legislation into law. The two lawsuits—*American Civil Liberties Union (ACLU) v. Reno* and *American Library Association v. United States Department of Justice*—were combined, with Janet Reno named as the defendant in her official capacity as U.S. attorney general.

Opponents of the CDA felt that the legislation improperly restricted adults' right to free speech and that the statute was both overly vague and too broad in that it did not set specific standards or give adequate notice of what was unlawful. Proponents of the act, including the government and family rights' advocates, maintained that the law protected children from readily available offensive material on the Internet. In addition, the courts had previously decided cases in favor of laws that protected minors from speech that was not obscene by adult standards in other communication technologies.

Each form of communicative technology (radio, telephone, television, and cable) is treated somewhat differently in applying the First Amendment. Factors that the courts have considered when setting the standards for measuring the constitutionality of governmental limitations include ease of access by children, extent of intrusion into the privacy of people's homes, possibility of constant risk of exposure to offensive material (pervasive nature of the medium), and degree of other types of regulation by the government.

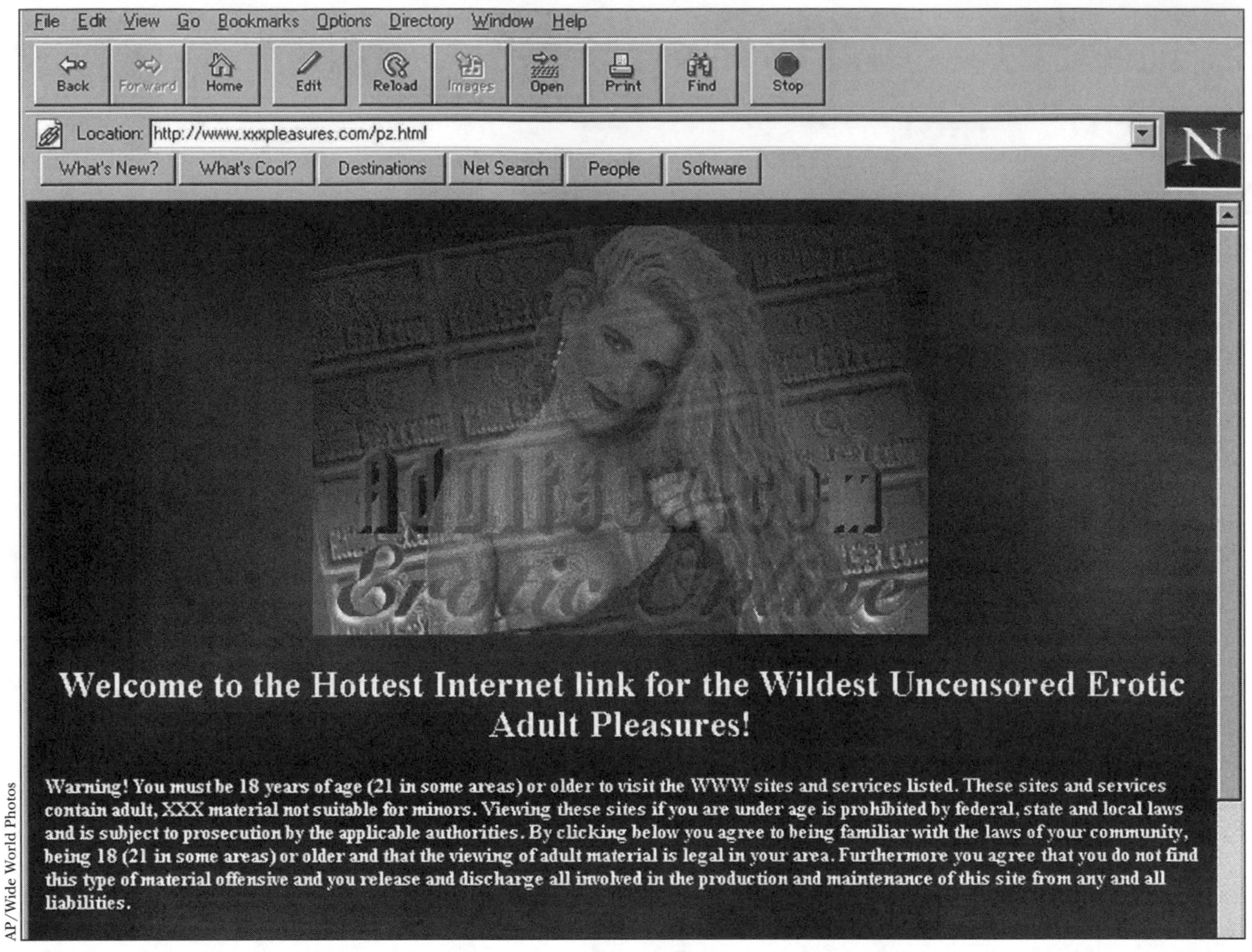

AP/Wide World Photos

A law banning sexually explicit material such as that pictured here from the Internet where children could access it was ruled unconstitutional by the Supreme Court.

Statutes preventing broadcast media (radio, television, and cable companies) from distributing pornography are more likely to be upheld by courts because of the pervasive and invasive nature of these three forms of media. With any of these, a user could come across sexually explicit material by accident. Also, government regulation of radio and television stations and cable companies by licensing and other methods is a long-recognized and well-established practice. Laws that restrict obscene commercial communications over the telephone are judged by a stricter standard because callers must take active steps to receive such messages.

Reno v. American Civil Liberties Union

The CDA itself provided for expedited review of its constitutionality. A three-judge panel of the District Court of the Eastern District of Pennsylvania conducted the initial inquiry and found the statute to be unconstitutional in the case of *American Civil Liberties Union v. Reno* (1996). The case then moved by direct appeal to the U.S. Supreme Court.

The questions facing the Supreme Court in *Reno v. American Civil Liberties Union* (1997) were the selection of the appropriate standard for constitutionally regulating speech on the Internet and the validity of the CDA. Justice John Paul Stevens lauded the congressional goal of protecting minors from harmful influences in a medium of incredibly diverse content; however, the justices found the act to be unconstitutional. They selected the strict scrutiny test, which requires a compelling state interest and the least restrictive methods, for determining the constitutionality of the statute. They felt that the Internet is less invasive than television or radio, and users must initiate the contact and rarely reach a site accidently. The Court noted that while an imprecise search could lead the user to obscene material, most commercial sites have warnings or require verification for viewing. In addition, commercial software that allows parents to screen or limit access can easily be installed on home computers so that children are not likely to stumble on this type of site while exploring the Internet.

Already existing laws prohibit the publication or distribution of obscenity to minors. The CDA's broad statutory language effectively censored or infringed on the speech of adults and thus was not drawn narrowly enough to pass constitutional muster.

Consequences

In *Reno v. American Civil Liberties Union,* the Court imposed a very high constitutional standard for the government to meet before legislatively limiting the content of the Internet. The unanimity of the justices on this point sent a strong message that they felt that government censorship had a chilling effect on the free exchange of information, but the dissenting justices indicated that other methods of regulating content might survive constitutional scrutiny. Therefore, while the case set the tenor for discussions of constitutionality of such regulation, it certainly did not permanently resolve the issue, and litigation and debate over the First Amendment, government censorship, and the Internet are likely to continue.

Susan Coleman

Hong Kong Reverts to Chinese Sovereignty

A century and a half after losing control over Hong Kong to Great Britain, China regained sovereignty over the prosperous port city, which faced the uncertainties of merging its capitalist economy and democratic institutions into China's authoritarian communist system.

What: International relations; National politics; Social change
When: July 1, 1997
Where: Hong Kong, China
Who:
Margaret Thatcher (1925-), prime minister of Great Britain from 1979 to 1990
Deng Xiaoping (1904-1997), the paramount leader of the Chinese Communist Party in the 1980's and de facto ruler of China from 1978 to 1997

The Road to Restoration

Through most of the twentieth century, Hong Kong was a British crown colony. However, it always remained an almost purely Chinese city. As late as the 1990's, 99 percent of its people were Chinese. The city's colonial roots go back to 1842, when the Chinese government permanently ceded the island of Hong Kong to Great Britain after losing the Opium War. More than 90 percent of what became British crown colony lands was added later. The large hinterland tracts known as the New Territories were leased to Great Britain in 1989 for a period of ninety-nine years. That lease was due to expire in 1997.

Ninety-nine years is a long time, and the question of what would become of Hong Kong after the lease expired in 1997 seems to have been forgotten as the city prospered through the twentieth century. Finally, in September, 1982, British prime minister Margaret Thatcher visited the capital of China, Beijing, to negotiate with the Chinese Communist Party leader Deng Xiaoping to continue British rule in Hong Kong after 1997. However, Deng insisted on returning Hong Kong to China and persuaded Thatcher to sign a joint declaration in 1984 that China would resume sovereignty over Hong Kong on July 1, 1997.

Although both Thatcher and Deng repeatedly made public assurances of the political stability and economic prosperity of Hong Kong, the coming change led to four vastly different interpretations of what this change would mean in the context of Hong Kong's history and future.

Differing Interpretations

One interpretation was widely expressed in the American mass media, which adopted what may be called a "communist repression" perspective, which assumed that Hong Kong faced a difficult future. The American media frequently recalled the negative images of China's Tiananmen Square massacre of 1989. From this viewpoint, the Chinese government not only lacked the capability to govern Hong Kong's capitalist economy, but it wished to suppress Hong Kong's democratization. The media frequently called attention to the large numbers of Hong Kong residents who were emigrating before the changeover occurred.

British policy makers tended to endorse the American viewpoint, adding what may be called a "glorious colonialism" perspective. The British claimed that they had done a great service to Hong Kong by transforming it from a nineteenth century fishing village to a modern, twentieth century global city. Were it not for the British rule of law, efficient colonial government, and sound laissez-faire British policy, Hong Kong's economy would have remained as backward as that of most of communist China. The British accused the Chinese government of not living up to its promises to promote democratic reforms in

AP/Wide World Photos

President Rita Fan directs the first plenary meeting of Hong Kong's provisional legislative council under Chinese sovereignty.

Hong Kong and felt that their government they had a moral responsibility to supervise Hong Kong's post-1997 government.

By contrast, the Chinese government presented a "national reunification" perspective. July 1, 1997 was an event for celebration because it was the day to eradicate the national shame that Great Britain imposed upon China after winning the Opium War. The Chinese government insisted it would faithfully keep its "one country, two systems" promise, allowing Hong Kong to run its own affairs. From the viewpoint of the Chinese government, most Hong Kong Chinese welcomed the return of sovereignty to China, as evident by the facts that Hong Kong's stock market prices, real estate prices, and business confidence were at record heights in mid-1997. The Beijing government foresaw a bright future for Hong Kong due to the latter's closer economic integration with the Chinese economy after 1997.

What is generally missing in the 1997 literature, however, is a "class conflict" perspective. Hong Kong was a capitalist paradise because the Hong Kong business leaders exerted a hegemonic domination over the government, economy, and society. During the 1997 transition, however, many middle-class professionals—such as social workers, teachers, lawyers, and journalists—were politicized by the democratic opening. They won elections, entered the legislature, and turned themselves into a strong political force to challenge the Hong Kong government. The democrats won elections not only because they defended human rights, but also because they proposed to protect the environment, to regulate soaring housing prices, and to give more welfare to the grassroots population. It was this welfare challenge from the middle class that prompted the Hong Kong business leaders to develop an "unholy alliance" with the Chinese communist government in order to slow down democratic reforms in Hong Kong.

Consequences

After 1997, the decolonization process gradually faded away, as the British no longer played an important role. However, other structural trends that affected Hong Kong continued into the twenty-first century. The unholy alliance between Hong Kong's business leaders, welfare politics, democracy struggles, and U.S. concern of the human rights situation in Hong Kong intensified after the 1997 transition.

Since Hong Kong's post-1997 political structure was defined by the Basic Law (Hong Kong's mini-constitution), and since Hong Kong's economic integration with mainland China promised to stimulate a bright future for its own economy, it seemed that the peaceful public demonstrations of the democrats would not lead Hong Kong into political chaos or economic instability. On the other hand, although the pace of Hong Kong's democratic development might not have been as fast as the democrats want, Hong Kong might not continue as a paradise for the business leaders or a timid Special Administrative Region for the Chinese government in the early twenty-first century.

Alvin Y. So

U.S. Spacecraft Mars Pathfinder Lands on Mars

A spacecraft that had been launched from the Kennedy Space Center in Florida on December 4, 1996, landed safely on Mars after a flight lasting seven months.

What: Space and aviation
When: July 4, 1997
Where: Mars
Who:
Matthew Golombek (1954-), Mars Pathfinder project scientist at the Jet Propulsion Laboratory (JPL), Pasadena, California
Robert Manning, flight system chief engineer at JPL
Brian Muirhead (1952-), Pathfinder project manager at JPL
Donna Shirley (1941-), Mars exploration project manager at JPL

Launch and Voyage to Mars

On December 4, 1996, a Delta II rocket blasted into space from the Kennedy Space Center in Florida, carrying the Mars Pathfinder into Earth orbit on the first leg of its journey to the red planet. Pathfinder carried no crew but was designed to deliver a lander called Sojourner safely to the surface of Mars. The name Sojourner commemorates Sojourner Truth, a civil rights pioneer, and was chosen as a result of a competition among school students.

The seven-month journey through space covered 310 million miles (500 million kilometers) on a curved path between the orbits of the two planets. While in flight, Pathfinder was in constant radio contact with the control center on earth, located at the Jet Propulsion Laboratory (JPL) in Pasadena, California. The JPL scientists most concerned with the Pathfinder mission were Matthew Golombek, Pathfinder project scientist; Robert Manning, flight system chief engineer; Brian Muirhead, project manager; and Donna Shirley, Mars exploration project manager. Four trajectory-correction maneuvers helped keep the mission on course and to ensure the most favorable angle of descent into the thin Martian atmosphere.

Landing on Mars

Pathfinder entered the outer fringes of the Martian atmosphere at a speed of 5.4 miles per second (7.3 kilometers per second) and an altitude of 81 miles (130 kilometers). Entry took place at an angle of about 14 degrees, and the heat shield afforded protection as the spacecraft decelerated to a speed of 1,312 feet per second (400 meters per second) over a period of about three minutes. At that point, a 3.3-foot-diameter (1-meter-diameter) parachute was opened, allowing the speed to drop to 197 feet per second (60 meters per second) after about 20 seconds.

In the final phase of the landing, the heat shield and parachute were jettisoned, and a 22.3-foot-diameter (5.8-meter-diameter) cluster of air bags was inflated, surrounding the payload on all sides. Guided by an onboard radar altimeter, Pathfinder was slowed and moved to the side by solid fuel rockets, then made a sliding, bouncing landing on a vast, rocky plain known as Ares Vallis. When the lander came to a stop, the air bags were deflated, and the sides of the lander folded down like flower petals, revealing the rover and the three solar panels that would provide power for data transmission. The landing site was named the Carl Sagan Memorial Station, after the late astronomer and planetary scientist.

The successful landing made it possible for scientists to explore the site with four instrument modules designed to record different types of information. Visual observations were made by the Imager for Mars Pathfinder (IMP), which provided stereo and color pictures of the surroundings and could be elevated on a mast. In addition to images of the terrain, the IMP gave information about the atmosphere: dust content, water content, and wind conditions (using wind socks).

AP/Wide World Photos

The Mars Pathfinder rover Sojourner crawls along the surface of Mars.

An instrument called an alpha proton X-ray spectrophotometer (APXS) used alpha particles (helium nuclei) from radioactive curium to determine the elemental composition (all elements except hydrogen) of the soil and rocks. The alpha particles in the APXS interact in various ways with samples to produce characteristic patterns of scattering and proton or X-ray emissions. The spectrometer measures the energy of these patterns and uses the information to determine the elements present in the sample and their concentrations. The APXS was mounted on Sojourner so that its sensor came in contact with the soil or rock to be examined. A robot manipulator allowed the sensor to assume different elevations and angles in relation to the sample. This instrument was the latest refinement in a series of spectrometers that have traveled on U.S. and Russian Mars missions.

The Atmospheric Structure Instrument and Meteorology Package (ASI/MET) measured the temperature and density of the atmosphere at various elevations during entry and descent. In addition, the ASI/MET measured the atmospheric pressure and provided wind profiles.

Sojourner was a six-wheeled vehicle designed to move over the planetary surface, moving the APXS and the IMP to appropriate locations. Sojourner weighed 24 pounds (11 kilograms) and crept along at speeds of up to 1.9 feet per minute (0.6 meters per minute). The rover was built with a flexible chassis for stability on rocky, uneven terrain. Each of the six aluminum wheels was 5 inches (13 centimeters) in diameter and could move up or down independently of the others. Steering the rover was a painstaking task because radio signals took twenty minutes to make a round trip to Earth. Information from the instruments on Sojourner was sent to the lander, then relayed to Earth.

Consequences

An important achievement of Pathfinder was to demonstrate a low-cost method of getting to Mars and to prove the effectiveness of the novel combination of parachute, retrorockets, and air bags used for entering the atmosphere and landing on Mars. Sojourner and its instruments provided the first close scrutiny and elemental analyses of Martian rocks. Pathfinder tended to confirm and strengthen the picture of Mars provided by the Viking and Mariner missions. The planet is cold, dusty, and dry, with a thin atmosphere consisting largely of carbon dioxide. The ASI/MET instruments detected atmospheric dust devils—swirling masses of dust and gas. These probably exert a sandblasting effect on surface rocks and could explain the striated appearance of some of the rocks.

Dust at Ares Vallis was shown to be attracted to a magnet, as expected for certain iron oxides.

The physical characteristics of the rocks were consistent with the presence of much water in the Martian past. The APXS results showed a high silicon content in Martian rocks, similar to the terrestrial mineral andesite. Measurements of the precession (wobbling) of the planet's axis of rotation enabled the planetary core to be estimated at 810 to 1,240 miles (1,300 to 2,000 kilometers) in radius.

After the panoramic color images of Mars were published on the Internet, they became a source of fascination to all viewers back on Earth. Pathfinder sent back 2.6 billion bits of data, 16,000 images, and fifteen rock and soil analyses before communications with the mission ceased on September 27, 1997. On December 10, 1997, the U.S. Postal Service issued a commemorative three-dollar stamp in honor of Pathfinder, and a toy manufacturer began to sell miniature models of Sojourner.

John R. Phillips

Khmer Rouge Try Longtime Cambodian Leader Pol Pot

The Khmer Rouge, a radical guerrilla group that ruled Cambodia from 1975 to 1979, held a public trial of its leader, Pol Pot, and sentenced him to life imprisonment.

What: Civil war; Human rights; International law
When: July 25, 1997
Where: Anlong Veng, Cambodia
Who:
POL POT (1925-1998), leader of the Khmer Rouge, a radical political movement in Cambodia
SON SEN (1930-1997), Pol Pot's top lieutenant
NATE THAYER, an American journalist who videotaped Pol Pot's trial

A Surprising Trial

On July 26, 1997, the hidden radio station of the Cambodian Khmer Rouge guerrilla army announced that Pol Pot had been arrested and sentenced to life imprisonment the day before. Many Cambodians and many international observers were skeptical. Pol Pot had been the leader of the Khmer Rouge since 1963, and its radical soldiers had supported him during the late 1970's, when more than one million Cambodians died under the brutal regime of the Khmer Rouge. Since 1975, these soldiers had followed him through more than twenty years of guerrilla warfare against the Vietnamese, who invaded Cambodia in late 1978, and against the Cambodian government that took power after Vietnamese forces withdrew from Cambodia in 1988. Some observers argued that Pol Pot continued to hold power over his group and that the trial was just a ploy to cleanse the reputation of the Khmer Rouge. Most of these arguments were disproved, though, when an American journalist's video of the trial reached the international media.

Pol Pot's Fall

Saloth Sar, who would later become known as Pol Pot, was born in the village of Prek Sbauv, about ninety miles north of Phnom Penh, the capital of Cambodia. In 1949, Saloth Sar was awarded a scholarship to study in France. Cambodia had been a French colony since 1887, and many young Cambodians who believed in independence for their country were drawn to communism as a way to break away from the French and strengthen Cambodia. After returning to Cambodia in January, 1953, Saloth Sar joined the Indochina Communist Party, which had Vietnamese, Cambodian, and Laotian members. Although the king of Cambodia, Norodom Sihanouk, negotiated Cambodian independence from France in 1953, Cambodia continued to have a small ruling class and a population that consisted mainly of impoverished farmers. In 1960, Saloth Sar became a member of the Central Committee of the Cambodian Communist Party, which had been organized under the Indochina Communist Party, and about two years later, he became the party's general secretary. When Sihanouk, who had given up the title of king to become leader of the Cambodian government, cracked down on the leftists in the early 1960's, Saloth Sar and his comrades fled into the jungle. There, Saloth Sar took the name Pol Pot and the Cambodian communists became known as the Khmer Rouge (Red Khmers), a name given them by Sihanouk.

In 1970, Sihanouk was overthrown by the Cambodian army, and he allied himself with his former adversaries in the Khmer Rouge. The Cambodian countryside had been devastated by U.S. bombing of Vietnamese communist forces taking refuge in Cambodia, and the Khmer Rouge had become even more radical in their plans to completely remake their country. U.S. forces left Cambodia, Vietnam, and Laos in 1973, and the Khmer Rouge marched into Phnom Penh on April 17, 1975. Estimations of the numbers of people who died from executions, starvation, or disease while the Khmer Rouge were in power usually range from 1 million to 3 million.

The Khmer Rouge wanted to retake areas in Vietnam that had been under Cambodian rule, and fighting between the two groups led the Vietnamese to invade Cambodia on December 25, 1978. When the Vietnamese left in 1988, they left a Cambodian government in power that continued to fight the Khmer Rouge guerrillas. However, after 1994, many Khmer Rouge supporters began to go over to the government side, and the guerrilla leaders began to distrust each other. On June 9, 1997, Pol Pot ordered his supporters to kill Son Sen, one of his top lieutenants. Other Khmer Rouge leaders began to fear for their lives, and another Khmer Rouge official, Ta Mok, took Pol Pot captive.

In late July, 1997, Pol Pot was put on trial in the village of Anlong Veng for the murder of Son Sen and for other crimes. The Khmer Rouge invited the American journalist Nate Thayer, accompanied by a cameraman, to attend the trial. Surrounded by about five hundred onlookers, who chanted slogans against him, Pol Pot sat silently while he was accused and convicted of murder and treason. In a later interview with Thayer, in October, 1997, Pol Pot denied any wrongdoing and declared that his conscience was clear. On April 15, 1998, a Voice of America broadcast announced that, according to Thayer, the Khmer Rouge had decided to turn Pol Pot over to international authorities to stand trial for crimes against humanity. However, the old guerrilla leader died of an apparent heart attack on April 16, and his body was cremated.

AP/Wide World Photos

Khmer Rouge soldiers taken prisoner, like those looking at a newsphoto of Pol Pot here, were "re-educated" for five days then allowed to return home.

Consequences

The trial of Pol Pot did not end Cambodia's troubles. Disagreements among the followers of Prime Minister Hun Sen, Sihanouk's son Norodom Ranariddh, and other factions continued to plague the country. Nevertheless, it did appear that the Khmer Rouge would continue to decline and that prospects for peace and stability would grow.

Debates over how to respond to the deeds of the Khmer Rouge also did not end with Pol Pot's trial and death. Although Pol Pot was found guilty of murdering Son Sen and of other wrongdoing in the 1990's, he was never charged with any crimes for the years when the Khmer Rouge ruled Cambodia. Many major figures in the Khmer Rouge, who also participated in the misrule of the 1970's, continued to live freely in Cambodia. Cambodians and international observers faced the question of whether there should be efforts to bring high officials of the Khmer Rouge to justice or an agreement to let these officials live undisturbed in order to achieve peace for a troubled nation.

Carl L. Bankston III

Nuclear Submarine Helps Find Ancient Roman Shipwrecks in Mediterranean

The discovery of these shipwrecks in the open sea confirms the theory of the researcher that ancient sailors did not simply hug the coast as they were engaging in trade across the Mediterranean.

What: Archaeology
When: July 30, 1997
Where: 80 miles off the coast of Sicily, in the open sea
Who:
Robert Ballard (1942-   ), a deep-sea archaeologist

The Discovery

Deep-sea archaeologist Robert Ballard, who located the remains of the *Titanic* and the *Bismarck*, announced on July 30, 1997, that he had discovered five Roman shipwrecks, dating from about 100 B.C.E. to 400 C.E., as well as three much later wrecks some seventy miles off the coast of Sicily. The submerged ships were discovered by Ballard and his archaeology team while they were aboard the U.S. Navy nuclear submarine NR-1. The ships were explored further by the remote submarine *Jason*, which was able to recover more than 150 artifacts from the floor of the Mediterranean.

Technology Lends a Hand

The use of nuclear submarines changed the face of underwater archaeology. Although surface sonar cannot reveal what lies on the ocean floor in much detail, the type of sonar employed by these submarines detects even single artifacts. Ballard and his team of archaeologists found that the sonar operators were providing more information than they could handle and had to ask them to focus on large clusters of objects such as the shipwrecks themselves. Even after cutting back their search, the operators were uncovering an average of a shipwreck every other day.

The submarine was equipped with green thallium iodide lighting, which penetrates farther and provides much clearer images than normal lighting. In addition, the submarine was equipped with television and still cameras, providing a number of angles for observation of the seafloor. The submarine also had a retractable arm that enabled the viewers to remove objects for closer viewing or for seeing what was beneath them.

In comparison with human divers, a submarine not only can locate deeper sites but also can explore them in a fraction of the time. A submarine can map out the overall boundaries of a site in a matter of hours, while it would take humans weeks, if not months. Also the submarine's special lighting provides a far better glimpse of the wreck than what can be seen by the naked human eye.

Once the submarine established the sites that were to bc further explored, the remote-controlled submarine *Jason* was released from a support ship on the surface. *Jason* was equipped with special visual recording equipment and had robotic arms that enabled it to remove artifacts from underwater sites. It had a special netted hand that could scoop up fragile objects (such as glass) from the seafloor.

Jason descended to the seafloor attached to the surface ship by a tether, through which the archaeologists controlled its actions. Sonar units were placed on the seafloor to help locate *Jason* in relation to other objects. A floating platform, fitted with lights and monitoring cameras and connected by a tether, hovered above *Jason*. The team also lowered an elevator onto which *Jason* could load artifacts, allowing the submarine to spend more time exploring the site. *Jason* mapped each site in detail and retrieved selected artifacts.

Consequences

Before the use of nuclear submarine technology for archaeology, most underwater exploration was done by scuba divers, who can reach depths of only about 200 feet (61 meters). The wrecks found by Ballard and his team were at a much lower depth, 2,800 feet (853 meters). Without the technology provided by the nuclear submarines, the area in which these ships were found, as well as 97 percent of the world's ocean floor, is unavailable for exploration, being at depths greater than 200 feet. This technology opened up a completely new avenue for recovery of artifacts from ancient cultures.

The wrecks found by Ballard helped confirm his earlier hypothesis that not all ancient sailors hugged the coast as they engaged in trade across the Mediterranean but rather sailed across the open sea. These ships were found along a possible open-sea trade route between Rome and modern-day Tunisia, near the ruined city of Carthage, which the Romans destroyed in the second century B.C.E. This open-sea route passed through an area that is notorious for having dangerous currents. Ancient ships traveling along the route apparently tried to avoid sinking by throwing their cargoes overboard, as evidenced by debris trails found by the submarine (many ships must have succeeded, as wrecks were not found at the end of the debris trails).

Ballard's open-sea sailing hypothesis was further bolstered by his discovery in 1999 of Phoenician shipwrecks from the eighth century B.C.E. some thirty miles from the coast at a depth of 1,200 feet (366 meters). These wrecks were in the open Mediterranean, not along the coast but rather along the most direct route from the Phoenician coast to the Nile Delta. Additional confirmation came from the February, 2001, announcement of a discovery made in 1999 of Greek shipwrecks at least two hundred years older than Ballard's oldest wreck in the eastern Mediterranean between the Egyptian port of Alexandria and Rhodes, at a depth of more than 10,000 feet (3,048 meters).

Shipwrecks at these great depths preserve a number of artifacts better than open-air sites because of the lessened oxygen content. For example, if the shipwreck is encased in mud, even the wood is well preserved. In addition, deep-sea sites are more likely than sites near the surface to have nearly intact cargoes. Sites near the surface are subject to the force of waves and currents as well as to coral formations, none of which are deep-sea phenomena. Wrecks at these depths also have not been looted, unlike sites close to the surface or on land. Being able to see an undisturbed site makes it easier for archaeologists and historians to understand the evidence. Also, archaeologists can leave the site comparatively undisturbed (without fear of looting), meaning the sites can serve as virtual museums for future archaeologists and other viewers.

Mark Anthony Phelps

Microsoft and Apple Form Partnership

Apple Computers and Microsoft shock Apple loyalists everywhere when they announce that Microsoft is buying $150 million of Apple's stock.

What: Computer science; Technology; Business
When: August 3, 1997
Where: Macworld Expo, San Francisco, California
Who:
Steve Jobs (1955-), cofounder of Apple Computers
Bill Gates (1955-), chairman of Microsoft

The War Ends

Everybody cheered at the *Macworld* Expo on August 3, 1997, when Steve Jobs, cofounder of Apple Computers, appeared on stage to speak for the company for the first time in twelve years. The enthusiasm was real, but by the end of Jobs's presentation, the crowd was booing. One reason was what appeared behind Jobs's shoulder as he made the presentation: an image of Microsoft chairman Bill Gates on a large video screen. After years of feuding and costly legal battles over who owned the Graphic User Interface (GUI) that Apple pioneered, Apple and Microsoft were coming to an agreement. Microsoft would invest $150 million in Apple and would continue to develop software for the company, saving it from almost certain financial death.

Windows, developed by Microsoft and Gates, had become the leading computer Operating System (OS) software because of its ease of use, a feature Apple had prided itself on with its Macintosh computers. Apple computers, originally built by two friends in a garage (Jobs and Steve Wozniak), were designed for those who viewed computers as something too difficult to be used by anyone other than scientists. At one time, Jobs and Gates had shared a friendly rivalry, but the relationship between the two companies changed when Microsoft began marketing its Windows OS in January, 1988. With the release of Windows version 2.0, Apple saw that Microsoft had duplicated much of the Macintosh format, both in the appearance of the interface and in the use of the computer mouse.

By 1997, Apple was losing market share and suffering financially. It rehired original cofounder and visionary Jobs, who had been fired from Apple in 1985. Within six months, Apple had fired Chief Executive Officer (CEO) Gil Amelio and named Jobs interim CEO. Apple loyalists saw the return of Jobs as just what the company needed, but when Jobs spoke Gates's name at the *Macworld* Expo, Apple fans thought he had signed a deal with the devil himself.

The Compact

Despite how it might have appeared to Apple enthusiasts, Jobs had good reasons for striking a deal with Gates. By the time of the deal, Apple had largely resigned itself to being the computer supplier for graphic designers, a small market in which it was still supreme. Jobs was beginning to think that Apple and Macintosh would never be as popular as Windows. However, he felt that Apple still had a place in the computer market and offered a number of ideas for bringing customers back. First, he had to save Apple from possible financial disaster, even if that meant signing a deal with Gates.

Gates had problems of his own. The U.S. Department of Justice had been investigating Microsoft to determine whether the company was illegally monopolizing the market for computer operating systems. Apple controlled perhaps 8 percent of the operating system market, and Microsoft the other 92 percent. No other company had been able to introduce a different operating system. Any company that tried, including IBM with its OS/2 operating system, eventually failed and went back to using Microsoft's Windows. Apple was marketing the only alternative to Windows, and it seemed Apple was going bankrupt and turning to Microsoft for help.

However, Apple's situation was not as severe as it had appeared. Microsoft needed Apple as much as Apple needed Microsoft. Microsoft had been making a steady profit from Apple versions of its software for twenty years. As part of the deal between Microsoft and Apple, Gates promised Jobs that Microsoft would continue to write Apple software for at least five more years. Gates hoped that by keeping Apple alive, he would prove that Microsoft was not a monopoly. The crowd at the *Macworld* Expo booed, but Apple's stock price rose immediately, and soon afterward, Apple started a three-year return to profits.

AP/Wide World Photos

Apple Computer's Steve Jobs stands at a podium in front of a video screen on which Microsoft's Bill Gates appears at a MacWorld convention.

Consequences

Apple, with Jobs as its leader, once again shook up the computer industry as it had with its introduction of the Macintosh in 1984. In May, 1998, Jobs presented the iMac, an all-in-one computer that was so simple, Jobs claimed, it could go from in the box to up and running in under ten minutes. Television commercials showed the two steps required to set it up: Plug in the computer and plug in the telephone line. In addition, the iMac came in a variety of colors to demonstrate that computers were about style as well as substance. It also harkened back to Jobs' original all-in-one Macintosh, built more than fourteen years earlier.

Following on the heels of the iMac, Apple released the iBook, an all-in-one portable computer that students could carry in their backpacks or by its built-in handle. Priced at under $1,600, it made Apple competitive in the portable computer market. It also came in many colors. In 2000, Apple produced the world's first supercomputer, the G4, and in 2001, Apple introduced both a PowerBook portable computer made from titanium and Mac OS X, the first radical change to the Macintosh's OS since 1984.

Apple had seemed to be on the brink of bankruptcy almost since Jobs first left Apple in 1985, with every week seemingly its last. Jobs's reappointment as CEO and the subsequent boost from Microsoft seemed to be what Apple needed to stay alive. Apple did not have the majority of the computer marketplace, but year after year, its new products represented innovative breakthroughs in the industry. Though Apple never achieved the market share that Microsoft had, Jobs demonstrated the Macintosh's viability as an alternative computer system.

Kelly Rothenberg

117,000-Year-Old Human Footprints Are Found Near South African Lagoon

Scientists discovered ancient fossil footprints left by a woman who walked on the shores of a South African lagoon 117,000 years ago.

What: Anthropology; Archaeology
When: August 14, 1997
Where: Langebaan Lagoon, South Africa
Who:
David Roberts, geologist, Council for Geoscience, Bellville, South Africa
Lee Berger (1965-), paleoanthropologist, University of Witwatersrand, Johannesburg, South Africa
Stephan Woodborne, archaeologist, Council on Scientific and Industrial Research, South Africa

Footprints in the Sand

About 117,000 years ago, a human being walked in the sand beside the shores of Langebaan Lagoon, located on the western seacoast of South Africa about 60 miles (97 kilometers) north of Capetown. The sand was wet and soft because of a recent rainstorm, and as the human walked, the individual's feet created a right-left-right pattern of footprints on the sand dune. By comparing the size and shape of these footprints with those of modern South African native peoples, scientists estimate that the ancient walker was about 5 feet to 5 feet, 4 inches (156-160 centimeters) tall and probably a woman. She was walking from the sea toward the lagoon, perhaps hunting for mussels or looking for what the storm had washed up.

Through a rare geological process, some of the footprints were fossilized. Many layers of sand were blown by the wind, covering the footprints and preserving their shape. Over a long period of time, shell fragments in the sand were dissolved by water, releasing calcium carbonate and turning the sand into hard, cementlike rock. Many thousands of years later, erosion caused this rock to break apart and once again expose three footprints to the open air.

Discovery and Preservation Efforts

In 1997, the fossilized footprints were discovered by David Roberts, a geologist with the Council for Geoscience in Bellville, South Africa, who was exploring nearby rock formations. Roberts noticed that there were many small pieces of rock that looked as if they had been chipped out by ancient humans while they were making stone tools or weapons. He also saw fossilized animal tracks and began looking to see if any human fossil tracks might be found in the rock slabs on the shore. On August 14, 1997, he reported that he had found three ancient human footprints that were surprisingly well preserved. The prints are 8.5 inches (21.5 centimeters) long, and one of them gives a very clear imprint of the complete human foot (the big toe, ball, arch, and heel). Another footprint is less clearly marked, and only part of the third footprint remains.

Although scientists could not date the footprints themselves, the rock surrounding the prints underwent a series of tests by Stephan Woodborne, a South African archaeologist with the Council on Scientific and Industrial Research. Woodborne estimated that the fossils were about 117,000 years old. The footprints are very fragile, and in June of 1998, a resin cast was made to preserve a permanent copy of them. The footprints themselves were then removed and later placed on display at the South African Museum in Cape Town.

Roberts worked with Lee Berger, a paleoanthropologist at the University of Witwatersrand in Johannesburg, South Africa, to further study the footprints. Paleoanthropology involves the study of how humans lived many thousands of years ago, based on the fossils and objects that ancient people have left behind. Berger theorized about how the woman who left the footprints might have lived. People living in South Africa 100,000 years ago did not hunt or fish in complex ways but moved around from spot to spot gathering fruit and scavenging small animals and shellfish from the ocean's edge. They used stone tools but did not have bows and arrows. In the rock lying underneath the fossils, Roberts discovered Stone Age tools that he believed were made by the people who left the footprints. These included blades for scraping and cutting, a spear point, and a large stone core from which other tools were chipped. The woman's people would have lived in caves in small family groups, and they knew how to make fire. Although they did not create art, they would have had rituals that involved dancing and painting their bodies with ocher pigments. Clothing would have been made from animal skins, but no jewelry would have been worn. Scientists do not know if these people could speak or not.

Consequences

The significance of the fossil footprints lies partly in their age and partly in how rare they are. Only four sets of fossil footprints have ever been found in Africa. Even more important, the Langebaan footprints date from the point in history when modern humans were evolving. This is the period that saw the emergence of *Homo sapiens*—the first human species that looked anatomically like modern humans. Scientist have discovered very few fossil remains, perhaps only thirty-six for all Africa, from this time period. In contrast, there are 3,000 to 4,000 fossils that have been found from much earlier periods of human history (before 2 million years ago).

Berger has suggested that the footprints might be those of "Eve," an individual female living in South Africa between 100,000 to 300,000 years ago from whom scientists believe that all modern humans are descended. Berger believes that the southern tip of Africa was an ideal place for new species of humans to evolve, because it is isolated from the rest of the continent by geographical barriers (mountains and deserts). This isolation would have allowed the ancient ancestors of modern humans to evolve and change separately from other human species in the world, until *Homo sapiens* finally emerged. Berger's theory has not been accepted by other scientists, however. Based on deoxyribonucleic acid (DNA) evidence, paleoanthropologists agree that a common female ancestor existed, but there is no way of knowing whether she was the woman who left the Langebaan footprints

The Langebaan footprints do not help scientists understand in detail how ancient humans lived or how they looked. They are of interest simply because they look much like human tracks that might have been left on a shoreline only hours ago, although they were made by a human ancestor who lived more than 60,000 human generations ago.

Helen Salmon

Great Britain's Princess Diana Dies in Paris Car Crash

Princess Diana died in an automobile accident in Paris a year after her divorce on August 28, 1996, from Prince Charles of Great Britain. Her suitor, Dodi Fayed, and their driver also perished.

What: National politics
When: August 31, 1997
Where: Paris, France
Who:
Diana (1961-1997), princess of Wales, former wife of Charles, prince of Wales
Charles (1948-), prince of Wales, heir to the British throne
Emad al-Fayed, also known as Dodi Fayed (1955-1997), Diana's suitor, son of an Egyptian billionaire
Trevor Rees-Jones (1968-), Dodi Fayed's bodyguard, sole survivor of the crash

A Divorce and a Romance

On July 15, 1996, Charles and Diana, prince and princess of Wales, received a preliminary divorce decree, made final on August 28. Princess Diana, who retained the title princess of Wales, set about rebuilding her life, continuing her humanitarian efforts, particularly her work with acquired immunodeficiency syndrome (AIDS) patients. A public that adored her looked scornfully on Prince Charles, often depicted as detached and unfeeling.

As rumors of Diana's romances circulated, it became evident that she was unlikely to remain single for long. She continued her involvement with humanitarian causes, focusing on aiding victims of land mines throughout the world and spearheading a worldwide initiative to dismantle land mines. These efforts led to a three-day visit to Bosnia on August 8, 1997.

Shortly before this trip, Diana and her sons were guests aboard the *Jonikal*, Mohamed al-Fayed's $32 million yacht. Here she developed a close relationship with Mohamed al-Fayed's son, Dodi, whom she had previously known only casually. Love quickly blossomed between the two. The bands of photographers who perpetually followed Diana increased as word of her new infatuation spread. Using telephoto lenses, the paparazzi snapped pictures of Diana and Dodi Fayed kissing and embracing like two people in love.

When the cruise ended on July 20, Diana and Fayed parted, arranging another cruise from Nice, France, to Sardinia, Italy, for the first five days of August. Returning to London on August 6, they were mobbed by paparazzi, who pursued Diana to Bosnia. She sought to escape them by taking a short Aegean cruise with her friend Rose Monckton. She and Fayed planned another nine-day cruise beginning on August 21. Diana admitted to many people that she was in love with Fayed and hinted that she might soon marry him.

Despite the intrusive press, Diana was relaxed on the *Jonikal* and reassured in her feelings for Fayed. On August 30, the couple returned to Paris, where they planned to spend the night in Fayed's apartment. He had purchased a ring that cost $205,400, paid for by his father. He intended to propose marriage that night and to give Diana the ring.

Fayed made a dinner reservation at Benoît. As they approached the restaurant, however, they found hordes of paparazzi waiting. They retreated to the Ritz, owned by Fayed's billionaire father, for dinner at L'Espadon. They took a secluded table and ordered their meals, but when someone dining near them reached for a camera, they fled, asking that their dinners be served in the Imperial Suite.

The Fatal Crash

Leaving the Ritz for Fayed's apartment shortly after midnight, they found more than thirty photographers gathered outside. Fayed tried to thwart the paparazzi by dispatching a car with one of his bodyguards, giving the impression that he and Diana were spending the night at the Ritz. Meanwhile, he had summoned Henri Paul, a driver he trusted, who was off duty at the time. He arranged for Paul to bring a Mercedes to a rear service door of the hotel to collect the couple and Fayed's bodyguard, Trevor Rees-Jones. This ploy failed. The paparazzi, some on motorcycles and some in automobiles, were waiting.

Paul took off as fast as he could, deciding that rather than taking his usual route down the Champs Élysées to Fayed's apartment, he would take an alternate route through the Alma Tunnel, at whose entrance there is a curve followed by a sharp dip. By the time Paul reached the entrance, he was traveling about 80 miles per hour (128 kilometers per hour), trying to outrun pursuing photographers. Only Rees-Jones was wearing a seatbelt.

The Alma Tunnel is supported by reinforced steel pillars 15 feet (4.6 meters) tall and set a few feet apart, separating the opposing lanes. A small white car was ahead of the Mercedes. Paul had no alternative but to overtake it, clipping it as he passed. The Mercedes veered out of control into the right cement wall then crashed head-on into the thirteenth pillar. Both Paul and Fayed died on impact. Rees-Jones was badly injured, most of his face torn off. He eventually recovered.

The impact threw Princess Diana onto the floor. She lay facing the rear of the car, trapped in the foot well, bleeding from her mouth, nose, and ears. Her face had only small gashes on the forehead and upper lip. She was barely alive. Diana received immediate assistance from a doctor who happened upon the scene of the 12:24 A.M. crash. An ambulance arrived at 12:33 but did not leave the Alma Tunnel until 1:45, after attendants attempted to minister to the princess.

Diana arrived at La Pitié-Salpêtrèire hospital shortly after 2 A.M. Emergency personnel worked feverishly to save her. At 3:55, her vital signs failed, and she was declared dead at 4 A.M. The immediate cause of her death was a torn main artery to the heart.

Consequences

Subsequent investigations revealed that Paul, who was taking two powerful antidepressants, Prozac and Tiapridel, had been drinking before the accident and was legally drunk. Shortly after the crash occurred, nine photographers and one

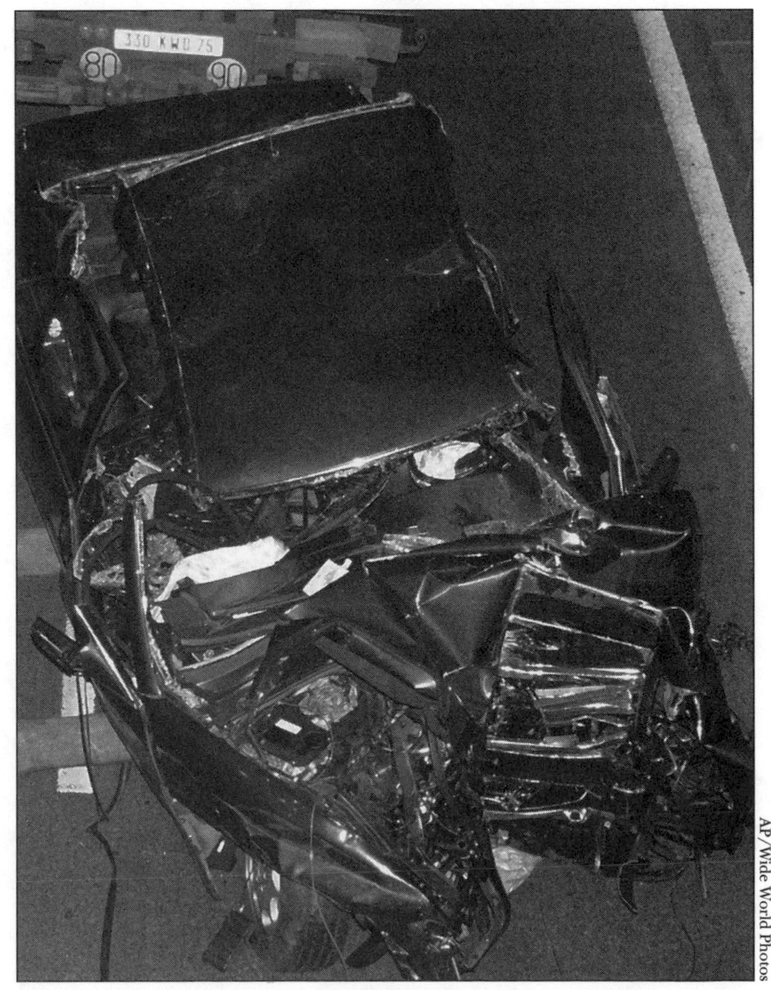

AP/Wide World Photos

The wrecked car in which Princess Diana was riding.

motorcyclist, members of the paparazzi, were arrested and charged with manslaughter and negligence. On September 3, 1999, all charges against these defendants were dismissed. Mohamed al-Fayed appealed this decision, claiming that the accident had been deliberately planned to prevent Diana from marrying his son, a Muslim. His appeal was dismissed on October 31, 2000.

On September, 1997, the only survivor, Rees-Jones, sued the Paris Ritz and Etoile Limousine for negligence. His case was settled out of court, as was Diana's family's suit against Mohamed al-Fayed, for $12 million.

One important consequence of the accident was that Queen Elizabeth II, who had been criticized for her seeming indifference to Diana's death, took positive steps to become a more human figure, going so far as to appoint a communications director, thirty-nine-year-old Simon Lewis, to help her reshape her austere public image.

R. Baird Shuman

Scots Vote for First Separate Legislature Since 1707

A referendum was held in Scotland as to whether the Scots wished to have a separate national assembly to govern them, and if so, whether it should be able to raise or lower taxes.

What: Government; Political independence
When: September 11, 1997
Where: Scotland
Who:
Tony Blair (1953-), British prime minister from 1997
Donald Dewar (1937-2000), Scottish secretary in Tony Blair's cabinet, subsequently first minister in the Scottish assembly

Scotland and the United Kingdom

The United Kingdom is made up of four countries: England, Scotland, Wales, and Northern Ireland. Its Parliament is held at Westminster in the British capital city of London, England. At one time, Scotland was a fully separate country, with its own sovereign and its own parliament. On the death of the English queen Elizabeth I in 1603, the Scottish king James VI inherited the English throne, and from then on, the two countries shared the same sovereign. Then, in 1707 in the Act of Union, the kingdom of Scotland and its parliament were united with the equivalent English institutions, and the seat of government moved to London. Despite these mergers, Scotland still retained its own legal, educational, and banking systems.

Until the mid-twentieth century this status quo was accepted by most Scots. Even though the Scottish members of parliament (MPs) were greatly outnumbered by the English, the benefits of trade and membership in the empire were seen as sufficient compensation. However, beginning in 1945, the emergent Scottish National Party (SNP) began to campaign actively for complete independence for Scotland, and the Liberal Party (later to become the Liberal Democrats) started to advocate devolution, or partial independence in internal affairs, though still retaining the Union.

A referendum (a vote by which an electorate can indicate its wishes on a single issue) was held in 1979 in Wales and Scotland by the Labour Government under Prime Minister James Callaghan to ascertain the wishes of the Welsh and Scots toward devolution. Although a small majority of voters was obtained for devolution in Scotland only, this was deemed insufficient to move ahead on the issue.

The 1997 Referendum

In the 1980's and early 1990's, a desire for regional devolution grew all over Europe. In Scotland, this was added to by newfound prosperity from the exploitation of the North Sea oil fields. In a change of heart, the Labour Party made devolution part of its policy manifesto for the 1997 elections, in which it was returned to power with a huge majority, especially in Scotland. The new Labour Government consequently made it one of its top priorities to set up referenda for Wales and Scotland and also to reinstate the Northern Ireland assembly. The date for Scotland was set for September 11, 1997.

On July 24, 1997, the government produced a white paper, a document outlining the legislative proposals. The Scottish assembly, or parliament, would consist of 129 seats, with the voting system being a mixture of "first past the post" for 73 of the seats (based on the existing Westminster constituencies); the remaining 56 seats were to be allocated on a system of proportional representation. Voters would have two votes: the first, whether they wanted to have an assembly at all; the second, whether they wanted it to have powers to vary the rate of income tax set by the Westminster Parliament by up to 3 percent (the so-called Tartan Tax).

The new assembly, if approved, would have responsibility for all Scotland's internal affairs, from forest management to health issues. The responsibility would extend to powers to regulate local council finances. The assembly would elect a first minister who would then appoint his own cabinet.

Electioneering was slow to start and was delayed by the tragic death of Princess Diana on August 31. It was decided to suspend campaigning until after her funeral on September 6, effectively leaving only four days for the campaign.

The only real proponent for a "no-no" vote was the Conservative Party, which had been totally defeated in Scotland in the earlier British elections. The main debate focused on the tax-raising powers. The governor of the Bank of Scotland predicted Scottish taxpayers could find themselves worse off by $450 per year. The Labour Government under Prime Minister Tony Blair (himself a Scot) took the position that voters should vote for such powers, although he promised a Labour assembly would not use such powers for the first four-year term of office. Opponents also objected to Scottish MPs still being able to vote at Westminster in purely English affairs, while English MPs would have no say in Scottish issues.

In the end, 74.3 percent of those voting (45 percent of the total electorate) cast their votes in favor of an assembly, and 63.5 percent voted for tax-varying powers. The heaviest votes in support of the referendum issues came from the urban areas of central Scotland and the remote North West Highlands. Only the Orkney Islands produced a majority vote against either issue.

Consequences

The election results were declared on September 12, 1997, the seven hundredth anniversary of William Wallace's defeat of the English. Donald Dewar, the Scottish secretary in Blair's cabinet, declared victory.

The referendum had technically been advisory and needed legal backing, so legislation was passed in late 1997 at Westminster. Elections were then set up for the first part of 1999, in which, as expected, the Labour Party emerged as the largest party but still in need of the support of the Liberal Democrats to form a coalition to govern. The Scottish National Party, who saw the referendum as a stepping stone toward full independence, became the main opposition, together with the Conservative Party, who had managed to gain a few seats under the proportional representation system.

The Assembly was opened by Queen Elizabeth II, the British monarch, early in 2000, at Edinburgh, the traditional Scottish capital. Its first minister was Dewar, the energetic Scottish secre-

AP/Wide World Photos

After casting his vote in the Scottish devolution referendum, Rory Murray leaves the polling station in Bo'ness.

tary who had campaigned for many years to change Labour policy in favor of devolution. However, he died within a year of taking office. Lord David Steele, a former Liberal Party leader, became the first speaker.

The part of the British budget that was formerly earmarked for Scotland, some $21 billion, was now administered from Edinburgh. The new administration, true to its election promise, did not use its tax-raising powers during its first term. Large numbers of civil servants moved up from London to administer the new departments, which dealt with transport, health, the environment, economic development, sport and the arts, public records, and agriculture, and which joined the existing departments of education and the law. The largest constitutional change to the Union since 1707 was effected remarkably smoothly.

David Barratt

Diet Drugs Are Pulled from Market

Fen-phen and Redux, popular prescription diet drugs, were withdrawn from the market after serious questions were raised concerning their link to heart-valve problems.

What: Medicine; Health
When: September 15, 1997
Where: United States
Who:
FOOD AND DRUG ADMINISTRATION (FDA), the government agency that oversees the approval and use of prescription drugs in the United States
AMERICAN HOME PRODUCTS CORPORATION (AHP), the company that manufactured Fen-phen and Redux

Dangerous Diet Drugs

On September 15, 1997, the Food and Drug Administration (FDA) asked the manufacturers of the popular prescription diet drug combination commonly known as Fen-phen, to voluntarily withdraw one of the two drugs from the market. They also asked that a chemically related drug marketed as Redux be withdrawn from the market. They made these recommendations after evidence emerged that heart-valve damage was a significant potential side effect of these drugs. Doctors were advised to take their patients off these drugs immediately.

Fen-phen, the popular name for a combination of two different drugs, was used in the treatment of obesity. Each of these two drugs was approved separately by the FDA for short-term use as an appetite suppressant (a substance that keeps an individual from feeling hungry). Doctors started prescribing these two drugs to be used in combination for long periods of time. They seemed to work together quite effectively in combating hunger. However, the FDA had no studies regarding the possible side effects or safety of using these two drugs together.

The first drug, fenfluramine, had been approved by the FDA for use as a short-term appetite suppressant. It was marketed under the brand name Pondimin. This drug affects serotonin production in the human brain. Serotonin is a brain chemical that contributes to feelings of satisfaction and well-being in an individual. Pondimin was designed to suppress cravings for specific foods. Heart-valve damage is believed to have occurred because Pondimin alters the way the body metabolizes serotonin.

The second drug, phentermine, was also approved by the FDA for short-term use. It is an amphetamine-like drug that is marketed under such brand names as Fastin and Ionamin. There are also several generic forms of phentermine. Amphetamines speed up the metabolism as well as suppress the appetite. Phentermine has not been withdrawn from the market because it is believed that this drug is not responsible for the heart damage.

Dexfenfluramine, commonly known as Redux, is a single pill that combines both types of appetite suppressants and works much as Fen-phen does. It is chemically related to fenfluramine. This drug was withdrawn because of the potential of heart-valve damage in patients taking it.

To the Heart of the Matter

Heart valvular disease is caused by a malfunction or leak in the valves of the heart. There are four major valves controlling the flow of blood in and out of the heart and also between each of the four chambers of the heart. Leaking valves cause the heart to work ineffectively and may produce severe heart and lung disease. Extremely ineffective valves may eventually lead to heart failure.

The first indication of potential problems occurred when doctors noticed an increase in the number of new heart murmurs in their patients taking the drugs. Heart murmurs are abnormal sounds created as blood flows over the heart valves. They can be heard through a stethoscope, the instrument a physician uses to listen to the heart. The valve damage could then be confirmed by the use of an echocardiogram, a machine used by a heart specialist to test heart function. Some patients exhibited more severe symptoms related to valvular disease such as shortness of breath, loss of tolerance of physical activity, and fluid retention in the legs and lungs. The most severe cases showed signs of primary pulmonary hypertension disease. This is an extremely dangerous disease caused by the narrowing of the blood vessels in the lungs. This disorder occurred in about one out of every twenty-five thousand patients using appetite suppressants for at least three months. Statistically, approximately 40 percent of those suffering from pulmonary hypertension die within four years.

Consequences

New guidelines were published in 1998 following the annual meeting of the American Heart Association (AHA) concerning those people who previously took Fen-phen or Redux. It was advised that anyone who took the drugs should be examined by his or her doctor, and if any of the symptoms of valvular heart disease were present, the patient should be further tested to determine the extent of the damage. In some cases, valvular heart disease can be controlled with medication. In others, surgery may be necessary to replace the damaged valves with artificial valves.

American Home Products Corporation (AHP), the company that manufactured Pondimin and Redux agreed to pay as much as $3.75 billion in damages to settle lawsuits arising from the use of these drugs. More than six million patients took the drugs, and studies have shown that as many as one-third of them suffered from some form of valvular heart disease. In early 2001, studies were released that showed that in most patients who stopped using the diet drugs, the valve damage did not worsen over the years, and in many cases, the patients actually improved.

The use and withdrawal of Fen-phen and Redux was yet another failure in the fight in the battle against obesity. Most of those taking the drug were successfully losing weight before the complications were detected. These drugs seemed like a dream come true, but as is often the case, they were too good to be true. Once again, it was proved that there are no quick fixes for obesity and that diet and exercise are still the best solution.

Mary Ellen Campion

Welsh Vote to Establish Own Assembly

By a small margin, voters in Wales approved of the establishment of an assembly that would govern Welsh matters while leaving national concerns, such as defense and taxation, in the hands of the government of Great Britain.

What: Political independence; Political reform
When: September 18, 1997
Where: Wales
Who:
TONY BLAIR (1953-), British prime minister from 1997

Wales and the United Kingdom

Like Scotland and Northern Ireland, Wales is a province of the United Kingdom. It is noted for its beautiful, rugged appearance, its old Celtic culture, and its remoteness from England. It has been united with England since the sixteenth century, when England took over administration and control of the area. Welsh nationalism and the desire for Welsh control of Wales has flared up from time to time over the centuries, and it gathered considerable strength in the final decades of the twentieth century. For example, the pride of many Welsh people in being able to speak the ancient Welsh language has increased considerably. A nationalist party, Plaid Cymru, has gathered the votes of the most committed Welsh nationalists. A few violent extremists have been active in the past, attacking some British public and private property in Wales.

Another trend over the past two centuries has been the immigration of many English people into Wales. Many English moved to Wales to help exploit the coal, iron, and slate industries in the nineteenth and early twentieth centuries. Since then, high-tech industries, financial service centers, and the lure of summer cottages in the beautiful countryside have brought more English immigrants. These people, in general, have not fully appreciated or understood Welsh nationalism. Therefore, two separate cultures with different sets of aspirations have coexisted in Wales for some time.

The Labour Party has always been strong in Wales, beginning with the strong devotion of Welsh coal miners when the party originated in the early twentieth century. In the election of 1997, the Labour Party enjoyed a landslide victory in Wales. Promises of devolution (partial independence in internal affairs) and the creation of a Welsh assembly were an important part of Labour's election platform. The Labour Party's promise of devolution involved giving power over local affairs to both the Welsh and the Scottish population through a Welsh assembly and a Scottish parliament, respectively. To fulfill this promise, the Labour Party under Prime Minister Tony Blair swiftly proposed legislation to promote Welsh and Scottish devolution. For Wales, new legislature was seen as the means by which the province could rule itself, or have home rule, for all matters pertaining exclusively to Wales.

In some ways, devolution is easier for Americans to understand than it is for many British people. In essence, devolution proposed something of a federal system, with Wales and Scotland gaining the powers of local representation and local control similar to those enjoyed in each of the fifty American states. For Wales, this meant the creation of a sixty-member assembly. Meanwhile, the central British government at Westminster, where Parliament is located, would continue to retain control over defense, foreign policy, and fiscal matters, including taxation levels.

The Referendum

A measure to give local control of Wales to a Welsh assembly passed in the British Parliament soon after Labour's victory in the 1997 election.

However, this issue was so fundamental that the voters of Wales had to register their approval or disapproval of what Parliament had proposed. If they voted against the referendum, Wales would continue to be administered by Britain; if they voted for it, Wales would have an assembly. A referendum on the question of devolution for Wales had come up once before, in 1979, and it had been defeated by a resounding majority of nearly four to one.

The second referendum on this question took place on September 18, 1997, and it passed, but only by a small margin, providing a "wafer-thin" majority in favor of devolution. Only 50.3 percent of the 2.2 million eligible voters in Wales cast a ballot, and the measure passed by 50.3 percent to 49.7 percent, a scant 0.6 percent, or 7,000 votes.

Much speculation has ensued regarding why the vote was so close and why so many people stayed home instead of voting. In the western part of Wales, Welsh nationalists guaranteed a vote in favor of the referendum. Hundreds of thousands of Welsh speakers were passionate about securing greater freedom from the English Parliament. However, the northeast and the heavily populated southeast voted against the referendum. These are the regions that contain more anglicized Welsh, fewer Welsh speakers, and more people originally from England. Many of these people commute to England on a daily basis or are attached to English commercial interests.

Many who voted against the referendum saw devolution as the slippery slope leading to separation from England and argued that, unlike the Scots, the people of Wales did not have a unified culture and history.

Consequences

The Welsh Assembly came into existence in Cardiff shortly after the passage of the referendum. The tasks put before the assembly have involved education, health, agriculture, the envi-

AP/Wide World Photos

A group of Welsh voters in Cardiff celebrate the passage of the referendum.

ronment, transportation, and planning, the same areas previously handled by a cabinet member of the British government with the title of secretary of state for Wales. Powers over what are called home affairs, police and the judiciary in particular, are retained by the British government at Westminster.

The assembly's primary responsibility was to take charge of and set priorities for a grant of £7 billion (or $12 billion), the annual budget. This money was formerly the responsibility of the Welsh office of the national government, which was to be considerably reduced in size. Similarly, a large number of quasi-autonomous organizations pertaining to Wales faced abolition. Scores of committees controlling local and regional matters in Wales came under the management of the assembly. The assembly functions as a miniature parliament, with the members of the leading party forming a committee to carry out delegated executive powers.

Henry G. Weisser

Million Woman March Draws Many Thousands to Philadelphia

The Million Woman March, the second in a series of activist marches in the late 1990's targeting empowerment and unity in the African American community, brought hundreds of thousands of black women together from all over the world.

What: Civil rights and liberties; Gender issues; Social reform
When: October 25, 1997
Where: Philadelphia, Pennsylvania
Who:
PHILÉ CHIONESU, founder and cochair, Million Woman March
ASIA CONEY, national cochair, Million Woman March
WINNIE MADIKIZELA-MANDELA (1934-), former wife of South African president Nelson Mandela
MAXINE WATERS (1938-), Democratic congresswoman from California

Inspiration

The Million Woman March took place on October 25, 1997, in Philadelphia, Pennsylvania. Cofounded by Philé Chionesu, the owner of an African artifacts store in Philadelphia, and Asia Coney, president of the Tasker Tenant Improvement Council in Philadelphia, the march created a network of African American women community activists. These women took on positions of leadership such as national cochair and regional coordinator in order to rally African American women to demonstrate their commitment to solving the problems that faced African Americans at the end of the twentieth century.

Although the march was not organized through a religious organization, it was inspired by the October 16, 1995, Million Man March organized by Minister Louis Farrakhan of the Nation of Islam. Philadelphia was selected as the destination because it is where the Declaration of Independence was signed. Organizers wanted African American women to make their own declaration of independence from poverty, discrimination, enslavement, and abuse. The march used grassroots organization and publicity, relying on word of mouth, the Internet, and black media sources instead of corporate sponsorship and mainstream television and print media.

Two years earlier, many African American women had supported the Million Man March, deferring to Farrakhan's request that women refrain from attending the march to avoid unnecessary distraction. However, many African American women were eager to have their turn to demonstrate their commitment to bettering the lives of black people, and the Million Woman March provided just such an opportunity.

A Day for Women

A complete program of events was designed to reinforce the motto of the October 25 march: Great Grandmother Taught Grandmother. Grandmother Taught Mother. Mother Taught Me. I Will Teach You. The march began with a 6 A.M. spiritual ceremony at Penn's Landing, a site on the waterfront of the Delaware River regarded as sacred by some African Americans because it is where Africans were bought and sold after reaching the colony of Philadelphia. Accompanied by the traditional rhythms of an African drum procession, participants marched two miles from Penn's Landing to the steps of the Philadelphia Museum of Art.

The Million Woman March addressed a wide variety of themes and issues, including sisterhood, positive relationships with men, domestic violence, women's health, incarceration of women, family, independent African American schools, leadership, global human rights, and the Central Intelligence Agency's possible role in crack cocaine trafficking in the inner cities. The march's mission statement highlighted these concerns, as well as a reaffirmation of women's roles as mothers, as nurturers, and as protectors of life. The march also addressed the African American woman's role in rebuilding deteriorated African American neighborhoods.

The seven-hour-long official program featured a diverse range of prominent black women in fields ranging from politics to religion, to music and the arts, and to activism. These speakers included Congresswoman Maxine Waters; Ilyassah and Camilla Shabazz, daughters of 1960's activist Malcolm X; actress Jada Pinkett; rapper and social activist Sister Souljah; singer-songwriter Faith Evans; Afeni Shakur, the mother of hip-hop artist Tupac Shakur; Pam and Ramona Africa of Philadelphia's MOVE organization; and Khadijah Farrakhan, wife of Louis Farrakhan. South African political activist Winnie Mandela gave the keynote address. Her international fame as the politically active former wife of South African president Nelson Mandela lent an air of internationalism to the march and added to its significance. Not all march participants were black women: BettyeMae Jumper of the Native American Tribal Council gave the program's prayer of unity. Organizers estimated that 2.1 million people convened in Philadelphia to support the march. Philadelphia police estimates range from 300,000 to one million attendees.

AP/Wide World Photos

Participants in the Million Woman March fill the Benjamin Franklin Parkway in Philadelphia.

Consequences

Many participants attended the march because they wanted to become a part of history. Women also attended the event to meet new people, exchange ideas, network, and see the faces of the hundreds of thousands of women whose daily lives are occupied with solving problems that affect the African American community. It is difficult to measure the exact impact of the march in real terms, but it is likely that the march at least served as a symbol to the world that African American women are leaders who are actively concerned with the progress of their communities. March leaders hoped that the image of hundreds of thousands of African American women gathered together to express their political power sent a positive message to the world that helped eliminate negative stereotypes of African American women. They also hoped it would heighten awareness of the difficulties of battling both sexism and racism.

The utility of march events such as the Million Woman March is the subject of debate. Critics question whether the money spent on travel, hotels, and souvenirs and the time involved in planning and attending the event could be put toward more tangible gains. For example, officials from the Philadelphia Convention and Visitors Bureau said that the march generated $21.7 million dollars for the city. What would happen, they ask, if the participants donated the equivalent of these expenses toward an African American social, political, educational, or economic cause instead of marching? The most consistent criticism is that it is difficult for a march to produce concrete achievements.

The Internet has provided one of the best means for maintaining the spirit of the Million Woman March and the networks it generated. African American women created forums, newsletters, and other opportunities for discussion of the march. In addition, African American women supported similar events such as the African American-centered 1998 Million Youth March in Harlem, New York; the unified, multiracial Million Mom March held in May, 2000, in Washington, D.C.; and the Nation of Islam-sponsored Million Family March in Washington, D.C., held in October, 2000.

Christel N. Temple

World Nations Approve Kyoto Plan for Reducing Harmful Emissions

An international agreement in Kyoto, Japan, to reduce industrial gas emissions provided hope that an organized global plan to combat human-caused global warming would be implemented.

What: Environment; Earth science
When: December 11, 1997
Where: Kyoto, Japan
Who:
Bill Clinton (1946-   ), president of the United States from 1993 to 2001
Robert T. Watson (1922-   ), chairman of the United Nations Intergovernmental Panel on Climate Change

Global Climate Change

Before the twentieth century, the direct impact of human activities on Earth's overall climate was relatively minor. In the latter part of the twentieth century, however, concern arose that human activities might be modifying Earth's surface and atmosphere sufficiently to create significant changes in global temperatures. In December, 1995, the United Nations Intergovernmental Panel on Climate Change, chaired by Robert T. Watson, issued a report concluding that "the balance of evidence suggests that there is a discernible human influence on the global climate." It projected that human activities may cause global temperatures, which are currently increasing, to rise an additional 1.8 to 6.3 degrees Fahrenheit (1 to 3.5 degrees Celsius) within the next century. Many scientists concur with the panel's assessment, and the threat of pronounced global warming has become one of the world's foremost environmental concerns.

The anthropogenic (human-caused) global warming potential stems chiefly from increases in atmospheric carbon dioxide (CO_2) levels, attributed primarily to the large-scale burning of fossil fuels. Another major cause of increasing carbon dioxide levels is deforestation, especially within the tropical rain forests. Carbon dioxide emissions, which are believed responsible for about 64 percent of the global warming resulting from human-released gases, have risen nearly fourfold since 1950. Several other anthropogenic gases, notably methane (CH_4), are also believed to be important contributors to global warming.

The current rate of global warming is indicated by climate records showing that worldwide mean temperatures rose by approximately 1 degree Fahrenheit (0.6 degrees Celsius) during the twentieth century. This amount of warming may not seem particularly large, but it is noteworthy that global temperatures have not varied more than 2.7 degrees Fahrenheit (1.5 degrees Celsius) over the past ten thousand years. In addition, most computer projections indicate that the rate of warming is likely to accelerate in the near future. By the end of the twenty-first century, global mean temperatures could be 2 to 6 degrees Fahrenheit (1 to 3.5 degrees Celsius) warmer than they were at the end of the twentieth century.

Because Earth's climatic elements are linked, a rise in global temperatures will affect other aspects of climate as well. The atmosphere's ability to hold water vapor rises with temperature; therefore, a general increase in global precipitation is likely. An increase in the intensity and severity of storms, especially hurricanes, appears probable. Evaporation rates will also become higher as temperatures rise, effectively producing drier climates in areas that do not receive enhanced precipitation. This will probably lead to more frequent droughts for many regions. As the climate becomes warmer, sea levels will rise because of melting glaciers and the thermal expansion of sea water. This is likely to produce greatly increased rates of beach erosion, damage to coastal ecosystems, and saltwater contamination of coastal water supplies. Continued global warming should also accelerate the already occurring migration of plant and animal species to higher latitudes and elevations and may result in the large-scale extinctions of less adaptable species.

The Kyoto Protocol

Concern over global warming has spurred the international community to hold a number of conferences in an attempt to regulate the release of atmospheric greenhouse gases. In December, 1997, a conference was held in Kyoto, Japan, that was attended by representatives of 171 countries. As a result of the conference, an agreement, the Kyoto Protocol, was reached on December 11 in which the Western industrial nations agreed to set specific goals for reduction of their emissions of carbon dioxide and five other greenhouse gases, including methane. During the period between 2008 and 2012, the United States pledged to achieve an average annual greenhouse gas output of 7 percent below 1990 levels; the Western European nations agreed to 8 percent cuts for this period, and Japan agreed to a 6 percent reduction. To help achieve these levels, countries can factor in greenhouse gas reductions brought about by changes in land use, including the planting of trees. Emissions trading—the purchase of emissions credits by one country from another country that has exceeded its specified emissions target—would be allowed.

To become legally binding, the Kyoto agreement must be ratified by at least fifty-five countries that account for 55 percent of the 1990 global emissions of carbon dioxide. A controversial aspect of the agreement is that developing nations, whose output of greenhouse gases is rapidly rising and will probably soon exceed those of the developed nations, are exempted from the necessity of making cuts. Although the agreement was strongly supported by President Bill Clinton, the U.S. Senate indicated that it would not ratify any agreement that would exempt developing countries that are economic competitors of the United States.

Consequences

Although it is an important agreement, the Kyoto Protocol is not in itself the answer to the problem of anthropogenic global warming. As of 2001, ratification of the agreement seemed uncertain at best, and even if it is approved, only a modest reduction in greenhouse gas emissions would probably occur; certainly not enough to stop further global warming. Another basic problem is that the Kyoto goals are for the near term. No goals have yet been set for the longer term (after the period 2008-2012) to maintain reduced emission levels. Much research and development work needs to be accomplished on alternative sources of energy such as solar,

AP/Wide World Photos

European Union's environment commissioner Ritt Bjerregaard speaks to reporters about the agreement on controlling global warming.

wind, tidal, and nuclear power to make them cost-competitive.

To work toward clarification and implementation of the agreement, several subsequent international meetings have been held, including one in Rio de Janeiro, Brazil, in 1998, and another in The Hague, Netherlands, in November, 2000. The key issues, which were not fully resolved at these meetings, were the development of an international emissions trading system, rules for counting emissions from "carbon sinks" such as forests, how to include developing nations without unduly restricting their economic growth, and enforcement mechanisms for any agreements reached. Probably the main achievement of all these meetings has been international recognition of the seriousness of the problem and the evidence of a desire on the part of almost all the participants to reach an effective solution. These meetings may lead to the creation of an international agreement that can be ratified in the near future. If the Kyoto Protocol or a similar agreement is ratified, its effectiveness in addressing long-term global warming will not be known for decades.

Ralph C. Scott

Pope John Paul II Visits Cuba

Pope John Paul II's first visit to Cuba signaled a thaw in long-chilly relations between the Vatican and Fidel Castro's communist-ruled Caribbean nation, but its impact on Cuba remained unclear.

What: Religion; International relations
When: January 21, 1998
Where: Cuba
Who:
John Paul II (Karol Wojtyła, 1920-), supreme pontiff of the Roman Catholic Church from 1978
Fidel Castro (1926 or 1927-), premier of Cuba until 1976, when he became president of the Council of State and Council of Ministers
Raúl Castro (1931-), Fidel's brother and commander of Cuba's armed forces
Pedro Meurice Estiú, archbishop of Santiago

A Holy State Visit

On the morning of January 21, 1998, Pope John Paul II arrived in Havana, Cuba, where he was cordially received by President Fidel Castro. The two men, who were in their seventies, were playing on different historical stages. The pope represented the oldest continually existing institution of Western culture, and Castro was the first and only communist leader still in power in the Western Hemisphere. In his welcoming address, Castro militantly defined himself by condemning the historical behavior of the Church, while also expressing admiration for the initiatives of the Vatican II Council (1962-1965), which had explicitly rejected all forms of racism and genocide.

Castro was honing his message for his communist constituency, which had just reelected him president and was ill at ease with the pontiff's visit. On the other hand, the pope's own message was one of reconciliation, saying that one had to love and be devoted to sacrifice and forgiveness. Further, he proclaimed that "this land should offer everyone an atmosphere of liberty, reciprocal confidence in social justice and lasting peace."

The site chosen for the first Mass in Cuba was Santa Clara, where in 1959 Castro's victory over government forces had brought him to power. Before the pope arrived there, the official newspaper of the Communist Party's central committee published a front-page article that declared there would be "no transition in Cuba to another type of government!" The government then braced itself for the pope's visit, unsure what he might publicly say.

Cuba's Invitation to the Pope

The primary motive behind Cuba's inviting John Paul II to visit appears to have been its desire to reduce its isolation from the outside world in a fashion that the United States could not prevent. The government had been angling for a papal visit since Castro visited Chile's President Salvador Allende in 1972, when he met with Chilean nuns and heralded their work as revolutionary. This was international propaganda and contrary to what was occurring inside Cuba, where churches were being persecuted. Nevertheless, it was not until the interview and publication of *Fidel and Religion: Conversations with Frei Betto* in 1986 that Cuba began in earnest to woo the pope to visit. Negotiations dragged on for nearly ten years, during which Castro visited the Vatican and had private audience with the pope. In a continuing effort to appease the Vatican, the Cuban government dropped its constitutional declaration of atheism.

During his stay in Cuba, the pope spoke of the Third Way—a path for national development that lies between the human degradation that capitalism can inflict and the stifling of the will associated with communism. He directly con-

AP/Wide World Photos

Cuban president Fidel Castro (right) and Pope John Paul II.

demned the United States' economic isolation of Cuba, because those who suffer most are the powerless, while also condemning the isolation that Cuba's own revolutionary elite had imposed on its people.

At the pope's first Mass, celebrated in Santa Clara, the scene was festive and both Catholic pilgrims and communist militants lowered their sense of distrust. Many believers in Afro-Cuban religions, such as Santería and spiritualist cults, attended. The next day the pope celebrated his second Mass in Camagüey, using the occasion to attack Afro-Cuban religions. He admonished Cubans to return to their Cuban and Christian roots. Afterward, he canceled an unscheduled visit with Santería believers and instead met only with leaders of mainline religious groups, Church-related institutions, and academics at Havana University.

The pope's most political Mass occurred on January 24. Before he spoke, he was preceded by Pedro Meurice Estiú, archbishop of Santiago, the birthplace of both Castro's revolution and of Castro himself and his brother, Raúl, commander of Cuba's armed forces—who both attended the Mass. The archbishop spoke harshly of Cuba's government, which he charged with having "confused their country for a party" and "the nation lives here and in the diaspora"—the latter a reference to Cubans living abroad. Many political leaders found his remarks ill-timed and offensive. By permitting the archbishop, a Cuban national, publicly to criticize the government signaled a new era in which the Church stood independently and publicly critical of Cuba's civil society.

The pope concluded his visit with a Mass at the Plaza of the Revolution in Havana on January 25. Not since Castro took power in 1959 had anyone other than a revolutionary leader spoken there. Many older people were emotionally overcome by the sight of the pope, but some Communists cried in frustration seeing the papal visit as yet another political setback. The pope was introduced by Cardinal Jaime Ortega, who was constantly interrupted by loud chants of "Cuba loves the Cardinal."

The pope's homily served two purposes: It emphasized the Church's message of love, and it stated that "a modern state cannot make atheism or religion one of its political functions." Castro was again in attendance, and the pope made an offhand remark that his attendance was unexpected but welcome. On this occasion, Castro forsook his customary green military fatigues for a conservative business suit and tie.

The Mass at the plaza took place in front of a five-story portrait of Ernesto "Che" Guevara, who had been Castro's right-hand man and military and philosophical strategist of the Cuban Revolution, and directly in front of the offices of the Central Committee of the Communist Party. Directly across from this also stood a five-story-tall poster of Jesus Christ, the likes of which had not been seen in Cuba for thirty-five years. At every point, the pope continued with his message of reconciliation, and mutual respect avoiding useless confrontations.

Consequences

It was obvious that the pope would neither follow his visit with permitting Catholics to join the Communist Party, nor did he highlight any of the social justice projects of the Cuban Revolution. When asked, officials traveling with the pope reminded individuals that the pontiff was there to plant a seed, which could blossom in the future, and possibly, in an unexpected manner.

The pilgrimage was a major break with the U.S. policy of isolation and another step toward ending the blockade. As the pilgrims left the island and returned to homes throughout the world, a Cuban radio announcer proclaimed the pope's visit a success because it had provided the state treasuries with nearly one billion dollars in revenue. Communists felt vindicated because inviting the pope was a means to an end: the continued survival of the only socialist revolution in the Western Hemisphere.

Andrés I. Pérez y Mena

Texas Executes First Woman Since Civil War

The execution of Karla Faye Tucker in Texas provoked protest around the world, even among supporters of the death penalty.

What: Crime; Law
When: February 3, 1998
Where: Texas
Who:
Karla Faye Tucker (1959-1998), convicted murderer
George W. Bush (1946-), governor of Texas from 1994 to 2001 who denied Tucker clemency

Crime and Punishment

On February 3, 1998, the state of Texas executed convicted murderer Karla Faye Tucker. She was the first woman executed in Texas since the Civil War and the second woman executed in the United States since 1976. Although Tucker admitted her guilt, many protested her execution, including some supporters of capital punishment such as television evangelist Pat Robertson and Speaker of the House Newt Gingrich. Others who opposed her execution, such as Pope John Paul II, Amnesty International, and Ron Cole, brother of Deborah Thornton, one of Tucker's victims, were death penalty opponents.

The major networks, Court TV, and Larry King of Cable News Network (CNN) provided extensive coverage of Tucker's case in the weeks before her execution. Coverage concentrated on Tucker herself and on the pace of executions in Texas. Tucker, a devout Christian, was young and attractive. Texas, whose governor, George W. Bush, was a presidential candidate, was described as the death penalty capital of the United States.

The Issues

On June 13, 1983, Tucker, Daniel Garrett, and Jimmy Liebrant went to the Houston apartment of Jerry Dean. Dean, a biker, had beaten his wife, who was Tucker's best friend, breaking her nose. Tucker, Garrett, and Liebrant plotted revenge for hours while drinking and taking a wide assortment of drugs. Finally, they decided that stealing Dean's motorcycle was an appropriate punishment. While Liebrant kept guard outside, Tucker and Garrett attacked Dean and Deborah Thornton, his lover, hacking them to death with a pickax. Afterward, Tucker and Garrett bragged about the murders to friends.

Eight months later, Garrett's brother and Tucker's sister reported them to authorities. Garrett and Tucker were tried and convicted of murdering Jerry Dean, although they were never tried for Thornton's murder. On April 23, 1984, they were sentenced to die. Garrett died in prison of liver disease in 1993.

Under Texas law, a death sentence requires a premeditated murder and a continuing threat to society. Tucker's supporters claimed that her case met neither standard. First, the murders occurred when both Tucker and Garrett were high from drugs and alcohol. They went to Dean's apartment to steal his motorcycle, not to murder the victims. Although they took guns to Dean's apartment, they found the murder weapon at Dean's apartment. Liebrant had testified at the trial that the murder was premeditated. In 1992, he changed his story, claiming he lied to receive a shorter sentence. Second, Tucker was no longer a threat to society. They claimed that when Tucker brutally murdered Dean and Thornton fifteen years ago, she was high on drugs and alcohol. She had a long history of drug abuse. At age eight, she began smoking marijuana, and by age eleven, she was shooting heroin. Jailed and removed from the drugs that dominated her early life, Tucker became a devout Christian. In 1995, she married Dana Brown, a prison chaplain. Her supporters argued that she was completely rehabilitated.

Tucker's gender was another important issue. Executions of women are relatively rare compared with those of men. Tucker would be the first woman executed in Texas since Chipita Rodriguez was hung in 1863 for killing John Savage. Ironically, one hundred years later, a unanimous Texas legislature voted that Rodriguez had not received a fair trial. Tucker would be only the second woman executed in the United States since 1976. The first was Velma Barfield. In 1984, North Carolina executed Barfield for poisoning her fiancé. Like Tucker, Barfield was a drug addict who became a born-again Christian in prison.

The number of executions in Texas disturbed some people. Although Texas has less than 10 percent of the population of the United States, it has held more than one-third of the executions. Between 1976 and April 4, 2001, Texas executed 245 prisoners. The rapid pace of executions was in part because of the clemency process. In Texas, the governor can reduce a sentence only with the recommendation of the State Board of Pardons and Paroles. The eighteen members of the board, appointed by the governor himself, seldom meet. They decide by mail if people they have never met will live or die. As of 1998, the board had never recommended clemency for any death penalty cases. Tucker's lawyers appealed the clemency process. The lawyers claimed that by never meeting Tucker, the board denied her due process. On January 28, 1998, the Texas Court of Criminal Appeals denied their appeal.

Although Tucker had many supporters, others bitterly opposed clemency. Among the most vocal opponents were Richard Thornton, the widower of Deborah Thornton, and Thornton's children. Opponents typically raised three objections: First, because Texas does not have a life term without parole, Tucker would become eligible for parole in 2003. Second, jailhouse religious conversions are common and often phony. Finally, two innocent people had been brutally murdered; Tucker was guilty of a vicious crime and deserved to die. On February 2, 1998, the Board of Pardons and Paroles rejected Tucker's clemency bid. On February 3, the U.S. Supreme Court refused to halt her execution, and Governor Bush rejected requests for a one-time thirty-day reprieve.

Consequences

On February 2, Tucker was flown from the state prison at Gatesville to Huntsville. She met with friends and family throughout that Monday and again Tuesday morning. At 6:01 P.M. on February 3, Tucker was led to the death chamber and strapped to a gurney. After her final statement, the lethal injection began. She was declared dead at 6:45 P.M.

Tucker's execution raised awareness of the issues surrounding the death penalty. The execution of a pretty young woman attracted extensive media coverage and contributed to declining support for capital punishment. Although most Americans supported the death penalty, in the late 1990's, support declined about 15 percent. The debate over the Tucker's execution did not slow the execution rate in Texas, however. In the three years after Tucker's death, Texas executed ninety-four more prisoners.

Virginia Thompson

U.S. Military Plane Cuts Italian Ski Cable, Killing Twenty People

A military jet plane flying at low altitude on a training mission in northern Italy cut through the cable of a ski lift gondola. The cable car fell, killing all twenty passengers.

What: Disasters; Military
When: February 3, 1998
Where: Cavalese ski resort in the Dolomite Mountains of northern Italy
Who:
Richard Ashby (1967-), pilot of the jet plane
Joseph Schweitzer (1967-), navigator
Peter Pace (1945-), Marine Corps Forces, Atlantic

The Accident

On February 3, 1998, a military airplane, part of the North Atlantic Treaty Organization (NATO) forces patrolling the skies over Bosnia, departed on a training flight from a U.S. airbase at Aviano, Italy. The twin-engine jet, an EA-6B "Prowler," was designed to fly very close to the ground to avoid detection by hostile antiaircraft installations. It is used for surveillance and for electronic warfare, such as jamming enemy radar to protect high-flying bombers.

The plane, manned by the pilot, a navigator, and two electronic countermeasure officers, was flying at more than five hundred miles (eight hundred kilometers) per hour at a very low altitude near Cavalese, a popular ski resort in the Dolomite Mountains of northern Italy. One wing of the plane sheared the steel cable near the ascending gondola car, causing it to crash to the ground. All twenty passengers in the gondola were killed by the impact. The descending gondola remained suspended by the cable, held by a safety mechanism, until its passengers could be rescued. The plane returned to its base without the pilot realizing that he had caused an accident.

The twenty victims who died included seven friends from one town in eastern Germany, a Polish mother and her twelve-year-old son, three Italians, five Belgians including a woman and her fiancé, two Austrians, and one Dutch woman. Families of the victims and political leaders from Europe called for a full investigation of the cause of the accident. Madeleine Albright, U.S. secretary of state, expressed condolence to the families and promised a full inquiry into the incident.

An Interview with the Pilot

The pilot of the airplane in the fatal accident was Captain Richard Ashby, an eight-year veteran of the Marine Corps. Richard A. Serrano, a reporter for the *Los Angeles Times*, interviewed Ashby, who grew up in Orange County, not far from Los Angeles. According to the interview, published May 3, Ashby had developed an enthusiasm for aviation during his boyhood by seeing and hearing the jet planes flying out of the nearby El Toro Marine Corps Air Station. He and his father had attended air shows at El Toro, and the young Ashby dreamed of becoming a pilot.

After the February accident, Ashby was portrayed in some of the European media as a show-off pilot, deliberately reckless and negligent about safety. In the *Los Angeles Times* interview, Ashby recalled their criticism, stating, "They were starting to call me Rambo and saying that we were trying to fly under the wire. . . . They said we were even betting beers." He insisted that nothing was done intentionally and expressed his sorrow for the victims of the tragedy. He recalled the loss that he felt when his own father died unexpectedly in a traffic accident.

Consequences

Lieutenant General Peter Pace, commander

of Marine Corps Forces in the Atlantic region, ordered Ashby and his navigator, Captain Joseph Schweitzer, to return to Camp Lejeune in North Carolina to stand trial on charges of negligent homicide, reckless endangerment, and obstruction of justice. The trial took place in February, 1998, and lasted about three weeks. The prosecution and defense attorneys presented their arguments to a jury of eight Marine Corps officers. Relatives of the victims and journalists came from Europe to observe the proceedings.

The prosecution argued that Ashby should be held responsible for the twenty deaths because his plane was flying too low and too fast, especially when operating in a populated area. In rebuttal, the defense attorney showed that the ski area at Cavalese was not marked on the flight map that Ashby had been given. Furthermore, the mountainous terrain apparently fooled the pilot into thinking that he was still above the minimum altitude of 500 feet (152 meters). The jury deliberated for two days and then announced its verdict: It found the pilot not guilty of recklessness or negligent homicide. After the decision in the pilot's case, the prosecutor decided to drop charges against the navigator.

The prime ministers of Italy and Belgium expressed shock and amazement when the verdict was announced. One Italian newspaper went so far as to demand that U.S. airmen be banned from all NATO bases in Italy. However, the Polish man who lost his wife and son wrote to Ashby that it was "a tragic accident" and that he "holds no ill will against the pilots, and the pilots are in our prayers."

A separate trial was held on the obstruction of justice charge. Ashby and Schweitzer admitted that they had destroyed a videotape that they took from the plane after it returned to the air base. They claimed that the tape recorder was off during the accident period but that the tape contained some footage from an earlier maneuver that they felt would be misconstrued by the media if it became public. Prosecutors argued that the tape would have shown whether the jet plane was operated recklessly. Both Ashby and Schweitzer were found guilty of obstruction of justice. They were dismissed from the Marine Corps, and Ashby was sentenced to a jail term of six months.

Hans G. Graetzer

AP/Wide World Photos

Workers examine the remains of the Mount Cermis cable car that fell after a U.S. Marine jet severed its cables.

Disabled Golfer Wins Right to Use Golf Cart in Tournaments

Golfer Casey Martin, disabled with Klippel-Trenaunay-Weber syndrome, successfully sued the Professional Golf Association (PGA) Tour to use a golf cart in tournament play.

What: Civil rights and liberties; Law; Sports
When: February 11, 1998
Where: Portland, Oregon
Who:
Casey Martin (1973-), professional golfer

Tee Time

Casey Martin began playing golf when he was six years old at his home in Oregon. He became good enough to be offered a golf scholarship to play, as a teammate of Tiger Woods, on the Stanford University golf team. After completing school, Martin decided on a professional golf career.

Unlike most golfers, Martin has Klippel-Trenaunay-Weber syndrome, a rare circulatory disease that limits the amount of blood that can circulate through his right leg. The lack of blood circulating causes the leg to be very weak. Martin walks with a pronounced limp and has indicated that any significant cut, bruise, or a fracture could require the amputation of his leg, which would mean the end of his professional golf career. Many days, because of the pain caused by his disability, he is unable to walk the entire eighteen-hole golf course. Martin asked the Professional Golf Association (PGA) Tour to allow him to ride in a golf cart instead of walking while playing tour events. The PGA responded that walking while playing golf in tour events was an integral part of the game. Therefore, they refused to allow Martin to use a golf cart for tour events.

The PGA also stated that as a private association, its rules should not be subject to review by an outside body. In 1997, Martin filed suit in an Oregon court against the PGA Tour to allow him to use a golf cart in tour events. Martin's attorneys cited provisions of the American with Disabilities Act (ADA) as the basis for his case. The ADA states that allowances must be made in public places for individuals with disabilities. They argued that Martin's circulatory disease, Klippel-Trenaunay-Weber syndrome, qualified Martin for this special treatment by the PGA.

Hole in One

On February 11, 1998, the U.S. District Court in Oregon handed down a ruling in favor of Martin, allowing him to ride a golf cart during PGA tour events and to continue to play professional golf. The court stated in its decision that the ADA applied to the PGA Tour and Martin's disease qualified him for coverage under the act. PGA members were divided on the issue. Some publically supported Martin, and others supported the PGA, saying that walking was an integral part of professional golf. The PGA argued that it was inappropriate for the court to attempt to change the rules of the game of golf as the PGA has established them.

Witnesses called during the trial indicated that they felt Martin riding in a cart while everyone else had to walk would give Martin an unfair advantage in tour competition. Martin stated that he did not believe this to be true. He added that because of the almost constant pain in his leg, he is very limited in both the time he can practice and play. After long sessions at the practice range or a difficult round on the golf course, he said, he is unable to sleep because of the pain and may play poorly for a number of days afterward.

In February of 1998, following the verdict against the PGA Tour, Magistrate Judge Thomas

Coffin issued a permanent injunction against the PGA Tour, requiring it to allow Martin to use a golf cart for all tour events. After the February, 1998, verdict and injunction by the district court in Oregon, the PGA Tour appealed the decision to the U.S. Ninth Circuit Court of Appeals in San Francisco, which hears appeals from all the district courts in nine Western states, including Oregon. In March of 2000, Judge William C. Canby, Jr., wrote a decision favoring Martin and upholding the decision of the district court. In the decision, he said that walking is not an integral part of the game and that hitting shots was what the game of golf is about.

The PGA Tour decided to appeal the case to the U.S. Supreme Court. In September of 2000, the Supreme Court agreed to hear arguments from both sides in this case. In January of 2001, attorneys for both Martin and the PGA Tour argued their cases to the nine justices of the Supreme Court.

AP/Wide World Photos

Casey Martin takes a shot during practice for the 1998 U.S. Open Championship in San Francisco.

Consequences

On May 29, 2001, the Supreme Court ruled 7-2 in favor of Martin. In the majority opinion, Justice John Paul Stevens said that organizations such as the PGA should "carefully weigh the purpose, as well as the letter" of the law in determining whether they should waive their requirements for disabled people. The Court ruled that the ADA, which requires "reasonable modifications" for disabled people unless such changes would fundamentally alter the event or place, applies to professional sports. According to the Court, walking the eighteen holes of golf is not essential to the game.

Martin, who continues to play professional golf using a cart, plays well in some tournaments and not as well in others, just like many of his competitors. His scores have not been either significantly better nor worse after he began using a golf cart.

A separate issue in the Martin case is the application of the ADA statutes to professional sports other than golf. One of the organizations supporting the PGA Tour was the men's professional tennis organization, the ATP Tour. Many critics of the Supreme Court decision fear that it will diminish the level of competition in other sports as special arrangements are made for disabled individuals in those sports.

Robert J. Stewart

U.S. Supreme Court Rejects Challenge to New Jersey's Megan's Law

The U.S. Supreme Court refused to hear an appeal, upholding a lower court's approval of a New Jersey statute known as Megan's Law, which required that convicted sex offenders register with the authorities and that communities be notified of their presence in the neighborhood.

What: Human rights; Law
When: February 23, 1998
Where: New Jersey
Who:
MEGAN KANKA, seven-year-old child raped and murdered by a neighbor who was a convicted sex offender
JOHN J. GIBBONS, lawyer for the unnamed sex offenders in the Supreme Court case

Megan's Laws

In 1994, Megan Kanka, a seven-year-old girl, was raped and killed by her neighbor, a convicted sex offender. Her parents argued that Megan would not have died had they been aware of their neighbor's history and been able to take precautionary measures. As a result, New Jersey and a number of other states passed Megan's Laws, laws designed to enable parents to know when convicted sex offenders were living in their neighborhoods.

Under the 1994 New Jersey law, the first of its kind, if sex offenders were registered as Tier II offenders, which meant that they represented a demonstrated risk to children in the community, authorities were required to send notification of their presence in the neighborhood to the affected schools, agencies, and community organizations. The law also required convicted sex offenders to register with the authorities. A group of sex offenders who had committed their crimes before the enactment of the law sued on the grounds that the harassment they suffered after their names and addresses were made public was a second punishment for their crimes, violating the constitutional guarantee against double jeopardy. The group also argued that this second punishment was beyond what was punishment under law when they committed their crimes and, therefore, that the Registration and Community Notification Laws, commonly known as Megan's Law, violated the constitutional guarantee against ex post facto laws.

Legal Analysis

On July 25, 1995, in *Doe v. Poritz,* the New Jersey state supreme court upheld the constitutionality of Megan's Law. However, the legal challenges to the law did not end. Alexander Artway, who had completed his sentence for sodomy and said the law prevented him from returning to New Jersey, challenged the legality of the statute, particularly its notification provisions, in federal court. U.S. District Judge Nicholas H. Politan found the notification portion of Megan's Law to be unconstitutional and issued an injunction preventing its enforcement. However, in April, 1996, Judge Edward R. Becker of the Third Circuit Court of Appeals overturned Politan's injunction against the enforcement of the notification part of the law and upheld the requirement that convicted sex offenders register with local authorities. On July 1, 1996, U.S. District Judge John W. Bissell lifted a temporary ban on the community notification provision of Megan's Law, reversing other judges' rulings against the requirement that communities be notified of the presence of a convicted sex offender. The legal battles continued. On June 30, 1997, U.S. District Judge Alfred J. Lechner ruled that Megan's Law was constitutional, clearing the way for it to go into effect. However, on July 3, the Third Circuit Court of Appeals issued an order that stayed implementation of the law for at least thirty days. On August 20, that court ruled that Megan's Law was constitutional and stated that court orders blocking its implementation would expire on September 10. On September 22, the Third Circuit refused to hear any additional constitutional challenges to Megan's Law but agreed to delay community notifications by ninety days while an appeal was made to the U.S. Supreme Court.

On December 8, 1997, U.S. Supreme Court justice David Souter lifted an injunction that had prevented community notification from being made. On February 24, 1998, the U.S. Supreme Court refused to hear an appeal of the Third Circuit court's ruling that Megan's Law, including the notification provision, was constitutional. The Court also declined to hear a challenge to rulings that upheld notification and registration provisions established by a New York version of Megan's Law.

Consequences

John J. Gibbons, the lawyer for the unnamed sex offenders in the U.S. Supreme Court case, argued that the notification policy made them subject to community harassment and violence and therefore constituted a second punishment for their crimes, violating the constitutional guarantee against double jeopardy. He also argued that the law increased the effective punishment for those convicted before Megan's Law was enacted. This retroactive effect, he said, violated the guarantee against ex post facto laws.

The Court, by letting the ruling of the U.S. Court of Appeals stand, did not prevent other states from offering legal challenges to other versions of Megan's Law. However, the refusal to hear the case made it appear likely that the Court would rule favorably on the constitutionality of other laws requiring local residents to be notified when a convicted sex offender moves into their area. President Bill Clinton expressed his approval of the Supreme Court's decision, calling the law a crucial tool to protect children.

A recent extension of Megan's Law is the Campus Sex Crimes Prevention Act. Congress passed the act as part of the Victims of Trafficking and Violence Protection Act of 2000. President Clinton signed the act into law on October 11, 2000. Under the Campus Sex Crimes Prevention Act, sexual offenders must notify state officials if they are enrolling or working at an institute of higher education. State officials then give the information to the campus police or the agency with jurisdiction over the campus area.

The success of Megan's Law in the United States led to a desire for a similar law in Great Britain. On July 1, 2000, schoolgirl Sarah Payne was abducted from a field near her grandparents' home in Hersham, Surrey, and murdered. Her body was found dumped in a field near Pulborough, West Essex, sixteen days later. Afterward, pressure was placed on Parliament to enact a "Sarah's Law" similar to New Jersey's Megan's Law; however, through 2001, no progress was made in that effort.

Dana P. McDermott

U.S. Scientists Report Discovery of Ice Crystals on Moon

Scientists announced that the preliminary findings from the Lunar Prospector mission indicated the presence of water ice in shadowed craters near the Moon's poles. They estimated that six billion metric tons of water ice might be buried beneath the lunar dust and could benefit future lunar colonization.

What: Space and aviation; Astronomy; Technology
When: March 5, 1998
Where: National Aeronautics and Space Administration (NASA)/Ames Research Center, Moffett Field, California
Who:
G. Scott Hubbard (1948-), Ames Research Center, Moffett Field, California
Alan Binder, Lunar Research Institute, Gilroy, California
William Feldman (1940-), Los Alamos National Laboratory, Los Alamos, New Mexico
Alex Konopliv, Jet Propulsion Laboratory, Pasadena, California

Water on the Moon

One of the more important discoveries made from the study of the lunar materials brought back by the Apollo astronauts was the apparent lack of water. There was no direct evidence, past or present, for the presence of water in any form on the Moon. Most geologists believed that water would at least be found in certain hydrated minerals, but this was not the case. On Earth, all rocks contain water in one form or another. The fact that scientists cannot reach a consensus on the absence of water has provided support for the giant impact theory for the origin of the Moon. The apparent absence of water on the Moon also creates significant problems for future lunar colonization.

An alternative theory to finding molecular water on the Moon suggested searching for water ice. This was first discussed as early as 1961 and later reconsidered after studying the data from the 1994 Clementine mission. It was theorized that cold traps could exist inside craters if certain conditions were met. First, water vapor from either volcanic eruptions or cometary impacts had to collect in craters near the poles. Second, the angle of the Sun had to be low enough to create shadowed areas that never saw sunlight and remained permanently frozen. Over time, the water ice and lunar dust could intermix to form layers somewhat similar to permafrost on Earth.

The Lunar Prospector Mission

The mission of the Lunar Prospector, managed by G. Scott Hubbard of the National Aeronautics and Space Administration's Ames Research Center, was to build on the data returned from the earlier Clementine mission and to make a comprehensive study of lunar surface material from a low polar orbit. Lunar Prospector was one of the first low-budget planetary missions designed to return maximum data from an essentially low-cost vehicle. The total mission cost was around $63 million. The spacecraft, built by Lockheed Martin Missiles, was a 1.28-meter-high by 1.3-meter-diameter (4.2-feet-high by 4.3-feet-diameter) graphite-epoxy drum with three attached radial instrument booms.

Lunar Prospector was launched on January 6, 1998, from Cape Canaveral, Florida. It took the spacecraft 105 hours to cover the distance from Earth to the Moon. Upon reaching the Moon, it was placed into a nearly circular 100-kilometer (62-mile) altitude polar orbit. It circled the

Moon every 118 minutes. Later in the mission, Lunar Prospector's orbit was gradually lowered to thirty kilometers (18.6 miles) and then to ten kilometers (6.2 miles) to increase data quality and resolution.

The proposed nineteen-month mission was designed primarily to construct a detailed map of the Moon's surface composition. Scientists hoped that this data could be used to gain a better understanding of the Moon's origin, evolution, current dynamic systems, and potential resources. The Lunar Prospector carried five science instruments—a gamma-ray spectrometer, neutron spectrometer, magnetometer, electron reflectometer, and alpha particle spectrometer—used to conduct experiments designed to search the lunar crust and atmosphere for potential mineral resources, water ice, and certain gases as well as to map the moon's magnetic fields and to characterize its core.

Alan Binder of the Lunar Research Institute functioned as principal investigator and managed the alpha particle spectrometer experiment, used to find outgassing events on the lunar surface and examine tectonic activity on the Moon. In addition, Alex Konopliv of the Jet Propulsion Laboratory designed a Doppler gravity experiment in which Doppler was used to map the Moon's gravitational fields. The information accumulated was meant to complement data returned from the Apollo and Clementine missions.

By the end of its mission, the Lunar Prospector had completed a high-quality gravity map of the Moon, compiled a global magnetic field map, and provided global absolute abundance data for eleven key elements, including hydrogen. William Feldman, of Los Alamos National Laboratory, used the craft's neutron spectrometer to measure hydrogen levels on the lunar surface. During its lunar orbit, the Lunar Prospector measured large amounts of hydrogen, which suggested the presence of water ice in the permanently shadowed areas of the Moon. At the conclusion of the mission, a decision was made to crash the spacecraft into a shadowed crater near the south pole in hopes of detecting evidence for the presence of water vapor released by the impact.

Consequences

To prove that the detected hydrogen was associated with water ice intermixed with the lunar regolith would require that mission scientists crash the spacecraft into a lunar crater near the south pole. Scientists hoped that the impact could produce sufficient energy to vaporize water ice that could then be detected from Earth-based instrumentation. Some of the early estimates suggested that as much as 192 pounds (87 kilograms) of water ice could be ejected and vaporized by the impact. It was hoped that the vapor plume would last for several minutes before the Sun's ultraviolet radiation would break up the water molecules. Scientists also believed that the resulting hydroxyl (OH) molecules could be detected from their ultraviolet emission lines for as long as fourteen hours after impact.

All preparations for the impact went as planned and the spacecraft crashed into the lunar surface on schedule. Unfortunately, no water signature was observed. Mission scientists of-

AP/Wide World Photos

A nearly full moon, photographed from the Apollo 8 spacecraft.

fered several explanations for the negative result. Some suggested that the spacecraft might have missed its intended target, or it may have hit a rock. Others thought that the impact did not release sufficient energy to vaporize enough water ice to be detected. There were other possibilities, but in the end most scientists agreed that initially the spacecraft probably detected pure hydrogen and not the inferred water ice.

Although the impact did not reveal water ice, it did not eliminate that possibility. In March, 2001, at a conference in Houston, Texas, Los Alamos's Feldman reported that a combination of the neutron spectrometer data with calculations of the sublimation processes of hydrogen compounds provided confirmation of the existence of water ice in the shaded regions of the lunar poles. Further research and analysis is necessary before the water ice theory can be confirmed or disproved.

Paul P. Sipiera

Astronomers See Light from Object 12.2 Billion Light-Years Away

Using the largest telescopes in the world, a team of astronomers extended their view back toward the beginning of the universe some 13 billion years ago.

What: Astronomy
When: March 13, 1998
Where: University of Hawaii, Honolulu, Hawaii
Who:
Esther M. Hu, team leader, University of Hawaii astronomer
Lennox L. Cowie (1950-), University of Hawaii astronomer
Richard McMahon, astronomer, Cambridge University, England

A Look Back

The universe has long been the subject of curiosity. To discover its shape and other attributes, scientists have been trying to determine the time of its birth, which according to the big bang theory, is about 13 billion years ago. This theory holds that the universe began with a big explosion and has been expanding ever since. On March 13, 1998, astronomers Esther M. Hu and Lennox L. Cowie of the University of Hawaii in Honolulu and Richard McMahon at Cambridge University in England, leaders of an international team, recorded seeing an infant galaxy, the most distant and, therefore, the oldest object ever viewed, at about 12.2 billion light-years away. The more distant a galaxy is from Earth, the higher is its redshift, the amount by which the expanding universe has shifted the light emitted by a galaxy into the red (longer) frequencies. For this galaxy, the team recorded a redshift of 5.64, which passed the earlier record of 5.34 and meant the galaxy it had found was older by about 60 million years.

The Birth of Light

The galaxy examined by Hu, Cowie, and McMahon, known as 0140+326RD1, lies some 12.2 billion light-years from Earth, which means the light that is reaching Earth from the galaxy left it about 800 million years after the galaxy came into being. Scientists believe that it may be a young galaxy, at the beginning of a long life of producing new stars. Light travels about 5.8 trillion miles per year in the vacuum of empty space, so the newly visible galaxy is far out on the edge of the universe. Astronomer McMahon said that astronomers hope to find galaxies that existed when the universe was only about 500 million years old

Larger, more accurate Earth-based telescopes enable scientists to see what was previously not visible. Even orbiting telescopes such as the Hubble Space Telescope, which produces crystal-clear views of many relatively close objects, cannot match the size and scope of the twin 10-meter (33-foot) Keck II telescopes atop the 14,000-foot (426-meter) extinct Mauna Kea volcano of Hawaii. The Keck telescopes can gather more visible light and detect weaker light than any other telescopes. Using these telescopes, scientists can use older methods of detecting distant objects that were developed for smaller telescopes to make numerous new discoveries.

One of the problems in detecting distant galaxies is that their mass and size are much less than those of galaxies such as the Milky Way, and therefore, they emit less light. One older method of searching for distant galaxies involved looking for Lyman-alpha emissions, an ultraviolet wavelength of light (121.6 nanometers) emitted by excited hydrogen atoms. These emissions, produced in relatively small quantities, are hard to find because they are absorbed by the dust generated as a galaxy ages. Although absorption proved a problem for detection with small telescopes, some scientists used this quality to create a ultraviolet dropout technique for detecting distant galaxies. Hydrogen gas, which is found throughout the universe, also absorbs ultraviolet light. These scientists reasoned that the farther away a galaxy lies, the more hydrogen gas between it and Earth, and the dimmer its ultraviolet emissions. To find distant galaxies, they look for those that vanish in the ultraviolet.

With the Keck telescopes, scientists are able to examine far more visible light, enabling them to better detect Lyman-alpha emissions to find a single redshifted length of light emitted by atoms of hydrogen, which reveals the presence of an object at the farthest reaches of time and space. Hu, Cowie, and McMahon stated a preference for this method of detection over the ultraviolet dropout technique because they believe it can detect older galaxies. For example, the dropout technique may be unable to see new galaxies with muted emissions because it relies on emissions over a broad range of colors for detection. Also, light from extremely distant galaxies begins to disappear from view at wavelengths in the visible range of the spectrum, which includes long-wavelength emissions from molecules in Earth's atmosphere. Lyman-alpha searches can eliminate this problem by avoiding the redshifted wavelengths at which Earth's atmosphere radiates. McMahon said that the team looks for galaxies where the sky is dark, in the dark spaces that occur between the emission lines that come from the sky.

Consequences

Scientists think they will be able to look back to within a million years of the birth of the universe. Scientists theorize that initially after the big bang, the universe was opaque, and all light was immediately absorbed by the dense cloud of initial elements. Then, the universe expanded into the empty clarity of nearly infinite space. Once the dust lightened up, it became possible to see. Astronomers would like to see back to that moment, and discoveries such as that made by Hu, Cowie, and McMahon bring that dream closer to reality by pushing back the time that the first stars and galaxies formed after the big bang.

Michael W. Simpson

Vatican Issues Document on Jewish Holocaust and Condemns Racism

The Vatican issued a remembrance on the Jewish Holocaust and the Nazi-inspired mass extermination of other peoples during World War II. The document condemned racism and expressed sorrow for the sins of "sons and daughters" of the Church who cooperated with or tolerated the evil, while recalling the heroism of those who assisted and protected Jews, including Pope Pius XII.

What: Religion
When: March 16, 1998
Where: Vatican City
Who:
JOHN PAUL II (KAROL WOJTYŁA, 1920-), pope of the Roman Catholic Church from 1978
EDWARD IDRIS CARDINAL CASSIDY (1924-), president of the Holy See's Commission for Religious Relations with the Jews

Background

During the course of World War II, the Nazi regime of Germany, headed by Adolf Hitler, engaged in a ruthless and brutal policy of extermination of about six million Jews in countries occupied by German forces. An equal number of non-Jews, mainly in Slavic countries of Eastern Europe, also perished in death camps, including thousands of Catholic clergy. The Nazi regime was both anti-Semitic and anti-Roman Catholic, though only Jews were eventually targeted for the final solution, or total extermination.

Although anti-Semitism had existed for many centuries in predominantly Christian countries in Europe, the Nazi ideology was unusual in the ferocity of its hatred. Nazism was also a racist ideology that viewed Aryan peoples as a superior race and regarded other peoples, such as Slavs, as inferior. Some Jews, anticipating the hostility of the Nazis, attempted to emigrate from Germany, and others were expelled by the Nazis during the 1930's. Although some Jews succeeded in establishing new homes, many others were turned away by governments in North and South America, including the United States.

As World War II began and the Nazi regime began to occupy territory outside Germany, Hitler's government engaged in forceful and brutal deportation of Jews. Protests by the Catholic Church and the international community had only a limited effect, and in some cases, they actually provoked the Nazis to intensify the scope and brutality of deportations. Most people, including the Jews and other victims of deportation, had little idea what awaited them in the camps and put up little resistance, an exception being the uprising and resistance by Jews in Warsaw in September, 1942. Reports of the exterminations began to surface after the Wannsee Conference in January, 1942, when the final solution was formally adopted and implemented, but the war was raging throughout Europe and Africa at the time, and the Nazis were able to execute their genocidal plan largely beyond international scrutiny and without opposition. The ultimate horror of the Nazi policy was not fully discovered and widely publicized until the end of World War II.

Many people, including Pope Pius XII, assisted and protected Jews, hid them from the Nazis and provided false baptismal certificates and travel documents to help Jews elude capture. However, many other people simply ignored the plight of the Jews, and others cooperated in their persecution. Some critics believe that too little was said and done by Christians to stop the Holocaust.

Motives for the Remembrance

With the new millennium approaching, Pope John Paul II called upon the Catholic Church to recount the events of the two thousand years since the birth of Jesus Christ. He hoped that by identifying wrongs committed in the name of the Church by Christians throughout the ages, Christians could enter the new millennium after a full cleansing of the conscience for all actions that fell short of Christian charity. The remembrance on the Shoah, or Holocaust, was just one of these efforts.

The document squarely places the lion's share of the blame of the Holocaust on the Nazi regime, where it appropriately belongs. It recounted, however, anti-Semitic tendencies in European practices that led to discrimination against Jews, attempts at forced conversion, and violent incidents in which Jews were treated by various populations as scapegoats for disasters. It also noted that while many Catholics and Christians came to the aid of the Jews during the Nazi regime, many others did not, and some actually cooperated with the Nazis.

The document noted that both Popes Pius XI and Pius XII had explicitly rejected racist policies before World War II and that other prominent Church leaders, at risk to themselves, had criticized Nazi anti-Semitic policy. Following the pronouncements of the Vatican II Council (1962-1965), the document explicitly rejected all forms of racism and genocide, calling Christians to never again allow such racial hatred to produce another Holocaust. It called on Christians to familiarize themselves with the Hebrew roots of Christianity and to recognize Jews as the elder brothers in faith to Christians. It expressed the Church's "deep sorrow for the failures of her sons and daughters in every age" and called for an "act of repentance" by all Christians and a renewed commitment to avoid anti-Semitism and to promote mutual respect between Christians and Jews. The remembrance did not, as some had hoped, specifically admit that the Church, by doing more or speaking out more forcefully, could have prevented the Holocaust.

Consequences

While the remembrance served as an important statement of concern and regret for the fail-

AP/Wide World Photos

Australian cardinal Edward Cassidy (right), president of the Commission for Religious Relations with the Jews, presents a document on the Church's role during the Holocaust.

ures of some Christians, it did not conclude that the Church itself was derelict in doing more for Jews and other victims of persecution during World War II. This and the fact that the document defended the record of Pius XII caused many Jewish activists and other critics of the Catholic Church to complain that the statement was an incomplete apology. However, the Church viewed the document not as an apology but rather as a remembrance of the moral tragedy of the Holocaust, which elicited both moral heroism and cowardice on the part of different people. The Church did not view itself as being in a position to prevent the Holocaust—an event that even the combined armies of the Allied nations failed to stop. The statement was, however, another step, among many taken by Pope John Paul II, to improve relations between the Catholic Church and the Jewish people.

Robert F. Gorman

Democratic Fund-Raiser Pleads Guilty to Making Illegal Contributions

The Justice Department and congressional investigations uncovered the use of foreign cash, other illegal contributions, and large amounts of unregulated money during the 1992 and 1996 elections, leading to suspicions of bribery and corruption.

What: Political reform
When: March 16, 1998
Where: United States
Who:
JOHNNY CHUNG, a Chinese American businessman and fund-raiser for the Democratic Party
JAMES RIADY, an Indonesian businessman and major contributor to Bill Clinton's 1992 campaign
BILL CLINTON (1946-   ), president of the United States from 1993 to 2001
AL GORE (1948-   ), vice president of the United States from 1993 to 2001

China Connection

On March 16, 1998, Chinese American businessman Johnny Chung struck a deal with government prosecutors. In exchange for a guilty plea to charges of bank fraud, tax evasion, and conspiracy in connection to $20,000 in illegal contributions to President Bill Clinton's 1996 reelection campaign, Chung agreed to cooperate with ongoing investigations of violations of campaign financing laws. His subsequent revelations were disturbing to the U.S. Justice Department. Chung admitted that the source of his funds was Liu Chaoying, the daughter of the most senior officer of the People's Liberation Army of Communist China. Family and professional ties further connected Liu to two other arms of Chinese national security: the China Ocean Shipping Company (COSCO) and the Commission on Science, Technology and Industry for National Defense (COSTIND). When a Clinton administration official later pressured the city of Long Beach, California, to lease port facilities to COSCO in preference to the U.S. Marine Corps, some Americans suspected that Liu's campaign cash had bought changes in U.S. policy favorable to China.

Campaign Financing Debates

Laws regulating campaign contributions and related activities have a long history in the United States. The Pendleton Act of 1883 prohibited fund-raising on federal property. Republican Richard M. Nixon's victories in the presidential elections of 1968 and 1972 served as the impetus for modern campaign financing legislation. Although historians and political scientists attribute Nixon's successes to divisions within the Democratic Party over the civil rights movement of the 1960's and the Vietnam War, liberals and Democrats at the time believed that Nixon won through massive campaign spending, funded by wealthy individuals and large corporations.

The result was the Campaign Finance Law of 1974, which limited individual contributions to particular candidates to $1,000 and attempted to cap spending in federal elections. Since then, liberals and Democrats have tended to favor further restrictive legislation as a way to limit the access of business interests to public officeholders and legislators. According to those favoring restrictions, business is able to buy legislation that is favorable to its needs and often detrimental to the public interest through campaign contributions. Conservatives and Republicans have argued that the vote of any given legislator is based primarily on the lawmaker's personal ideology, party affiliation, and constituent demands. According to Republicans, campaign contributors give to candidates who are already inclined to represent their viewpoints. Restricting political spending, conservatives maintain, is equivalent to restricting free speech and is thus a violation of the First Amendment of the U.S. Constitution. The U.S. Supreme Court partially agreed with the latter viewpoint when it struck down some campaign spending limits as unconstitutional in *Buckley v. Valeo* (1978).

In the 1990's, the debate over campaign financing focused on the raising of so-called soft money, funds that political parties spend for voter education on the issues. Although soft money was critical to Clinton's reelection in 1996, Democrats favored banning its use, while Republicans generally opposed any new campaign financing laws.

The greatest controversy of the 1990's, however, concerned Chinese attempts to influence U.S. elections. In early 2001, Indonesian James Riady pled guilty to numerous violations of campaign laws (during the 1992 campaign) and admitted that he had sought alterations in U.S. policy, including a highly favorable trade treaty for Communist China, which favored his family-owned megabusiness, the Lippo Group. Riady was successful in placing one of his employees, John Huang, in the U.S. Commerce Department, where the latter had continuous access to the Central Intelligence Agency's classified reports on worldwide business conditions. The Clinton administration had recommended Huang's employment.

AP/Wide World Photos

Johnny Chung.

In 1996, two American companies, Loral Space Systems and Hughes Electronics, illegally transferred missile technology to Communist China. This technology was used to improve China's commercial satellites, but it could be easily adapted to improve the types of rockets that carry nuclear bombs. When the Justice Department began investigating these companies in 1998, Clinton signed an executive order making such technology transfers legal. Between 1996 and 1998, Loral's chief executive officer, Bernard Schwartz, gave approximately $2 million to the Democratic Party. The U.S. Air Force and the bipartisan Cox Commission later concluded that the technology transfers had probably damaged U.S. security. Clinton fell under suspicion of having enabled the selling of military technology to a potential enemy in exchange for campaign cash. The charges seemed believable to some Americans because Clinton appeared to have reversed his position on China. Although Clinton had been extremely critical of his opponent, President George Bush, during the 1992 election for being too lenient with the Chinese, he had begun to speak of China as a strategic partner. Clinton's defenders pointed out that there was no evidence showing that the executive order was signed in exchange for money nor was there any evidence that Clinton was even aware of Schwartz's donations or the Chinese money in Democratic coffers. Even Clinton's severest critics admitted that such proof was needed to sustain charges of treason and bribery against him.

Clinton and Vice President Al Gore were continuously investigated beginning in 1997 for possible violations of the Pendleton Act. Their White House "coffees" for major donors seemed to violate the rules against fund-raising on federal property. Gore was also suspected of holding an illegal fund-raiser on tax-exempt property, a Buddhist temple in California, where thousands of dollars in donations exceeding the $1,000-per-individual limit were disguised as coming from the monks. Gore was also accused of lying under oath concerning these matters. Clinton's attorney general, Janet Reno, refused to appoint a special prosecutor to investigate these charges, despite a recommendation for such action by two Justice Department lead investigators and the head of the Federal Bureau of Investigation. Reno said that there was no specific or credible evidence that laws had been broken.

Consequences

Democrats and most Americans blamed these problems on the lack of restrictions and limits on campaign fund-raising and spending. Therefore, the movement for more campaign-funding legislation gained momentum in the early twenty-first century. Clinton's popularity remained high until he left office because the American economy was prospering; however, the perception that Gore was less than completely honest started with the fund-raising scandals and dogged him throughout his 2000 campaign for the presidency. It may have contributed to his ultimate defeat at the hands of Republican George W. Bush.

Michael S. Fitzgerald

Meteorite Is First Extraterrestrial Object Found to Contain Water

The Monahans meteorite was the first extraterrestrial object to provide a sample of liquid water from an asteroid. The water, trapped in salt crystals, demonstrates that liquid water existed early in the history of the solar system, and the association of water with salt crystals suggests that brine evaporated on or near the surface of the asteroid.

What: Astronomy
When: March 22, 1998
Where: Monahans, Texas, and National Aeronautics and Space Administration (NASA) Johnson Space Center, Houston, Texas
Who:
Alvaro Lyles, witness to the fall of the Monahans meteorite
Everett Gibson (1940-) and
Michael E. Zolensky, scientists at NASA Johnson Space Center

The Fall to Earth

Witnesses saw a brilliant fireball and heard a sonic boom over West Texas on March 22, 1998. Shortly afterward, two stones, weighing 2.9 pounds (1,344 grams) and 2.7 pounds (1,243 grams), fell to the ground in Monahans, Texas. These meteorites, believed to be small fragments broken from an asteroid, fell in front of the home of Orlando Lyles. One of the meteorites penetrated the asphalt of the city street in front of the Lyles's house. The fall was witnessed by seven boys who were playing basketball nearby, including eleven-year-old Alvaro Lyles. They reported the event, and within hours, scientists at the National Aeronautics and Space Administration (NASA) Johnson Space Center in Houston, Texas, were alerted.

The next day, Everett Gibson, a meteorite researcher at the Johnson Space Center, arrived in Monahans to examine the meteorites. Gibson persuaded the city of Monahans, which claimed the meteorites because they fell on the city roadway, to loan the rocks to NASA for scientific research. Within forty-eight hours of the fall, the Monahans meteorites had been taken to the Johnson Space Center for examination. This quick recovery of the Monahans meteorites reduced the possibility of alteration and contamination of the interior of the rocks by rain or other natural processing on Earth.

The Discovery of Water

The smaller of the two Monahans meteorites was broken open to allow inspection. When the NASA researchers cracked open this piece of the meteorite, they were surprised to see purple crystals on the otherwise gray-colored broken surface. These purple crystals were quickly identified as halite, or rock salt, a compound of sodium and chlorine. Although microscopic crystals of halite had been seen previously in other meteorites, the halite crystals found in the Monahans meteorite were much larger, up to 3 millimeters (0.1 inch) in size, than those found before.

The real surprise came when these halite crystals were examined under a microscope. These purple halite crystals contained microscopic inclusions of fluid and solids. The fluid was analyzed using a technique called Raman spectroscopy, in which an intense spot of light from a laser shines on the sample and light is emitted at specific energies characteristic of the particular molecules in the sample. Raman spectroscopy indicated the fluid was water. This was the first time that liquid water had ever been found in a meteorite. Many of the solid inclusions were identified as sylvite, a compound of potassium and chlorine.

The first important issue was to determine if this water was really from the asteroid, or if the water and the halite crystals had formed in the few days the meteorite was on Earth. Radioactive clocks, based on the amount of decay of one element into another, are frequently used to determine how long ago a crystal was formed. One of these clocks is based on the decay of an isotope of the element rubidium into an isotope of strontium. The rubidium and strontium isotope contents of a small fragment of halite from the Monahans meteorite were measured by mass spectrometry, a technique that separates different isotopes of an element based on their mass. From these measurements, the scientists determined how long rubidium had been decaying in this sample, and they concluded that the halite crystals formed about 4,700 million years ago, an age essentially identical to the age of the solar system. Therefore, the halite and the water inside it predated the fall of the meteorite, eliminating the possibility that this was water from Earth.

Shortly after water was found in the Monahans meteorite, halite crystals containing water were found in another meteorite, called Zag, that fell in Morocco in 1998. Scientists at the University of Manchester, England, and at the Natural History Museum in London used another radioactive dating technique, based on the decay of an isotope of iodine to an isotope of xenon, to determine that the halite in Zag formed within two million years of the birth of the solar system.

Consequences

Scientists had previously found evidence that water existed on asteroids, mainly by the presence of hydrated minerals such as clays that contain water in their structure. However, previous studies of these hydrated minerals could demonstrate only that water was present within 100 million years of the formation of the solar system. These measurements shortened that interval to only two million years, indicating that liquid water was present on some asteroids at the time the asteroids were forming.

Michael E. Zolensky, a scientist at the Johnson Space Center, and his coworkers speculated that the impact of icy comets might have carried water to the surface of the asteroids or that water from the interstellar medium, the space between star systems, might have been incorporated into the asteroids at the time they were forming.

Actual liquid water from the asteroids remained elusive before the fall of the Monahans and Zag meteorites. When asteroidal water became available for study in the laboratory, scientists had many questions about that water, including whether the water was a brine, a mixture of salts of potassium and sodium with water, and what its acidity was. The answers came from analogy to processes on Earth, where halite forms when sea water evaporates. The association of water with halite and sylvite, when seen on Earth, indicates formation by the evaporation of brine. Therefore, the researchers concluded that brine must have existed very early in the history of the asteroid that was the parent of the Monahans meteorites.

The presence of liquid water on some asteroids has important implications for understanding the geology of these small bodies of the solar system. It means that, early in the history of the solar system, asteroids were warm enough, from the heat produced by radioactive decay, for water to remain liquid rather than to form ice. Liquid water may also play an important role in volcanic activity on asteroids.

George J. Flynn

Food and Drug Administration Approves Sale of Viagra

Developed by a team of scientists of the Pfizer Company, Viagra was the first anti-impotence drug to be approved by the U.S. federal Food and Drug Administration.

What: Biology; Gender issues; Health; Medicine
When: March 26, 1998
Where: Washington, D.C.
Who:
Nicholas Terrett and
Peter Ellis, Pfizer Company researchers who had primary roles in the development of Viagra
Simon Campbell (1941-   ), a vice president of the Pfizer Company

The Invention of Viagra

As hundreds of persons were involved in the development of Viagra, the Pfizer Pharmaceutical Company avoids giving credit to any of their employees involved. In June, 1991, three researchers working at the Pfizer facility in Kent, England, discovered that a compound called suldenafil citrate was effective in treating heart problems such as angina. Of these workers, Nicholas Terrett is considered by some to be the inventor of Viagra. In 1994, he and his colleague Peter Ellis noted that during one of the trial studies of suldenafil citrate for heart problems among male patients, the compound increased the flow of blood to the penis. From this observation came the idea of using it to overcome male impotence, or erectile dysfunction (ED).

The American press gives principal credit to Simon Campbell, previously senior vice president of medical discovery at Pfizer. He was the department head under whose direction Viagra was developed. The British press recognizes two other Pfizer employees as having developed the nine-step chemical process by which Viagra is manufactured. They are Peter Dunn and Albert Wood, both of Kent, England. However, whoever receives primary credit, many were involved in the development of Viagra, the first anti-impotence pill to be approved by the U.S. Food and Drug Administration (FDA).

Aphrodisiacs and Their Action

Throughout human history, plants were used in the preparation of medicines. In many parts of the world, folk medicines relied upon plant and animal products for use as aphrodisiacs—medications intended to stimulate sexual desire (known as the libido) and male sexual performance (erection and orgasm), or both. However, many such products are known to be ineffective or even dangerous. The development of organic chemistry in the nineteenth century and its application to medicine led to the creation of entirely new, synthetic drugs. Among the most recent of these is Viagra.

An example of a classical aphrodisiac is the so-called Spanish fly, a drug made from an insect. When taken internally, it is highly irritable to the urinary and genital (reproductive) tracts, and can, in fact, often produce erections. However, serious damage to users' urogenital tracts often results, and some fatalities among users are known.

A prescription drug used in the late twentieth century to overcome impotence was papaverine, obtained from the opium poppy. When papaverine is injected by needle into the base of the penis about an hour before intercourse, it produces an erection that lasts from thirty minutes to more than an hour. However, not surprisingly, the method of papaverine's administration severely limited its popularity.

Before the development of Viagra, the leading prescription drug for the treatment of erectile dysfunction was a preparation called yohimbe.

Made from the bark of an African tree, yohimbe has been used as an aphrodisiac for centuries, but with only limited success. Tablets containing yohimbe hydrochloride to be taken orally have been manufactured, but clinical studies cast doubts on their effectiveness, and yohimbe has never been approved by the FDA.

Ginkgo, an herb extracted from the leaves of a Chinese tree, has been shown to be effective in the treatment of certain circulatory problems. For example, it has been used to increase circulation of blood to the brain, with the effect of alleviating certain memory problems, including Alzheimer's disease. Because erectile disfunction is also due primarily to insufficient circulation of blood, to the penis, ginkgo has also been used to treat it—though only with limited success.

It has long been known that the penis is under the control of the central nervous system (CNS), especially the brain. In recent decades, a more complete understanding of sexual functioning has been revealed by modern molecular biology. As a result, body functions can often be explained at the level of molecules within cells and tissues. When impulses from the brain are received by the penis, it responds by releasing special chemical messengers, especially nitric oxide and acetylcholine. These neurotransmitters cause the artery to the penis to relax which increases blood flow to the organ. As spongy chambers inside the organ fill with blood, the veins which otherwise drain blood away are compressed, thus trapping blood inside. Viagra works by slowing the breakdown of these chemicals responsible for an erection.

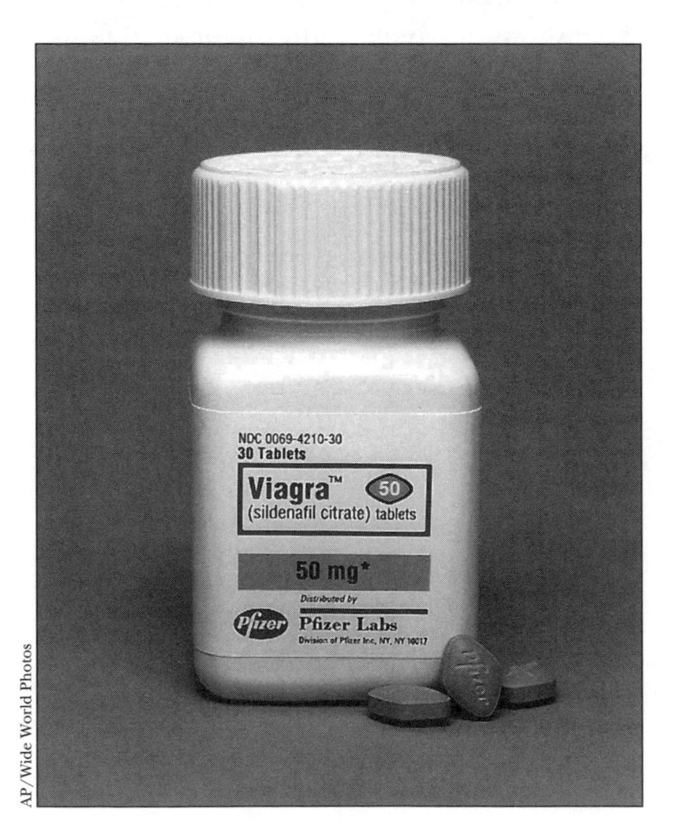

AP/Wide World Photos

It is known that Viagra, when taken by women, also produces an enhancement of sexual desire and performance. However, further studies are necessary before it can be recommended for use by women.

Consequences

Viagra is prescribed in the form of purple tablets of 25, 50, or 100 milligrams, with the specific dosage determined by physicians for individual patients. In men under age sixty-five, 50 milligrams taken about an hour before sexual activity has been found to be successful among 70 percent of patients. In men over that age, 25 milligram tablets are more often used. The retail prices for Viagra in the year 2001 ranged around ten dollars per pill, regardless of their strength. Since few health insurance plans cover the drug, its use has been limited to those who could most easily afford it.

The release of the drug generated considerable publicity and may have led many people—especially older men—to conclude that it was the answer to all impotence problems. Although purchase of the drug requires a physician's prescription, many people were able to acquire it through online and mail-order distributors without even being seen by doctors. Patients should, however, be screened carefully before prescriptions are written, particularly because Viagra should not be combined with certain other drugs and foods. As with all drugs, some of its users have experienced undesirable side effects, ranging from headaches, rashes, and indigestion to more serious problems.

By the year 2001, Viagra had clearly improved the sexual lives of many men and their partners. It was expected that continued studies and use on a larger scale would result in greater safety and lower retail costs.

Thomas E. Hemmerly

Judge Dismisses Sexual Harassment Lawsuit Against President Clinton

A federal judge ruled that the allegations of sexual harassment issued by Paula Jones against President Bill Clinton were without merit. She noted that Jones had not suffered on the job as a result of declining then Arkansas governor Clinton's alleged proposition.

What: Gender issues; Law
When: April 1, 1998
Where: Little Rock, Arkansas
Who:
Bill Clinton (1946-), president of the United States from 1993 to 2001
Danny Ferguson, Arkansas State Trooper
Paula Jones (1967-), former Arkansas state employee
Susan Webber Wright (1948-), U.S. District Court judge in Arkansas

Sexual Harassment

On April 1, 1998, Judge Susan Webber Wright of the U.S. District Court for the Eastern district of Arkansas threw out the Paula Jones sexual harassment lawsuit against President Bill Clinton. The president had moved for summary judgment because Jones could not show either direct retaliation or hostile work environment sexual harassment as required under Title 42 U.S. Code Section 1983 (Civil Rights Act of 1871, section 1). Section 1983 grants legal remedies for the deprivation of any rights, privileges, or immunities secured by the Constitution and laws of the United States by any individual acting under the "color of law" (any city, state, or federal official such as a police officer).

In any civil action in a court of law, the plaintiff must demonstrate that he or she has suffered an injury or damage. The record plainly demonstrated that Jones did not suffer any tangible job detriment, let alone one caused by her supposed rejection of Clinton's alleged sexual advances. Also, the alleged actions did not constitute severe or pervasive abusive conduct. The court further stated that if the sexual harassment claim was dismissed, it had to dismiss Jones's conspiracy claim as well. Jones had claimed conspiracy to deny her civil rights under Title 42 U.S.C. Section 1985(c); however, the plaintiff must show that he or she has suffered an injury for there to be a case. Jones did not show that any such conspiracy resulted in a deprivation of her rights, and the undisputed facts did not show any agreement between then Governor Clinton and Arkansas state trooper Danny Ferguson to deprive Jones of her constitutional rights.

Contempt and Impeachment

The political costs of the dismissed lawsuit were high. The court's discovery order of December 11, 1997, in the Jones civil lawsuit indicated that Jones, the plaintiff, was entitled to information regarding any state or federal employee whom the president had propositioned or with whom he had been sexually involved. On January 17, 1998, President Clinton gave a deposition to Judge Wright in which he testified in response to questioning from the plaintiff's counsel and his own attorney that he had no recollection of having ever been alone with Monica Lewinsky, who had been a White House intern. The president denied that he had engaged in an extramarital sexual affair or had ever been in a sexual relationship with Lewinsky. An affidavit submitted by Lewinsky in support of her motion to quash a subpoena for her testimony and made a part of the record of the president's deposition also denied that she and the president had engaged in a sexual relationship. During the president's deposition, Clinton said that Lewinsky's statement was absolutely true.

AP/Wide World Photos

Paula Jones.

On August 17, 1998, the president admitted in a televised address to the nation that he did have a relationship with a female employee (Lewinsky) that was inappropriate and wrong. The court considered the question of whether the president could be held in contempt of court and thereby sanctioned. The court ruled that the U.S. Constitution did not place the president's unofficial conduct beyond judicial scrutiny, and hence, the president was subject to the court's contempt power. On April 12, 1999, Judge Wright initiated contempt proceedings against the president for his willful failure to comply with the court's discovery orders.

As a result of all the legal proceedings in Arkansas, Kenneth Starr, an independent counsel in Washington, D.C., began an investigation into whether the president lied or obstructed justice in order to prevent other sexual relationships from being revealed. A Republican-dominated U.S. House of Representatives, acting as a grand jury under its constitutionally granted power, voted to impeach or indict President Clinton. The controversy moved to the Senate, which eventually declined to convict the president of the charges. On February 12, 2000, the Senate acquitted President Clinton on counts of perjury and obstruction of justice.

Consequences

On June 30, 2000, the Committee on Professional Conduct of the Arkansas State Supreme Court sued to disbar President Clinton, charging him with telling lies in his testimony relating to his relationship with former White House intern Lewinsky. In July, 2000, Pulaski County (Arkansas) Circuit judge Leon Johnson accepted the case for Clinton's disbarment. In August, 2000, independent counsel Robert Ray announced that he was proceeding with the possible indictment of Clinton for perjury and obstruction of justice. Ray eventually entered into discussions with David Kendall, Clinton's attorney.

On January 19, 2001, the last full day of his presidency, President Clinton accepted an extended plea agreement. He admitted guilt and signed an Agreed Order of Discipline, which states that he knowingly gave evasive and misleading answers in the Jones case deposition and he engaged in conduct that is prejudicial to the administration of justice. As part of the package, Clinton agreed to accept a five-year suspension from the practice of law in Arkansas.

The general political fallout from the Jones/Lewinsky furor continued even after President Clinton left office. Some questioned Clinton's last-minute pardon of fugitive financier Marc Rich. A few even suggested impeaching Clinton for that or reducing the former president's pension, office rent allowance, and other administrative expenses. The response to the last-minute pardons affected even the popularity ratings of the president's wife, Hillary Clinton, who won election to the U.S. Senate, representing the state of New York, in 2000.

Dana P. McDermott

India and Pakistan Conduct Nuclear Weapon Tests

Despite condemnation from the proponents of nuclear nonproliferation, India and Pakistan tested nuclear weapons and joined a select group of countries possessing nuclear weapons.

What: Military capability; Weapons technology; International relations
When: May, 1998
Where: Rajasthan, India, and Chagai, Baluchistan, Pakistan
Who:
Atal Bihari Vajpayee (1926-   ), prime minister of India from 1998
Nawaz Sharif (1948-   ), prime minister of Pakistan from 1997 to 1999
Abdul Kalam, head of the Indian nuclear program
Abdul Kadir Khan, head of the Pakistani nuclear program

Historical Divide

With the end of the Cold War in 1991, the ever-present menace of nuclear war was believed to have come to an end. However, the relief felt by many countries was short-lived, as India and Pakistan stunned the world by conducting a series of nuclear weapons tests in the month of May, 1998. Although it was widely perceived that the dangers of a nuclear war were receding fast, the actions of India and Pakistan reawakened the world to such a possibility.

When British colonial rule came to an end in 1947, India was divided into two parts. The predominantly Muslim areas of the east and the west became the new state of Pakistan, and the rest became a newly independent India. The partition of India into two countries resulted in a significant level of migration of people across the newly established borders. Hindus and Sikhs moved from the area given to Pakistan to India, and a large number of Muslims moved from India into Pakistan. This mass movement of people was coupled with large-scale religious and ethnic violence that left about a half a million people dead.

The birth of independent India and Pakistan was marred not only by the violence associated with this mass migration but by the problematic division of Kashmir, a territory in the north of India and south of Pakistan, over which India and Pakistan went to war in 1947. The war did not decide the conflict over Kashmir, and the territory remained the main point of disagreement between the two countries. India and Pakistan fought two more wars, in 1965 and in 1971. Traditional rivalry between India and Pakistan worsened during the Cold War because India supported the Soviet Union and Pakistan served as proxy of the United States. At the end of the Cold War, India and Pakistan still had substantial differences on a number of issues, leading them to compete with each other in developing military strength. The most urgent issue of difference between India and Pakistan remained Kashmir, which each country claims as part of its territory.

India and Pakistan Go Nuclear

In March, 1998, for the first time in the political history of India, a coalition government of the Hindu nationalist party, the Bharatyia Janata Party, assumed office with Atal Bihari Vajpayee as prime minister. The Hindu nationalists brought back the issue of conducting nuclear weapons tests to consolidate their support among the people of India. On May 11 and 14, without warning, the Indian government conducted five nuclear tests in the western desert state of Rajasthan.

The Indian government advanced a number of reasons for conducting the tests, including the possible threat posed by its neighbor China, which possesses nuclear weapons. India also argued that the tests, directed by Abdul Kalam, head of the Indian nuclear program, were a protest against the hypocrisy of the declared nuclear weapons countries—the United States, Russia, England, France, and China—which maintain their own arsenals of nuclear weapons and force other countries to sign the Comprehensive Test Ban Treaty and thereby give up the right to conduct nuclear tests.

In response to India's sudden nuclear tests, Pakistan, under Prime Minister Nawaz Sharif, responded with its own nuclear weapons tests in Chagai, Baluchistan. Arguing that its national security had been endangered by India's blast, Pakistan conducted an underground explosion with the power of 2 to 12 kilotons on May 28 and a second test on May 31, both under the direction of Abdul Kadir Khan, head of the Pakistani nuclear program.

Consequences

Both India's and Pakistan's tests were condemned worldwide. The leaders of the Group of Eight (Britain, Canada, France, Germany, Italy, Japan, Russia, and the United States), who were meeting in England, criticized the nuclear blasts in South Asia. A serious of punitive actions was announced, including the imposition of sanctions by the United States, the suspension of a $26 million annual grant by Japan, and the freezing of development aid by Germany and Canada. A number of other countries, including Sweden and Australia, also made small cuts to their developmental aid.

Although India's economy was strong enough to withstand the effect of these sanctions, Pakistan's economy was not and has deteriorated. Analysts believe that if the International Monetary Fund were to stop loaning money to Pakistan, it would face severe trouble in sustaining a working economy. Both the Indian and Pakistani govern-

AP/Wide World Photos

Czech members of Greenpeace protest against India's underground nuclear tests in front of the Indian embassy in Prague.

ments are under pressure to develop nuclear missiles and are increasingly diverting scarce resources meant to improve the education and health of the poorest sections of their societies.

Some experts predict that the presence of three nuclear powers—China, India, and Pakistan—will make the region more stable because a nuclear weapons attack would result in their mutually assured destruction. However, many others fear that conflicts over things such as the still unresolved Kashmir issue may lead to a nuclear war between India and Pakistan. In May, 1999, when Pakistan sent intruders into Kargil, a disputed part of Kashmir, India and Pakistan came close to launching a war. India was able to dislodge the intruders after a few months, but the incident reminded the world of how small, seemingly insignificant events might set the stage for a future nuclear confrontation.

Amandeep Sandhu

Europeans Agree on Euro Currency

The European monetary union created a single European currency, the euro, which links twelve countries in a powerful economic market whose policy is guided by the European Central Bank.

What: Business; Economics; International relations
When: May 3, 1998
Where: Brussels, Belgium
Who:
WIM DUISENBERG (1935-   ), first president of the European Central Bank
JEAN-CLAUDE TRICHET (1942-   ), president of the Bank of France

European Unification

In 1992, the fifteen member nations of the European Union ratified the Maastricht Treaty on European Union in response to inflation, unemployment, recession, and currency instability in the region. The Maastricht Treaty became the blueprint for monetary union in the European community, establishing an independent central bank for the region and a set of criteria countries must meet to become eligible for a single European currency. Within five years, or by 1997, nations who wished to join the European Monetary Union would need to demonstrate price stability and fiscal prudence in the areas of national debt and budget deficits, keep their own currency within normal fluctuation margins against other European currencies, and maintain interest rates in line with other European Union countries. Membership in the European Monetary Union would be determined in early 1998 and would determine whether a member state would be eligible for the single European currency, the Euro.

Establishment of a single European currency would link all participating nations in a union with strong trading power and a strong currency, one that rivaled the United States dollar or the Japanese yen. The European Central Bank would be in a position to set sound monetary policy for all the participating nations, and dealing with only one currency would benefit business and tourism by eliminating the costs of conversion from one national currency to another. European identity would be strengthened.

A Bank Begins

The fifteen members of the European Union met in Brussels, Belgium, on May 1-3, 1998, to decide on participating member states based on their performance under the Maastricht criteria and to appoint a head of the new European Central Bank. The conference was marked by a dispute over who would preside over the European Central Bank and take on the job of ushering in the euro. Most of the nations backed Wim Duisenberg, a strong advocate of conservative monetary policy who defined the role of a central bank as primarily to fight inflation with a policy of austerity. The dispute was mainly over nationality, with France insisting that the central bank's president be a Frenchman, the equally conservative and well-qualified president of the Bank of France, Jean-Claude Trichet. In the end, a compromise was reached in the form of a gentleman's agreement that Duisenberg would be appointed but would voluntarily retire half-way through his eight-year term, at which time Trichet would take over.

Eleven countries—Austria, Belgium, Finland, France, Germany, Ireland, Italy, Luxembourg, the Netherlands, Portugal, and Spain—were chosen for the first round of membership in the European Central Bank and the introduction of the euro. Britain, Sweden, and Denmark chose not to participate. Greece had not met the Maastricht criteria but subsequently qualified, joining the euro group as the twelfth member on January 1, 2001.

The new currency was introduced according to a time line. On January 1, 1999, the euro became the official currency of the eleven participating countries. Monetary policy began to be

AP/Wide World Photos

The faces of all Euro coins have the same map of Western Europe; designs on the reverse sides are unique to each member state.

set by the European Central Bank, headquartered in Frankfurt, Germany. Interbank commerce and stock exchange transactions among the participating countries began to be carried out in euros. National currencies technically became denominations of the euro, and would continue to be used as a matter of convenience only until 2002.

On January 1, 2002, circulation of euro banknotes and coins began. National currencies were to be completely replaced by the euro within six months after the introduction of the euro notes and coins. A dual circulation period was to last until February 28, 2002, at which time the legal tender status of national banknotes and coins would be canceled.

Consequences

The success of a stable and successful euro would depend on strict budgetary discipline within the participating countries. The experience of the European Union proved that abiding by a single monetary policy would not be easy. In some cases, meeting the Maastricht criteria required austerity measures that were not always popular. Those measures took a political and economic toll. Unemployment rose, and economic growth stalled. Deficit-reduction measures in France included an attempt to cut back retirement pensions, causing a national transportation strike and contributing to the fall of the Gaullist government. Britain withdrew from the system under the pressure of keeping interest rates and employment benefits in line with continental economies.

The European Monetary Union links the economies of the participating nations. Because there is only one Europe-wide interest rate, individual countries that increase their debt will raise interest rates in all other countries. Countries with very different economic needs are bound by one common economic policy. European Union countries with stronger economies may have to come to the aid of countries with weaker economies to preserve the value of the euro. Although this interdependence strengthens European identity and its economy as a whole, it can be seen as eroding national industry and identity. The French nationalist struggle for control of the European Central Bank in 1998 indicated that national conflicts would not easily fade.

The decision to create a single European currency also created the European Central Bank as a powerful force to set monetary policy and key interest rates. In terms of international financial clout, central banker Duisenberg has been compared to Alan Greenspan, the influential head of the U.S. Federal Reserve Bank. The European Monetary Union and the introduction of its currency, the euro, created a powerful financial market second only to the United States dollar.

Susan Butterworth

Indonesian Dictator Suharto Steps Down

Faced with massive corruption charges, economic catastrophe, and widespread antigovernment demonstrations, President Suharto of Indonesia was forced to resign after thirty-one years, leaving a legacy of repression, poverty, and corruption.

What: Civil strife; Government; Political reform
When: May 21, 1998
Where: Jakarta, Indonesia
Who:
SUHARTO (1921-), president of Indonesia from 1967 to 1998
BACHARUDDIN JUSUF HABIBIE (1937-), Suharto's longest-serving minister
MEGAWATI SUKARNOPUTRI (1947-), daughter of former president Sukarno and opposition leader

Storms in Indonesia

In the summer of 1997, the soaring economy of Indonesia was hit by a financial storm that also struck the economies of the other industrializing nations of Southeast Asia. Indonesia, whose barely regulated economy was controlled by a network of family members and friends, was hit hardest of all the so-called Asian tigers. Inflation soared out of control as living standards of the poor steeply declined. The International Monetary Fund (IMF) offered Indonesia a $43 billion loan, but only if stringent economic reforms were implemented, including the dismantling of the vast corporate empire controlled by Suharto's family and friends. This Suharto was unwilling to do until more than a year later.

By the end of 1997, sixteen major banks had collapsed, as had the political and physical health of Indonesia's aging leader. Meanwhile, natural disasters compounded human-made catastrophes. The country was in the midst of a dengue fever epidemic that received world attention as well as one of its worst droughts in modern history. Starvation threatened people in many parts of the formerly self-sufficient nation, and forest fires, blamed partly on the government's failure to regulated timber interests, raged out of control. Choking smoke blanketed much of Indonesia, home to the world's second largest rain forest. Loss of control of the natural world and the economy accelerated the forces aimed at wresting away Suharto's political control.

Riots, Protests, and Resignation

By January, 1998, price riots had erupted throughout Jakarta, accompanied by violent attacks against the city's wealthy ethnic Chinese residents and businesses associated with the Suharto family. Student protests aimed at overturning the corrupt Suharto regime grew in size and gained outside support from more tradition-oriented Muslim organizations. Protests were met by repression by an army and police force used to restricting dissent. Public rage grew after April 8, when Suharto announced steep cuts in government subsidies of fuel and food in order to secure an IMF loan.

On May 12, riot police opened fire on student protesters at Jakarta's Trisakti University, leaving six dead. This act did not discourage protests but rather created martyrs for the cause and increased the number and intensity of demonstrations. Over the next three days, rioters torched shopping malls, buildings, and automobiles, and wrecked the Chinatown section of North Jakarta. The end result was more than one thousand deaths and an estimated $1 billion in property damage. Suharto's promise on May 15 to restore fuel subsidies was too little and too late.

Protests came to a head on May 18, when a huge crowd gathered outside the parliament building as military units and riot police stood by helplessly. Thousands of students entered the parliament building to begin a sit-in. They demanded Suharto's immediate resignation. What they received the following day was a promise by Suharto, who had been reelected by parliament in 1998 to a five-year term as president, to hold early elections and not to run again. Suharto's offer was immediately rejected by the demonstrators. Resignation pressures were increased on May 20, when Speaker of the House Haji Harmoko delivered a letter to Suharto to resign or face impeachment proceedings in two days.

Bowing to the inevitable, Suharto resigned from the presidency on May 21. Eleven of Suharto's cabinet ministers also resigned. Power automatically was transferred to Vice President Bacharuddin Jusuf Habibie, a close political and business associate of Suharto. An aeronautical engineer, Habibie had been regarded as Suharto's adopted son since he was thirteen years old. He had served in cabinet posts for the past twenty years and was financially involved in at least forty Suharto business enterprises. Protesters claimed that all they had got for their actions was a photocopy of Suharto.

Lacking appeal and national confidence, Habibie did not produce stability during his presidency. Bloody clashes continued in eastern Indonesia between Muslim and Christian populations, and scandal after scandal was revealed, many of them about the Suharto family's plundering of national wealth. In 1999, early elections were held. On October 20, Abdurrahman Wahid, a Muslim theologian and religious leader who headed the National Awakening Party, became Indonesia's first freely elected president.

Consequences

Suharto's resignation removed from power one of the world's longest-serving heads of state. Suharto had used the military to seize power from Sukarno, president from 1945 to 1967, and became Indonesia's second president in 1968. Suharto steered Indonesia away from Sukarno's Cold War neutralism toward a strong pro-U.S. policy. He attacked leftists and communists, killing at least one-half million and imprisoning more than a million. Suharto also used the army

AP/Wide World Photos

Suharto (left) announces his resignation at Jakarta.

to annex East Timor in 1975, after Portugal vacated its former colony. Bloody fighting continued for the next twenty years, killing approximately 200,000 East Timorese.

Suharto's New Order economic policy was based on foreign investments and development of oil and timber resources. As world oil prices skyrocketed in 1974, the economy boomed. By the 1980's, stable authoritarian control over labor and foreign investments gave rise to manufacture of textiles, clothing, and footwear. By 1985, Indonesia achieved rice self-sufficiency and was no longer dependant on imports. In addition, Indonesia was a model nation regarding successful birth control policies; the average of 6 children per woman in 1965 had fallen to 3.3 by 1985. The stability of government in a complex nation composed of three thousand inhabited islands and an average economic growth rate of 7 percent a year for more than thirty years would be a difficult feat for Suharto's successors to duplicate.

On the other hand, unrestrained corruption made relatives, friends, and associates of Suharto fantastically wealthy while the working poor remained impoverished, earning at the time of Suharto's fall less than two U.S. dollars per day. The middle class became increasingly alienated, particularly after the expanding economy burst in the economic crisis sweeping Southeast Asia in 1997. As a lasting legacy, Suharto family members are so enmeshed in the monopolistic Indonesian economy, it will be difficult to eliminate them or the vast patronage system that they headed. The billions of dollars the family looted from the national wealth and put into foreign banks and overseas investments has proved to be hard to get back. Patterns of corrupt business practices, the legacy of rule through military power, and the callous destruction of native cultures and tropical rain forests will be difficult things for successor governments to reverse.

Irwin Halfond

California Voters Reject Bilingual Education

After thirty years of experimentation with bilingual education in California's public schools, voters decided it did not work and voted overwhelmingly to end it in a ballot initiative.

What: Education; Ethnic conflict; Politics
When: June, 1998
Where: California
Who:
RON UNZ (1961-), physicist, businessman, and political candidate who spearheaded the voter initiative against bilingual education
GLORIA MATTA TUCHMAN (1942-), schoolteacher and coauthor of the Unz ballot initiative
JAIME ESCALANTE (1930-), nationally known mathematics teacher

Bilingualism as an Ideal

Beginning the early education of schoolchildren in their own languages became a goal of California's bilingual education policy during the 1970's, when educators hoped that by giving children a strong educational start in their own languages, they would be better prepared to succeed after shifting over to English-language education. However, as time passed, that goal seemed impossible to achieve. California's schoolchildren speak an estimated 140 different languages in their homes. To teach each group of them in their own languages before teaching them in English was beyond the resources of California's massive education system.

In June, 1998 the issue of bilingualism was placed before the voters of California in a referendum. Their response was strong: "No Mas"—no more bilingual education in their public schools.

The liberal California of the 1960's appeared to have reprogrammed itself in latter years with regards to social issues. In 1994, for example, Californians voted against providing government benefits to undocumented immigrants. Next, they voted against affirmative action. Finally, in November, 1997, voter groups filed petitions for a movement called English for the Children—a ballot measure sponsored by Silicon Valley millionaire Ron Unz, whose mother was an immigrant from Russia. In 1994, the Republican Unz had unsuccessfully challenged Republican incumbent Pete Wilson in the primary election for governor.

California's bilingual education controversy was also developing around the same time that the board of the Oakland Unified School District made its widely ridiculed pronouncement that Ebonics—black English—should be regarded as a language separate from English. Ironically, just as African Americans were split on the Ebonics debate, California's Hispanic population was also splintered on the need for bilingual education.

The Unz Initiative

In 1987, just over 500,000 California children attended some type of bilingual classes. By 1997, the number had risen to nearly 1.4 million. According to a 1997 *U.S. News & World Report* story, California, with its burgeoning immigrant population, led the nation in proportion of its students who were not proficient in English, with a figure of 25 percent, compared with 6.7 percent of students nationally. By definition, these are students who cannot understand English well enough to keep up in school. Eighty-eight percent of California's public schools had at least one LEP student, and 71 percent had at least twenty limited English proficient (LEP) students. (In 1997 the acronym LEP was changed to EL, for English learners.)

Traditionally immigrants to the United States, speaking an assortment of languages, regard English as the language of upward mobility and want their children to learn it as quickly as possi-

AP/Wide World Photos

Proposition 227 author Ron Unz.

ble. This attitude was still held by many immigrant families in California in the late 1990's, and some of them were among the opponents to bilingual education. The largest non-English speaking groups in California were Latinos, especially Mexican Americans, 84 percent of whom indicated in late 1997 that they would support the Unz bilingual education initiative, according to a *Los Angeles Times* poll. That figure compared impressively with the 80 percent of white voters who indicated that they would back the initiative.

The Unz measure was cochaired by Gloria Matta Tuchman, a Mexican American teacher who had used English immersion to teach students for about fifteen years. The measure also benefited from a strong endorsement from Jaime Escalante, the Bolivian immigrant who taught calculus to urban Latino youths and became California's most famous schoolteacher—thanks, in part, to the 1988 film *Stand and Deliver.*

The Unz initiative called for a one-year English immersion program, which many educators said wouldn't prepare students for academic work in English, although it would allow them to speak more easily to their friends on the playground. Initially many state Republicans avoided the bilingual education debate, fearing that the Democrats would label supporters "racists." They also recalled that while many Latinos had earlier begun by supporting Proposition 187, the ballot initiative to deny benefits to undocumented aliens, only to turn against it later and vote against the measure's Republican supporters in the 1996 elections.

Critics of the Unz measure argued that it ignored important research data that demonstrated successes in bilingual education programs. Supporters of the measure countered that bilingual education created an educational ghetto by isolating non-English speaking students and preventing them from becoming successful members of society. They also accused politicians and educators of profiting from bilingual education. For example, they noted that some bilingual teachers were paid up to five thousand dollars a year extra, while school districts were receiving hundreds of millions of extra dollars simply for placing students in bilingual classes.

Opposition to the referendum was led primarily by major African American organizations, Democrats, the ethnic news media, several Asian American groups, bilingual education teachers, Latino activists, and organizations such as the Mexican American Legal Defense and Education Fund, the California PTA, the California School Boards Association, the California Teachers Association, and the California federation of Teachers, AFL-CIO. Several of these organizations publicly denounced the Unz measure and viewed it to be the third in a chain of anti-immigrant, anti-ALANA proposals that emerged in the mid-1990's. (ALANA stands for African, Latino, Asian and Native American.)

In March, 1998, three months before the state measure actually passed, California state board of education voted unanimously to discard its thirty-year-old bilingual education policy. Although California's basic bilingual education law had expired in 1987, state law still required native-language instruction when necessary to provide immigrant children with an equal chance for academic success.

Consequences

In June, 1998, Californians voted, by a margin of 61 percent to 39 percent, for Proposition 227, which placed major restrictions on bilingual education, limiting parent choice on the education programs for their children. Proposition 227 mandated a one-size-fits-all approach to instruction of English learners. Afterward, Californians Together, a round table of education and civil rights groups and organizations, analyzed selected schools still providing bilingual instruction to substantial numbers of students and determined that their students could equal or exceed the performances of students in English-immersion classes.

Additional discussions and findings on California's and the nation's bilingual education future will likely come from education organizations and think tanks such as the National Clearinghouse for Bilingual Education located at George Washington University, and the National Association for Bilingual Education, also headquartered in the nation's capital.

Keith Orlando Hilton

School Prayer Amendment Fails in House of Representatives

The U.S. House of Representatives rejected a controversial constitutional amendment that would have permitted organized prayer and religious instruction in public schools and displays of religion on public property.

What: Government; Religion; National politics
When: June 4, 1998
Where: Washington, D.C.
Who:
Ernest Istook (1950-), U.S. representative from Oklahoma
Newt Gingrich (1943-), Speaker of the House of Representatives
Bill Clinton (1946-), president of the United States from 1993 to 2001
Barney Frank (1940-), U.S. representative from Massachusetts

The School Prayer Debate

On June 4, 1998, an attempt by religious conservatives to amend the U.S. Constitution failed when the vote in the House of Representatives fell sixty-two votes short of the two-thirds majority needed to pass the bill, known as the Religious Freedom Amendment, to the state legislatures for ratification. The bill, also known as the Istook Amendment, would have allowed teachers and other school officials to lead organized prayers and other religious activities in public schools and permitted the government to fund various other religious activities. By weakening the separation of church and state as set forth in the First Amendment to the Constitution, the amendment would have altered the Bill of Rights (the first ten amendments to the Constitution) for the first time since its ratification in 1789.

The role of religion in public education had been a subject of debate for most of the twentieth century. Prayers and religious instruction were part of the daily routine at many American public schools during the eighteenth and nineteenth centuries. However, as the country became more religiously diverse during the twentieth century, many individuals and groups began to question the predominantly Christian religious activities that were taking place in some public schools, on the grounds that they violated the constitutional guarantee that religion and government would remain separate institutions.

Legal challenges to religious activity in public schools led to a series of court cases that culminated in a landmark decision by the U.S. Supreme Court in 1948 in the case of *McCollum v. Board of Education,* in which the Court ruled that religious instruction in public schools violated the First Amendment principle of separation of church and state. The Court further clarified its position in 1962 and 1963 by handing down decisions that prohibited school-sponsored prayer and Bible readings.

Although the Court insisted that organized religious activities be kept out of public schools to protect the religious beliefs of individuals, these decisions drew the anger of some people and groups who interpreted them as an effort to remove religion from society. The issue became a rallying point for conservative Christian politicians in the late 1970's and early 1980's, as leaders of the religious right blamed various social problems on the removal of religion from public schools and promised a constitutional amendment allowing school prayer and religious instruction in public schools. When conservative Republicans took control of Congress following the 1994 election, newly installed Speaker of the House Newt Gingrich promised that a school prayer amendment would be a priority of the new conservative House leadership.

The Istook Amendment

Few political observers expected a school prayer amendment to win the required two-thirds majority in the House and Senate; however, Republican leaders had promised religious conservatives, who had become a powerful influence within the Republican Party, that an amendment would be introduced. In 1995, as the campaign for the 1996 presidential election began, a bill sponsored by Republican representative Ernest Istook of Oklahoma was introduced in the House. The bill proposed a constitutional amendment that would guarantee individuals the "right to pray and to recognize their religious beliefs . . . on public property, including schools." Numerous conservative members of Congress rushed to support the bill, which was eventually cosponsored by 151 representatives.

The Istook Amendment prompted spirited debate among politicians and private citizens alike. Supporters of the bill argued that it was necessary to reverse decades of hostility toward religious expression in public schools by courts and school administrators and that restoring religious instruction to public schools would diminish problems of violence, drug use, and disobedience among school children by restoring a sense of morality to the nation's youth. Opponents of the amendment, which included President Bill Clinton and a broad coalition of religious groups from various denominations, stressed that individual students or groups of students who wished to pray at school on their own were already permitted to do so by the First Amendment and that the bill would allow teachers and administrators to impose their religious beliefs on students. Moreover, opponents argued, the bill would permit the government to provide tax money to religious organizations to fund activities that are clearly religious in nature and would give all religions—including followers of witchcraft, Satanism, and cults—access to public schools.

The Istook Amendment reached the floor of the House on June 4, 1998. After four hours of debate, the House voted 224-203 in favor of the bill, but according to the Constitution, a proposed amendment must be approved by two-thirds of both houses of Congress and then by the legislatures of three-fourths of the states to become part of the Constitution. The bill had fallen sixty-two votes short of moving on to the next step in the process.

Consequences

Both sides of the debate over religion in public schools claimed victory following the defeat of the Istook Amendment. Opponents of the amendment, led by President Clinton and Representative Barney Frank, called its defeat a victory for religious freedom and reminded people that public school students already had the right to pray and participate in student-initiated religious activities at school. Religious conservatives argued that bringing a school prayer amendment to the floor of the House for a vote was itself a victory and was simply the first step toward the eventual passage of a school prayer amendment.

Nevertheless, the defeat of the Istook Amendment appeared to stifle the momentum of school prayer supporters. A bill similar to the Istook Amendment was introduced in the Senate but received little support. Istook reintroduced the proposed amendment in the House in 1999, but the bill failed to reach the floor of the House for a second vote.

Michael H. Burchett

Mitsubishi Motors Pays Record Settlement in Sexual Harassment Suit

In a settlement with the U.S. Equal Employment Opportunity Commission, Mitsubishi Motor Manufacturing of America agreed to pay $34 million to hundreds of current and former female employees who had charged that they were victims of sexual harassment.

What: Gender issues; Law
When: June 11, 1998
Where: Normal, Illinois
Who:
Tsuneo Ohinouye (1932-), chief executive officer, Mitsubishi Motor Manufacturing of America (MMMA)
Patricia Benassi, attorney representing the female plant workers
Gary Schultz, vice president and general counsel of MMMA
Lynn Martin (1939-), former U.S. secretary of labor, hired by MMMA

A Problem at Work

On June 11, 1998, a settlement was announced in the lawsuit brought by the U.S. Equal Employment Opportunity Commission (EEOC), on behalf of women workers at the Normal, Illinois, factory of Mitsubishi Motor Manufacturing of America (MMMA), the U.S. subsidiary of the Japanese company Mitsubishi Motors Corporation. The EEOC is the federal agency that enforces civil rights laws.

The EEOC alleged that a large number of women were victims of sexual harassment at the plant, that male employees called female employees insulting names, made crude comments about them, subjected them to physical abuse, and vandalized their property. After the men learned that the women had registered complaints, some women reported that threats were made against their lives.

The plant management and the parent company, MMMA, would be liable for damages if complaints about sexual harassment were ignored. Their seeming indifference was the basis of a private lawsuit against MMMA filed in December, 1994, by Patricia Benassi, a Peoria, Illinois, attorney, representing twenty-nine current or former women workers at the Normal plant. The suit alleged that by refusing to take action after complaints were made, the plant management encouraged the abuse of women in the workplace.

Because federal laws prohibit discrimination on the basis of gender, the EEOC took action. On April 9, 1996, after a fifteen-month investigation that included interviews with more than one hundred women, the EEOC filed a class-action lawsuit against MMMA. Because this suit included all the women who had worked at the plant, each of whom might recover up to $300,000, millions of dollars were at stake.

The Battle

MMMA's counterattack was led by vice president Gary Schultz. As the company's general counsel, he insisted that there was no basis for the charges, claiming that the company had policies and procedures in place that eliminated sexual harassment. MMMA attorneys identified employees who would testify either that they had not experienced sexual harassment or that, if they had, their complaints to management were addressed promptly and adequately.

Employees were also pressured to telephone their elected representatives or the local media, voicing their support of the company. A free phone bank was set up at the plant for this purpose. Schultz also organized a demonstration outside the EEOC offices in Chicago. On April 22, 1996, more than two thousand workers were given a day off with pay, loaded onto buses chartered by the company, and transported into the city, where they waved signs and chanted protests.

However, the demonstration resulted in negative publicity for MMMA. Civil rights leader Jesse Jackson joined with the National Organization for Women in calling a national boycott of Mitsubishi products. The boycott was called off after a meeting between Jackson and Mitsubishi executives. Nevertheless, there was good reason to believe that the unfavorable publicity would cause a sharp decline in sales of Mitsubishi vehicles, especially to women.

A month after the demonstration, MMMA executives sent signals that they were ready to negotiate with the EEOC. They were also making important changes in policy and procedure. Lynn Martin, who had been both an Illinois congresswoman and a secretary of labor, was hired by MMMA to develop ways of dealing effectively with issues of sexual harassment, discrimination, and diversity.

In September, 1997, the private suit filed by Benassi was settled out of court for an undisclosed sum. In June, 1998, MMMA reached a settlement with the EEOC in which the MMMA agreed to pay a total of $34 million to current and former women workers who had been victims of sexual harassment. This was the largest monetary amount ever paid in a federal sexual harassment lawsuit.

Consequences

Changes began to take place at the Normal plant even before the negotiations were concluded. Some came out of Martin's recommendations, including new policy statements and a new training program targeting sexual harassment. Others involved personnel. For example, the executive director of human resources was transferred to Japan, and an Illinois State University affirmative action expert was recruited to deal with discrimination issues.

Personnel changes also were made at the cor-

AP/Wide World Photos

Mitsubishi Motors attorney Walter Connolly (left) confers with Mitsubishi executive vice president Kohei Ikuta (center) and Mitsubishi Motor Sales of America president Takashi Sonobe.

porate level. The same month that the EEOC settlement was announced, Takashi Sonobe took over as chairman of MMMA. He soon asked that the company create the position of chief operating officer at the factory. Longtime Ford plant manager Richard Gilligan was chosen for the position. Within a year, Sonobe was credited with having turned around the Normal plant. As a result of his success, when the president of Mitsubishi Motors Corporation was ousted in September, 2000, Sonobe was named to replace him.

That same month, monitors appointed by U.S. District judge Billy Joe McDade reported favorably on the progress made by the Normal plant in dealing with sexual harassment. The women workers had accomplished a great deal there. However, their action had repercussions far beyond Normal. For centuries, Japanese men had viewed women as inherently inferior and therefore subject to discrimination and harassment in the workplace. However, the MMMA case proved that such attitudes are not tenable in a global marketplace. As a direct result of the MMMA decision, Japan enacted a law banning sexual harassment and sex discrimination and began enforcing it vigorously. Thus, the actions of a few hundred American women were ultimately responsible for a significant change in Japanese attitudes toward male-female work relationships.

Rosemary M. Canfield Reisman

Dow Corning Agrees to Pay Women with Faulty Breast Implants

A tentative agreement between Dow Corning and patients with faulty breast implants ended a long-standing legal battle. The company agreed to a $3.2 billion settlement for thousands of women claiming damages from breast implants.

What: Consumer issues; Gender issues; Health; Medicine
When: July 8, 1998
Where: Midland, Michigan
Who:
Gary E. Anderson, president of Dow Corning
Francis McGovern, mediator in the settlement

Women's Self-Image and Social Pressures

The female breasts are mammary glands, organs designed to provide nourishment for the young. Paired organs that occur only in mammals, the breasts are usually fully developed in adult females. Although their biological role relates strictly to feeding, human female breasts have come to symbolize beauty, femininity, and sexuality in most Western societies, with shapeliness and large size generally regarded as desirable.

The notion that female breasts have become objects of adornment only in modern time is an error. Countless artists from earlier times captured beautiful female breasts on canvas, in clay, and in marble. Moreover, many centuries ago, wealthy Europeans occasionally employed "wet nurses" to suckle infants in order to preserve the aesthetic appearance of the natural mothers' breasts. Throughout history, women also sought to enhance their breasts with specially designed garments, such as tight corsets and other restraining garments. Both men and women of earlier times clearly perceived the breasts as symbols of feminine beauty.

However, for many twentieth century women, ill-fitting corsets and similar restraining garments were eventually discarded. Glamorous women of the 1920's, better known as flappers, strutted about, attempting to flatten their chests and to de-emphasize their breasts. As with most fads, the flat-chested look eventually passed away, and fashionable women once again acknowledged their bosoms. Moreover, women found additional ways to enhance their chests: bras with built-in padding, removable foam-rubber inserts ("falsies"), and supportive wiring sewn below bra cups—all these devices and others promised more beautiful and shapely breasts. Undoubtedly, women who used these "under cover props" looked and felt more attractive.

During the 1960's, many American women began exposing more of their bodies to public view, especially after the skimpy, two-piece French bikini became fashionable on beaches. Even though European women of all shapes and sizes had worn similarly scant swimming attire, American women frowned upon showing less-than-perfect bodies. However, enhancing breasts in bikinis and in the clinging street wear that was coming into vogue required more than artificial padding and wiring.

Breast augmentation

The creation of larger, firmer breasts by means of plastic surgery seemed to be the ideal solution for those who could afford the procedure. Between the 1960's and the 1980's, approximately two million American women resorted to surgically inserted breast implants. They were able to choose from different types of implants. These ranged from silicone gel-filled implants to silicone gel-filled implants with polyurethane coating, and inflatable saline-filled implants and double-lumen implants. Composed of polymers containing alternating silicon and oxygen atoms, silicone played a major role in shaping and redefining the appearance of women's breasts.

By the 1990's, approximately 100,000 American women chose to alter the shape of their breasts every year. Approximately 87 percent of them based their decision on a desire to have fuller and firmer breasts. Only a very small number of women who elected this surgery were motivated by a desire to reconstruct their breasts because of medical problems, such as surgery associated with the removal of malignant and nonmalignant tumors. Next to nose reshaping and liposuction (surgical withdrawal of excess fat from areas under the skin), breast augmentation had become the most popular plastic surgery procedure performed in the United States.

Adverse Effects

Unfortunately, many women who underwent breast augmentation experienced adverse effects which they and their physicians associated with silicone breast implants. These problems included painful, hardened breasts; infections and deformities; implant ruptures; allergic reactions to silicone; harmful silicone displacement throughout the body; swollen joints; and hardening of the body's connective tissues. It had also been shown that implant devices interfere with routine mammography screening for breast cancer.

These complaints and others forced a response from federal regulators. In 1989 the Food and Drug Administration acknowledged that certain studies proved that health risks were associated with breast implants. Nevertheless, women continued to subject themselves to the procedure, and the number of complaints continued to grow. Literally, thousands of women claimed injuries which appear to be associated with the implants.

On July 9, 1998 a tentative agreement was reached which ended one of the most publicized legal battles between consumers and a major corporation. Dow Corning Corporation, a major manufacturer of implants and attorneys representing numerous women reached an agreement. The corporation agreed to pay $3.2 billion to the plaintiffs. This agreement also allowed Dow Corning to emerge from Chapter 11 bankruptcy which the corporation had declared in 1995 to protect itself from the thousands of pending damage suits resulting from their product. Although the agreement was tentative, it was an important step in resolving a serious consumer health issue.

Consequences

The Dow Corning settlement did not resolve the problems associated with breast implants. The company wanted to act before a court-appointed panel of physicians released its findings concerning health risks associated with the implants. Many implant recipients who accepted the settlement did so without the benefit of having all available scientific evidence on the medical risks they faced. At the same time many consumer groups concerned with women's health believed that women have been and continue to be taken advantage of by major corporations concerned with profits rather than health. Described as "ticking bombs" by one women's health network, implants continue to be health risks. Health groups advise women with implants to have themselves checked thoroughly by their primary doctors. Meanwhile, it appeared that the search for less risky breast augmentation procedures would continue and that consumer groups would continue to monitor the implants still in use.

Betty L. Plummer

Congress Approves Overhaul of Internal Revenue Service

The Internal Revenue Service (IRS) Restructuring and Reform Act of 1998 mandated fairer treatment for taxpayers and replaced outmoded procedures and methods the IRS used in the conduct of its business.

What: Business; Government
When: July 9, 1998
Where: Washington, D.C.
Who:
Charles O. Rossotti (1941-), Internal Revenue Service commissioner
William V. Roth (1921), Republican senator from Delaware, chairman of the Senate Finance Committee
Bob Kerrey (1943), Democratic senator from Nebraska

The People Speak

On July 9, 1998, the Senate, by a vote of 402-8, passed the Internal Revenue Service (IRS) Restructuring and Reform Act of 1998, which the House of Representatives had approved, 96-2, on June 25. The bill, drafted in response to concerns about problems involving the IRS, represented a compromise between legislation passed in the House in the fall of 1997 and a similar bill passed by the Senate in the spring of 1998. Republican senator William V. Roth of Delaware, chairman of the Senate Finance Committee, and Democratic senator Bob Kerrey of Nebraska, chairman of a commission that spent a year studying the IRS, were largely responsible for the final form of the legislation. Newly appointed IRS commissioner Charles O. Rossotti promised immediate implementation of the reforms mandated by the legislation, which was signed into law on July 22 by President Bill Clinton.

The bill required that the IRS abandon its geographical structure of thirty-three district offices and ten service centers. Past performance goals, which ranked offices and centers based on revenue collected and enforcement results, were to be abandoned. In their stead, the act called for the creation of four departments, each dealing with the needs of specific segments of taxpayers. These departments were wage and income (for those whose income consists solely of wages and investment income), small business/supplemental (for those who file Schedules C, F, or E or Form 2106; partnerships; S corporations; and corporations with incomes less than $5 million), large corporate and middle market (for those corporations with incomes greater than $5 million); and tax exempt (for employee plans, exempt organizations, and state and local governmental entities).

Reforms

The act introduced a number of reforms designed to protect taxpayers. For example, the act stipulated that the IRS must provide taxpayers with notice of taxes due within eighteen months of the filing of the return or due date of the return. If notice was given after this eighteen-month period, interest and certain penalties were to be suspended until twenty-one days after notice was received by the taxpayer. The eighteen-month notification deadline was to be reduced to twelve months beginning in the year 2004. In addition, those who were the unknowing, innocent victims of misfiling and tax evasion by a spouse would no longer be responsible for the total tax bill incurred by the spouse. Instead, the innocent party would be held responsible only for the portion of taxes directly attributable to his or her income. Also, taxpayers were to be afforded protection from IRS collection on tax levies during the period in which an offer in compromise was being processed, for thirty days following rejection by the IRS of an offer in compromise, for any period during which appeal of the rejection of an offer in compromise was being considered, and for the period during which an installment agreement was pending.

The act also addressed legal issues. For example, attorney-client privilege was extended to certified public accountants or enrolled agents in addition to attorneys. Also, in the past, a taxpayer whose case reached the U.S. Tax Court was responsible for proving the case. The act shifted the burden of proof to the IRS. However, the taxpayer was required to cooperate by, among other things, keeping records and providing them to the IRS upon request and granting access to any possible witnesses.

Property seizures by the IRS were another area that the bill addressed. The act stipulated that the IRS would no longer be able to seize a residence or other nonrental property used as a residence by a person other than the owner/taxpayer to satisfy a tax liability of $5,000 or less, including penalties and interest. In addition, taxpayers would also have the right to recover civil damages for property that was improperly seized by the IRS or for the negligent activities of IRS employees. Damages of up to $100,000 could be awarded if employees knowingly disregarded provisions of the Internal Revenue tax code or U.S. treasury regulations in the collection of federal taxes. Damages of up to $1 million could be awarded for harm caused by an employee of the IRS who violated provisions of the bankruptcy code related to stays or discharges.

Consequences

The Internal Revenue Service Restructuring and Reform Act of 1998 was meant to make IRS officials and employees aware that they must treat taxpayers with fairness and courtesy. The IRS has made efforts to be more responsive to taxpayers, by creating problem-solving days in which taxpayers can meet directly with IRS employees in an effort to resolve problems and by opening local offices on Saturdays during the busiest part of the tax season to help taxpayers prepare their returns.

To ensure that taxpayers receive that fairness and courtesy, the act created an advocacy office for taxpayers and an oversight board for monitoring IRS activities. Under the provisions of the act, taxpayers no longer had to seek redress on their own for violations by IRS employees. The Office of Taxpayer Advocate, which replaced the Office of Taxpayer Ombudsman, was responsible for protecting the rights of taxpayers in disputes with the IRS. In the event of collection disputes, the advocate attempts to work out an equitable payment arrangement or compromise in the amount due. By law, the advocate can be overruled only by the IRS commissioner or deputy commissioner.

The act also provided monitoring of the IRS by a nine-member oversight board, consisting of the secretary of the treasury, a representative of the IRS employees union, the IRS commissioner, and six qualified citizens, who have expertise in business areas such as management and information systems.

Elizabeth Algren Shaw

AP/Wide World Photos

House Speaker Newt Gingrich (seated, left) and Senate Majority Leader Trent Lott (right) sign the IRS reform bill.

Food and Drug Administration Approves Thalidomide to Fight Leprosy

The U.S. government approves the use of a historically controversial drug to treat an age-old disease.

What: Biology; Chemistry; Health; Medicine
When: July 16, 1998
Where: Washington, D.C.
Who:
Celgene Corporation, pharmaceutical company
Food and Drug Administration, federal governmental control agency

Tragic History

For decades, the prescription drug thalidomide was associated with perhaps the worst pharmaceutical disaster in history. The word conjured up images of malformed infants and children who were born to mothers who had used thalidomide to combat severe morning sickness during their pregnancies. In the late 1990's, however, a new use was found for the infamous drug.

Thalidomide was initially released in Germany in 1956 under the name Contergan. First marketed as a tranquilizer and sleep aid, it was later widely used to help pregnant women who suffered from morning sickness but with devastating results. Within months, the first of what would become known as "thalidomide babies" were born. Children whose mothers had taken thalidomide during their pregnancies were born with severe abnormalities, including missing limbs and fingers. Many of these children had flipperlike appendages in the place of limbs or had severely underdeveloped limbs. Other birth defects involved internal organs such as the heart and kidneys or parts of the nervous system and digestive tract. Some thalidomide babies were miscarried or died within a few days of birth.

Though extensively tested on laboratory animals, particularly rats, thalidomide had not been tested on pregnant laboratory animals, and its effect on the fetus was not known before it was approved for use by humans. The drug was never approved for use in the United States but was eventually marketed in forty-eight other countries around the world. Just one dose of thalidomide taken in early pregnancy was shown to be toxic to the developing fetus. The drug was removed from the market in 1961 but not before between 10,000 to 12,000 children were born with disabilities because of its use.

New Uses Found

Thalidomide was removed from the market but remained on the shelves in some pharmacies and doctors' offices. It was eventually used again in countries outside the United States as a sleep aid and anti-inflammatory agent. After thalidomide was shown to alleviate many of the discomforts associated with erythema nodosum, skin lesions that developed in patients suffering from Hansen's disease, more commonly known as leprosy, it was used outside the United States to treat the disease.

In the early 1990's, Celgene Corporation, a pharmaceutical company, began testing the formulation on leprosy patients in the United States with remarkable results. Thalidomide, an anti-inflammatory agent, seemed effective in diminishing the size and number of skin lesions. Celgene applied to the U.S. Food and Drug Administration (FDA) for review and approval of the drug for use in cases of Hansen's disease and also for severe wasting conditions in acquired immunodeficiency syndrome (AIDS) patients. On July 16, 1998, the FDA approved the use of thalidomide for the treatment of leprosy. Thalidomide would be distributed by Celgene under the brand name Thalomid.

Consequences

Though the FDA approved the use of Thalomid, no one had forgotten the devastating effect of the drug on the children whose mothers had taken it while pregnant. Consequently, strict guidelines were established for doctors, pharmacists, and patients in the drug's use. Together the FDA, Celgene, physicians, and pharmacists created an unprecedented safety system designed to prevent fetal exposure to the drug. In order to prescribe, dispense, or use the drug, physicians, pharmacists, and patients must adhere to strict guidelines established by the FDA and Celgene known as the System for Thalidomide Education and Prescribing Safety (STEPS).

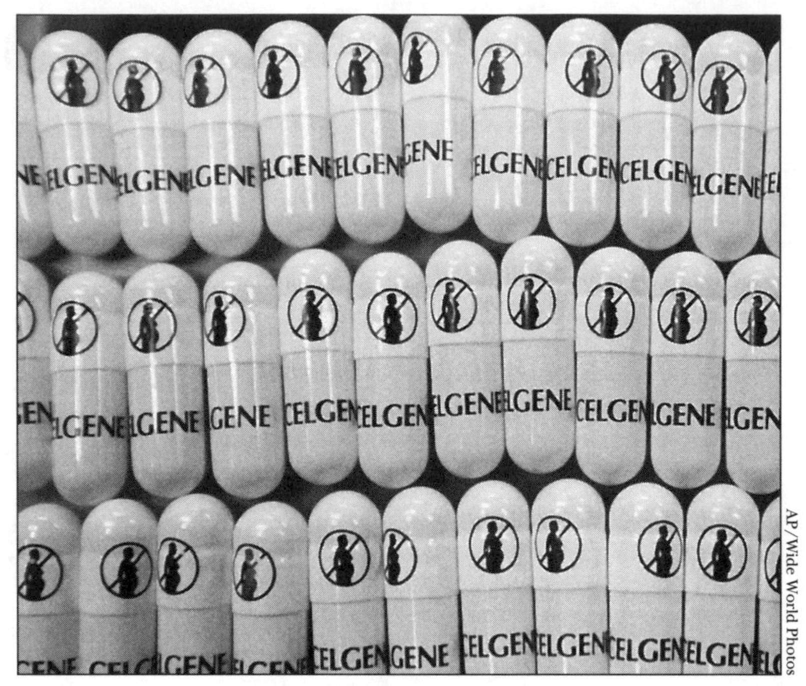

AP/Wide World Photos

Thalidomide capsules bear a symbol warning pregnant women against using them.

Under these guidelines, physicians who wish to prescribe Thalomid to their patients must first register with Celgene and participate in the drug company's training program. In addition, only those pharmacists who have registered with Celgene are allowed to dispense the drug. Physicians must monitor their Thalomid patients frequently and are allowed to issue prescriptions for no more than a twenty-eight-day supply of the drug. Extensive records must be kept by both the physician and pharmacist on each patient. Patients who are considering taking Thalomid are provided with educational material and must receive counseling from their physician explaining both the drug's effectiveness as well as the considerable risks involved in taking it. Patients are required to sign informed-consent forms for both their doctor and their pharmacist. Periodic patient surveys must be completed to determine how well patients are complying with the STEPS program.

All packages of the prescription drug prominently display warnings of the dangers of its consumption. Women of childbearing age are required to take a pregnancy test before receiving the drug and at least monthly while on it. They must use two different forms of birth control. Men taking Thalomid are required to use condoms during sex as long as they are on the drug and for one month after usage terminates. Thalomid patients are prohibited from donating blood or sperm.

Even with strict guidelines in place, many groups, including surviving victims of thalidomide, fear that patients or physicians will become lax and that more thalidomide babies will be born. Some fear that a black market or underground distribution of Thalomid will develop to supply desperate patients, creating uneducated and unmonitored consumption of the drug.

Thalomid shows promise in the treatment of other conditions. Testing continues in the use of thalidomide for AIDS patients and for treatment of those suffering from other autoimmune diseases including Crohn's disease, lupus, and rheumatoid arthritis. Thalomid is also being researched in the treatment of certain cancers in which the drug might block the blood supply to tumors and in connection with transplant surgery.

Leslie A. Stricker

International Monetary Fund Approves Rescue Plan for Russia

The International Monetary Fund provided substantial financial assistance to help the ailing Russian economy, but the results failed to solve the problems of the postcommunist period.

What: International relations; Economics
When: July 20, 1998
Where: Moscow and Washington, D.C.
Who:
Michel Camdessus (1933-   ), managing director of the International Monetary Fund
Boris Yeltsin (1931-   ), president of the Russian Federation from 1991 to 1999

The Russian Economy After 1991

Following the collapse of the communist system and the breakup of the Union of Soviet Socialist Republics in 1991, Russia failed to make a satisfactory transition to a free-enterprise capitalist system. Inflation grew at a rapid pace. Declines occurred in the production of foodstuffs and manufactured goods and the sale of oil on the international market, causing a reduction in government revenues. In 1995, President Boris Yeltsin and his government were forced to seek substantial financial assistance from the International Monetary Fund (IMF) to help meet budget expenditures and international loan obligations.

This international assistance, more than $10 billion in a February, 1996, agreement with the IMF, did not prove sufficient to solve Russia's economic difficulties. This led to further Russian negotiations with the IMF in 1998 to seek additional funding.

The timing of the July, 1998, IMF agreement was important for the Russian economy. The 1998 deficit was estimated to be 4.7 percent of the gross domestic product—the total value of goods and services—as compared with the 1997 level of 6.5 percent. However, in early 1998, the economy faced mounting challenges. Millions of workers and public employees had not been paid for months, leading to growing unrest and even strikes in such key occupations as coal mining. Low levels in tax collection, a sign of the Yeltsin regime's ineffectiveness, seemed no closer to solution despite the government's assurances to the IMF that it would collect a higher percentage of taxes owed to the government from Russian businesses.

Because of its continued need for revenue, the government borrowed heavily from banks and other Russian sources to meet its expenditures and cope with existing loan obligations. Wary of the government's ability to repay, these sources demanded extremely high interest rates on short-term loans. Some went as high as 150 percent or even 300 percent in 1998, a phenomenal level at which the interest was higher than the loan principal. At the same time, the stability of the currency fluctuated greatly, further reducing public confidence. The Russian Central Bank responded with the purchase of billions of dollars worth of rubles in mid-1998 in a major effort to stabilize the value of the nation's currency. This tactic had only limited success. Another problem was the severe decline in the Russian stock market, as investors sold shares to salvage what they could of their investments. From January to late May, 1998, the overall value of stock shares fell by 50 percent, with a major drop of 40 percent occurring in May alone.

The financial challenges intensified the Russian government's efforts to obtain IMF funds as rapidly as possible to shore up its shaky finances. Moscow officials in May confirmed the ongoing talks with the IMF and also with the World Bank to seek funds. The situation was complicated by IMF concerns about the weakness of the Russian economy, which resulted in

AP/Wide World Photos

In this May, 1998, photograph, an elderly Russian woman, one of many Russians struggling to get by, eats the remains of a McDonalds cake she found in the trash outside the restaurant.

postponing several $700 million installments. IMF officials also announced that the IMF would reevaluate the Russian government's spending plans and tax collection efforts before continuing financial assistance.

The Agreement

In July, 1998, Michel Camdessus, managing director of the IMF, announced that the IMF agreed to extend its prior Russian loan to Moscow by one year, with the final installment to be made in 2000.

Negotiations for additional IMF funds were completed by early summer, and Camdessus announced on July 20 a second major loan of $17.1 billion to the Russian government. The funds would be provided in several installments beginning in 1998, and the loan would be completed by the end of 1999. This large figure was in addition to the remainder of $5.5 billion that the IMF previously had approved for Russia in 1996. Thus, a total of $22.6 billion in loans had been guaranteed for the Yeltsin government over the next eighteen months.

As a condition of the 1998 loan, the IMF required Russian authorities to attain higher levels of tax collection. President Yeltsin also agreed to reductions in government expenditures to help balance the budget. To ensure compliance with these expectations, the IMF reduced the initial installment of $5.6 billion by $800 million until satisfactory progress had been made.

Consequences

The difficult conditions in the spring and summer of 1998 showed how volatile the currency and fiscal system had become. Rumors spread that the Russian government might default on international loans, including those owed to the IMF. President Yeltsin denied these rumors, but in August, 1998, the Russian government announced that it would implement a sweeping devaluation of the currency and that a number of loan obligations would not be paid on schedule. This was an extremely serious step for any national government to take. It meant that the Russian government could not be trusted to meet its fiscal obligations.

In the aftermath of the August crisis, the IMF suspended further loan installments until the situation greatly improved. Many IMF funds had already been spent, and IMF officials were reluctant to provide more aid that might be wasted. However, Camdessus agreed in April, 1999, that additional loans to Russia would be made but only if they were applied to payment of existing IMF obligations to avoid further defaults. No money would be sent to Russia directly. This suspension lasted nearly a year before the installments were resumed, compounding the economic difficulties of the Russian government in meeting its budget priorities.

The 1998 loan helped Russia in its budgetary and fiscal efforts, but IMF assistance could not solve the economic challenges faced by the country. Only more comprehensive and fundamental efforts by the government could create economic conditions that would aid the country in its uneven transition from communism to democracy and capitalism.

Taylor Stults

General Motors and Union End Production-Stopping Strike

A seven-week strike by two auto-parts plants in Flint, Michigan, which were part of a large network of factories, halted production at many General Motors factories. Although the strike involved local issues, it brought global issues of manufacturing and trade to the attention of Americans.

What: Economics
When: July 28, 1998
Where: Flint, Michigan
Who:
General Motors Corporation
United Auto Workers union
Paul Gadola (1929-), U.S. district judge

An Automaker and a Union

The 1998 strike involved two major, established organizations: a large motor vehicle manufacturer and a powerful union. The strike began at two local General Motors Corporation auto-parts plants in Flint, Michigan, but the lack of parts created by the strike forced other manufacturing plants around the nation to sit idle until the conflict was resolved. In the short term, the strike created economic hardship for the striking and laid-off workers, damaged local economies in areas in which work stoppages occurred, cost General Motors billions of dollars, and forced consumers to wait for their new cars or buy other brands. Over the long term, the strike forced American companies, workers, and consumers to consider the changes brought about by the globalization of the marketplace.

General Motors was founded by William C. Durant in 1908. Over the years, the company merged with other automobile and truck companies, including Chevrolet Motor Company, Olds Motor Vehicle Company, Oakland Motor Car Company (predecessor of Pontiac Motor Division), the Fisher Body Company, United Motors Corporation, and Allison Engineering, to become one of the three major automobile companies in the United States. During World War II, General Motors delivered more than $12.3 billion of war material—airplane engines, airplanes and parts, trucks, tanks, marine diesels, guns, and shells—to the U.S. Armed Forces. In 1971, the company worked with Boeing to develop the first moon vehicle, the lunar roving vehicle (LRV). In the 1980's, General Motors developed the Saturn Car Company, an automotive division incorporating new technology and ways of doing business.

The United Auto Workers (UAW) union was founded in 1935. It is part of a larger organization that includes the International Union, Aerospace Workers, and Agricultural Implement Workers of America. It is one of the largest and most diverse unions in North America, with members in virtually every sector of the economy. The UAW has approximately 750,000 active members and more than 500,000 retired members in the United States, Canada, and Puerto Rico, and has more than 1,000 local unions. The goal of the unions is to work with employers to ensure the health, safety, and rights of their members. The UAW has approximately 2,100 contracts with some 1,000 employers in the United States, Canada and Puerto Rico.

General Motors first recognized the UAW after a sit-down strike in Flint in 1937. The company has worked with the union since then to improve relations between employers and employees. In the 1990's, General Motors and the UAW signed a new labor agreement that emphasized improvements in competitiveness, quality of life at work, and job security. However, at the same time, things were not going well for General Motors. Since the 1970's, General Motors had lost nearly sixteen points of market share to its rivals and had begun to lag behind Chrysler, Ford, and the Japanese auto manufacturers. The company had been forced to cut its hourly payroll nearly in half to 218,000 employees. It completed a rationalization plan, reviewing which vehicles, plants, and future investments it would maintain. Financial analysts suggested that the company needed to cut nearly 50,000 jobs, close three assembly plants, and transfer more work to other plants outside the United States to remain competitive and compete in the twenty-first century.

News about possible job cuts added to the growing debate over unsafe working conditions that had become a point of concern for autoworkers in Flint and for all UAW members. Autoworkers had been complaining about working with hazardous materials and in unhealthy noise conditions for years.

The Strike

The strike began at a metal-stamping plant in Flint on June 5, after General Motors secretly slipped into the plant over the Memorial Day weekend to remove key metal-stamping dies for truck parts. That act was unacceptable to the workers. It reminded them of the 1936-1937 sit-down strike that forced General Motors to first recognize the UAW in that those strikes also began when General Motors attempted to remove dies from the plant. The June 5 strike was quickly followed by another strike on June 11 at the nearby Delphi East plant, which makes spark plugs, oil filters, air meters, fuel pumps, gauges, and instrument clusters.

Only the 9,200 UAW members employed at the two plants in Flint went on strike. These two factories, however, supplied sixteen assembly plants in the United States, Canada, and Mexico with fenders, doors, hoods, and other sheet-metal parts for most of General Motors full-size light trucks and some cars. General Motors operated on a just-in-time delivery schedule, which minimizes inventory, and therefore, the short-

AP/Wide World Photos

Workers from the General Motors auto plant at Orion, Michigan, turn out to support striking Flint Metal Center employees outside the Flint center.

age of parts from the Flint plants had a rapid effect on the system, forcing the company to close twenty-six of its twenty-nine North American assembly plants and many parts plants. The Flint strikes halted production at other General Motors factories within a few days, idling more than 200,000 workers and stopping General Motors' output by as many as 12,000 vehicles a day.

At the end of June, General Motors took the unusual step of filing a grievance claiming that the strikes were illegal. The company took the issue to a federal judge on July 15. On July 21, U.S. District judge Paul Gadola, issued an order for immediate arbitration. On July 28, 1998, a tentative agreement was announced between the striking UAW union and General Motors. The seven-week strike, which virtually shut down General Motor's production, cost the automaker more than $2 billion and knocked out nearly all of the automaker's North American production.

Consequences

The strike was the longest against General Motors since 1970, and it brought to the forefront the effect a seemingly local event could have on local, national, and global issues. The local issues in the strike involved health and safety, production standards, and job security. Flint had already lost more than 40,000 auto jobs in the previous few decades, and it hoped to hang on to as many jobs as possible. Although the strike sprung from local concerns, it caused problems in other parts of the nation and in Canada and Mexico when it idled other General Motors plants and diminished sales at car dealerships.

The strike also highlighted the global nature of the automobile industry in the late twentieth century. Faced by competition from domestic and international automakers, General Motors found it necessary to increase its productivity and had done so in part by using cheaper labor in Mexico. This global-level act had influenced job security, a concern at the local level in Flint. In the autoworkers' strike, the globally integrated character of the economy was shown to have an effect on companies and workers at the local, national, and global level.

Alison E. Feeney

Terrorists Attack U.S. Embassies in East Africa

Two U.S. embassies in Kenya and Tanzania were the subject of coordinated terrorist attacks. The bombings are attributed to terrorist leader Osama bin Laden and his associates.

What: Terrorism; International relations
When: August 7, 1998
Where: Nairobi, Kenya, and Dar es Salaam, Tanzania
Who:
OSAMA BIN LADEN (1957-), wealthy businessman turned leader of a terrorist organization
WADIH EL-HAGE, naturalized American citizen born in Lebanon, important underground agent in Kenya
HAROUN FAZIL, native of Comoros, a key participant in the terror bombing

Bin Laden and His Warriors

When the Soviets invaded Afghanistan in 1979, many Arabs decided to help Afghan guerrillas and fight for the Islamic cause. Osama bin Laden, whose fortune is estimated at $270 million, joined them. In 1987, he decided to expand the struggle beyond Afghanistan and start a holy war (jihad) against Muslim governments he considered secular, as well as the Western states supporting them. This war would make way for Islamist regimes run strictly according to the Qur'an. In 1988, he founded an organization to fight his jihad. Using his personal wealth, he brought warriors from around the world to his training camps in Afghanistan. In 1989, the Soviet Union withdrew from Afghanistan, a triumph for bin Laden's warriors and an incentive to fight elsewhere.

In 1991, when the United States sent forces to Saudi Arabia to liberate Kuwait, bin Laden was infuriated. He denounced the Saudi monarchy for letting in U.S. forces, whom he viewed as infidels. The Saudi government expelled him and, in 1994, deprived him of his citizenship. He moved to Sudan and, in 1996, to Afghanistan, where a new group of Islamic militants, the Taliban, was taking control. In February, 1998, from a camp in eastern Afghanistan, bin Laden publicly announced that he and several other leaders of radical groups had founded the International Islamic Front for Jihad Against Jews and Crusaders, an umbrella organization. The group issued a directive stating that it is the duty of every Muslim to kill Americans and their allies, both civil and military, in any country in which this is possible. Even before this date, bin Laden was suspected of working with those who carried out the attack on the World Trade Center in New York in 1993 and the bombing of U.S. military personnel in Saudi Arabia in 1995 and 1996. Evidence indicates that fighters trained in bin Laden's camps were sent to a variety of countries to create underground cells and carry out terrorist operations.

Terrorist Bombing of the Embassies

In the August 7, 1998, coordinated attacks, some 220 people, including 12 Americans, were killed in the Nairobi embassy and adjacent streets and buildings, and more than 5,000 were injured. In Dar es Salaam, eleven were killed and eighty injured. Shortly after the attacks, authorities made several arrests that provided evidence that the bombings had been orchestrated by bin Laden, and $5 million were offered for his capture. Among those arrested were Wadih el-Hage, Haroun Fazil, Ali A. Mohamed, and Mohamed Rashid Daoud al-'Owhali. El-Hage, a naturalized American citizen born in Lebanon and described as bin Laden's personal secretary, was reported to have been instrumental in setting up the terrorist cell in Nairobi. Fazil, a native of Comoros (an Indian Ocean island state), was accused of having played a leading role in the bombing in Nairobi. He was believed to have rented the villa in which the bomb was assembled and, on the day of the attack, to have driven a pickup to the embassy, thereby leading the way for the truck carrying the bomb. Mohamed, a California resident described as a close confidant of bin Laden, told the Federal Bureau of Investigation (FBI) that he knew who had carried out the attacks. Al-ʿOwhali was a twenty-four-year-old Saudi would-be suicide bomber and participant in the Nairobi attack. Some of those arrested faced trial in the United States; others were taken into custody in Tanzania and Pakistan. Reportedly, at the United States' urging, law enforcement agencies in twenty countries arrested some one hundred militants over a period of two years following the embassy attacks.

Consequences

At the time of the bombings, the United States had been making efforts to render its embassies less vulnerable to terrorist attacks and had even considered relocating some embassies. Before the attack, the embassy in Nairobi had informed the State Department of its utter vulnerability at its current site. These protective measures gained new weight and urgency as a result of the attacks.

On August 20, 1998, less than two weeks after the bombings, the United States carried out two cruise missile attacks against bin Laden's installations, the first against a terrorist training camp in Afghanistan and the second against a pharmaceutical plant in Sudan that was believed to be owned by bin Laden and used to store or produce nerve gas. The first attack, in which bin Laden was not harmed, did not produce much of an international reaction, as few countries feel much sympathy for the Taliban extremists.

The second attack, however, met with some criticism. The Sudanese government disavowed any knowledge of chemical weapons and insisted that the plant manufactured pharmaceuticals for civilian use. Sudan suggested that it might ask for an inspection by the United Nations Security Council, but no such investigation took place. Although supported by traditional U.S. allies such as Great Britain, Israel, Germany, and Australia, the attack was neither condemned nor endorsed by China, Japan, and Kenya. Muslim nations such as Iraq, Iran, Libya, and Pakistan protested the bombing of the plant, and anti-American demonstrations took place in many Muslim countries. Some stateside critics of President Clinton's actions suggested that the attacks had been ordered only to distract Americans from his involvement in a sex scandal.

The United States formally requested Afghanistan to surrender bin Laden to face trial. The Taliban refused, stating that he was a guest in their country. In November, 1998, following a unanimous vote in the Security Council, the United Nations imposed sanctions that were to freeze Afghanistan's economic assets abroad and curtail international flights by the country's

AP/Wide World Photos

The U.S. embassy in Tanzania, one day after the blast.

airline if bin Laden and one of his chief aides were not handed over to the United States. The sanctions were not expected to produce immediate results, and bin Laden and his associates around the world remained a major threat. For example, the October, 2000, bombing of the U.S. destroyer *Cole* in Aden, Yemen, is widely believed to have been carried out by the bin Laden organization. In January, 2001, the U.S. embassy in Rome was closed for several days after intelligence reports indicated it might be the terrorist group's next target. A few days later, U.S. Naval units in the Mediterranean were diverted from Naples, which they had been scheduled to visit.

Jean-Robert Leguey-Feilleux

Swiss Banks Agree on Compensation of Holocaust Victims

U.S. sanctions and class-action suits force Swiss banks to return pre-1945 deposits to Holocaust survivors or their heirs.

What: Business; International relations
When: August 12, 1998
Where: New York City
Who:
Edgar Bronfman, Sr. (1929-), president of World Jewish Congress
Alfonse M. D'Amato (1937-), senator from New York
Paul A. Volcker (1927-), chair of Committee of Eminent Persons
Edward Fagan, New York attorney

Withdrawals Thwarted

When European Jews were threatened with annihilation under German chancellor Adolf Hitler's regime (1933-1945), many placed their savings and valuable possessions in Swiss banks for safekeeping. Switzerland maintained neutrality, and its banks had developed a labyrinthine system of secrecy in their manner of conducting business. After the Allies defeated the Germans in 1945, it was determined that 6 million Jews had been killed by the Nazis. European Jews left behind many accounts in the Swiss banks, accounts whose worth the American Jewish Conference estimated to be in excess of $8 billion (U.S. dollars) in 1945.

The Swiss banks did not contact the owners of the dormant accounts or their heirs and paid no interest into the accounts, although they continued to deduct service charges. When the rightful heirs subsequently tried to claim their family fortunes, the banks demanded death certificates from the concentration camps, a type of documentation that did not exist. In addition, the banks charged each claimant 300 Swiss francs, nonrefundable. In 1949, the Swiss banks illegally liquidated the dormant accounts of Polish Jews for the purpose of reimbursing Swiss business concerns in Poland that had been nationalized by the Polish communists.

In 1962, under pressure from the United States, Britain, and France, the Swiss government approved the Disclosure Decree, which required the Swiss government to search its banks for the assets of Holocaust victims. Ten years later when the decree expired, the Swiss government had allegedly found only $2 million belonging to 1,048 owners. Only twenty-six of five hundred Swiss banks responded, and 7,000 people had their claims denied.

Legal Challenges and a Settlement

The collapse of the Soviet Union in 1991 enabled many Jewish people in Eastern Europe to seek redress from the Swiss banks for the first time. In 1992, the World Jewish Congress (WJC) took up the cause of the rejected heirs. In 1995, fifty years after the end of World War II, documents from the joint American and British intelligence operation Safehaven were declassified in the United States, providing details of bank accounts that the Swiss banks had previously failed to mention.

In December, 1995, WJC president Edgar Bronfman, Sr., enlisted the aid of Senator Alfonse M. D'Amato from New York. They reasoned that the Swiss banks, which do 25 percent of their business in the United States, could be pressured. On May 2, 1996, the Swiss Bankers Association and the World Jewish Congress jointly established the Committee of Eminent Persons, chaired by Paul Volcker, former chairman of the U.S. Federal Reserve Board, which would employ international auditors to track down unclaimed Jewish assets in Swiss banks.

AP/Wide World Photos

Holocaust survivor Margaret Zentner (left) hugs Senator Alfonse M. D'Amato in front of the U.S. District Court in New York.

As evidence of Swiss dishonesty continued to amass, New Yorker Gizella Weisshaus, whose father was killed in Auschwitz, filed a suit against Swiss banks in a New York court on October 3, 1996. Her attorney, Edward Fagan, soon attracted more clients and filed a class-action suit for $20 billion against three Swiss banks: Crédit Suisse, Swiss Bank Corporation (SBC), and Union Bank of Switzerland (UBS). Other class-action suits were quick to follow.

On December 13, 1996, the Swiss government passed a federal act prohibiting the destruction of bank records; however, the banks did not comply. On January 8, 1997, Christoph Meili, a night watchman, took documents pertaining to the Third Reich from the shredding room of UBS and passed them on to the Israelische Kulturgemeinde Zürich (Israeli Cultural Foundation of Zürich), which called a press conference.

In February of 1997, the New York City Council, which usually put $150 million in overnight investments in the Swiss banks, proposed prohibiting business with them. The City of Chicago and the state legislatures of New York, New Jersey, Rhode Island, and Illinois threatened the same. On October 9, 1997, the City of New York excluded UBS from bidding on a billion-dollar transaction. On October 14, California treasurer Matt Fong placed a moratorium on business with the Swiss banks. In March, 1998, New York State blocked the merger of SBC and UBS on the grounds that they had not returned deposits to their rightful owners and heirs.

On June 19, 1998, the Swiss banks offered $600 million to settle all the lawsuits. The Americans stated that they found that amount insultingly low and filed more suits against the banks. Nearly two months later, the two sides were brought together by Judge Edward R. Korman, U.S. District judge in Brooklyn. On August 12, 1998, the Swiss banks Crédit Suisse Group and UBS, the World Jewish Congress, and the lawyers representing the U.S. class-action plaintiffs agreed on a settlement of $1.25 billion, to be paid in four installments over a period of three years. The lump sum consisted of $530 million plus all dormant assets identified by the Volcker audits. The banks were also prepared to pay an additional amount "to avert the threat of sanctions as well as long and costly court proceedings."

Consequences

The $1.25 billion settlement represents only a fraction of the money the Swiss banks received from persecuted Jews during the Third Reich. The Volcker commission published lengthy lists of names of account holders but had no way of determining the ownership of anonymous accounts or of accounts set up for Jews by Swiss trustees. There was also no way of knowing how many accounts were illegally liquidated because the documentation had been destroyed. Still, the settlement had symbolic significance, and some

Holocaust survivors had their wealth returned to them, albeit decades after the war.

Class-action suits continue against Swiss insurance companies, and the banks of other countries are coming under close scrutiny. Some are busy shredding documents, and others are cooperating in the investigations. In addition, works of art that changed hands during and after the war are being scrutinized. As a result of the Swiss bank investigations, museums and galleries that acquired works of art that formerly belonged to Holocaust victims are realizing that they must seek out the rightful owners of these works or their heirs.

Only under the threat of heavy financial sanctions did the Swiss banks relinquish to victims of the Holocaust what was rightfully theirs. The landmark settlement on August 12, 1998, placed new emphasis on accountability in the struggle between monied interests and an emerging morality of international relations.

Jean M. Snook

Court Rules That Food and Drug Administration Cannot Regulate Tobacco

A panel of federal appeals court judges ruled that tobacco did not fall under the definition of substances that could be controlled by the Food and Drug Administration, preventing efforts by the administration of President Bill Clinton to crack down on tobacco sales and advertising.

What: Law; Government; Health
When: August 14, 1998
Where: Charleston, West Virginia
Who:
David Kessler (1951-), commissioner of the U.S. Food and Drug Administration
Bill Clinton (1946-), president of the United States from 1993 to 2001
William Osteen, Sr. (1931-), federal appeals court judge

Food, Drug, or Neither?

On August 14, 1998, a three-judge panel of the Fourth U.S. Circuit Court of Appeals ruled that the Food and Drug Administration (FDA) had no authority to regulate tobacco products on the grounds that tobacco did not fit the definition of either a food or a drug. The ruling effectively overturned recently enacted FDA regulations limiting the sale and advertising of tobacco products, derailing a campaign by the administration of President Bill Clinton to crack down on youth smoking and certain methods of marketing tobacco products.

The decision by the appeals court added a new chapter to a longstanding debate over the legal definition of tobacco products and who was responsible for controlling their sale and use. The effects of smoking on health had been the subject of debate for centuries. Tobacco was included in the 1890 edition of the *U.S. Pharmacopoeia*—an official catalog of drugs published by the U.S. government—but was dropped from later editions to gain the support of tobacco-producing states for the Food and Drug Act of 1906, the legislation that created the FDA.

During the first half of the twentieth century, as the long-term health risks of tobacco use became increasingly evident, a growing number of politicians and public health groups demanded increased government control over the manner in which tobacco products were packaged and advertised. In 1965, the Federal Trade Commission required tobacco companies to place warning labels on cigarette packages, and in 1970, the U.S. Congress voted to prohibit television and radio stations from airing cigarette commercials. During the 1980's, many states enacted restrictions on tobacco sales to minors. However, despite increased demands for public intervention, the FDA continued to insist that tobacco products were outside its authority because they did not fit the definition of a food or a drug.

The FDA Tests Its Authority

The FDA's position changed in 1996 in the light of new studies showing how nicotine affected the body and because of promises by President Clinton and Vice President Al Gore to reduce rates of teen smoking and incidents of health problems related to smoking. In August of 1996, the FDA, under the leadership of Commissioner David Kessler, declared that it had the authority to regulate tobacco on the grounds that the nicotine in tobacco products is a drug. Kessler announced a series of new regulations that would ban the sale of tobacco products to persons under the age of eighteen nationwide, require stores to check the identification of cigarette and smokeless-tobacco buyers who appeared to be under twenty-seven years of age, and prohibit cigarette vending machines except in bars and other adults-only establishments. In addition, the regulations banned billboards advertising tobacco products near schools and playgrounds, limited tobacco advertising in publications read by young people, and prohibited tobacco companies from sponsoring sporting events, concerts, and other public attractions.

These regulations brought fierce opposition from tobacco companies, who sought to challenge in court the FDA's authority to regulate their products. In April of 1997, the issue was brought before U.S. District judge William L. Osteen in North Carolina. Tobacco company lawyers had hoped that filing suit against the FDA in the heart of tobacco country would result in a favorable ruling; however, Judge Osteen upheld the agency's crackdown on tobacco sales to minors and vending machine restrictions, ruling that the FDA had overstepped its jurisdiction only when it attempted to restrict tobacco advertising, which Osteen deemed a violation of First Amendment guarantees of free speech.

The ruling did little to please either the Clinton administration or the tobacco industry. On June 11, 1997, the Department of Justice filed an appeal on behalf of the Clinton administration to the U.S. Court of Appeals for the Fourth Circuit, arguing that the FDA should have the authority to restrict tobacco advertising. The tobacco industry filed its own appeal two weeks later on the grounds that the FDA did not have any authority to regulate tobacco. A panel of three judges heard oral arguments on August and had planned to issue a ruling the following year; however, the case was reargued in June, 1998, because of the death of a member of the panel, Judge Donald Russell, in April, 1998. The panel of judges voted 2-1 to reverse Judge Osteen's decision, stating that the court was "of the opinion that Congress did not intend to delegate jurisdiction over tobacco products to the FDA" under the Pure Food and Drug Act of 1906.

Consequences

The ruling by the appeals court was a major setback to the Clinton administration's efforts to use government agencies to regulate tobacco products. Nevertheless, the administration persisted. In January of 1999, the U.S. solicitor general asked the Supreme Court to hear argument that the Fourth Circuit ruling should be overturned. With the support of more than forty public health and children's organizations—including the American Medical Association, the American Heart Association, and the American Cancer Society—attorneys for the federal government argued the FDA's case before the Supreme Court in December, 1999, against attorneys from the five major tobacco manufacturers.

On March 21, 2000, the Supreme Court upheld the appeals court ruling that the FDA could not regulate tobacco by a 5-4 vote, stating that Congress had repeatedly rejected proposals to allow the FDA to regulate the nicotine in tobacco as a drug despite evidence that "tobacco use, particularly among children and adolescents, poses perhaps the most significant threat to public health in the United States." The Court's ruling effectively put an end to efforts by the Clinton administration to restrict tobacco advertising and marketing through regulatory means. Nevertheless, some of the questions argued before the Court were already irrelevant by the time it handed down its ruling, as tobacco companies had already agreed to advertising restrictions and efforts to reduce teen smoking as a condition of a multibillion-dollar settlement with state governments who had sued them in federal court.

Michael H. Burchett

U.S. Missiles Hit Suspected Terrorist Bases in Afghanistan and Sudan

U.S. military forces conducted coordinated cruise missile attacks on what were identified as terrorist facilities in Afghanistan and the Sudan. These facilities belonged to the terrorist group reputed to be behind the earlier bombing of two U.S. embassies in East Africa.

What: International law; International relations; Military; Terrorism
When: August 20, 1998
Where: Reputed terrorist facilities in Afghanistan and Khartoum, Sudan
Who:
Bill Clinton (1946-), president of the United States from 1993 to 2001
William Cohen (1940-), U.S. secretary of defense
Hugh Shelton (1942-), U.S. Army general and chairman of the Joint Chiefs of Staff
Osama bin Laden (1957-), millionaire Saudi Arabian suspected of funding terrorist bombings of the U.S. embassies in Kenya and Tanzania

Rationale

On August 7, 1998, two bombs exploded at the U.S. embassies in Kenya and Tanzania, killing more than 250 people and injuring more than 5,000. An Islamic terrorist group identified as the International Islamic Front for Jihad Against Jews and Crusaders indirectly claimed responsibility for the bombings and threatened additional terrorist attacks on American targets worldwide. This terrorist group was reportedly founded and financed by a renegade Saudi Arabian millionaire, Osama bin Laden, previously linked to other attacks by terrorist groups. U.S. intelligence and military officials, reviewing the available circumstantial evidence, concluded that bin Laden's group was responsible for these two embassy bombings.

Intelligence reports also indicated that this group of terrorists was in the process of developing chemical weapons in the Sudan and concurrently training additional Islamic terrorists in Afghanistan. President Bill Clinton, on advice from Defense Secretary William Cohen and General Hugh Shelton, chairman of the Joint Chiefs of Staff, determined that an immediate surprise military response would appropriately punish the perpetrators. The president also concluded that such a military strike would delay, and possibly prevent, additional terrorist attacks by this group. Detailed military planning for the strikes commenced immediately and took about one week to complete.

Missile Strikes in Retaliation

To avoid losses, U.S. armed forces used only cruise missiles rather than manned aircraft to attack what had been identified as terrorist-related facilities in Afghanistan and the Sudan. Cruise missiles, fired from U.S. Navy ships in the Red Sea and the Arabian Sea, simultaneously struck the reported terrorist facilities. The missiles hit six individual targets in Afghanistan, most of which were located along the border with Pakistan. Specific targets were in Khost, south of Afghanistan's capital city of Kabul, and Jalalabad, east of the capital. Collectively, these targets constituted a suspected terrorist training complex in which bin Laden's group reportedly trained hundreds of other terrorists.

In the Sudan, the target of the U.S. missiles was the El Shifa Pharmaceutical Industries plant. This factory, located in Khartoum, Sudan's capital, was alleged to be storing chemical weapons for later use by the terrorists. The precise number of casualties in Afghanistan and the Sudan as a direct result of the missile strikes was never established with certainty. No U.S. casualties occurred in any of the attacks.

Consequences

Following the missile strikes, U.S. spokespeople conceded that the strikes would not eliminate the problem of state-sponsored terrorism but said that they would clearly convey that there would be no safe haven for terrorists who chose to attack the United States or its embassies worldwide. Bin Laden and his closest associates were all reported to have survived the attacks unscathed. Damage to the facilities themselves was extensive, however, and military planners labeled both attacks as successful. Already under some international pressure for harboring bin Laden, a fugitive, the government of Afghanistan showed no sign of reconsidering its policy because of the U.S. strike. Instead, Afghan spokespeople interpreted the American missile strike not as an attack on bin Laden or terrorists but instead as an attack on the Afghan people.

The Sudanese government immediately condemned the attack on its nation and disavowed any knowledge of chemical weapons on Sudanese territory, claiming that the plant manufactured pharmaceuticals for civilian use. The Sudanese government also suggested that it might request an inspection by the United Nations (U.N.) Security Council to firmly disprove the allegations of chemical weapons at the destroyed plant. No U.N. inspection was subsequently conducted.

American politicians, regardless of political affiliation, almost universally supported the need for the attacks, as well as the manner in which the missile strikes were conducted. The China, Japan, Kenya, and several other countries

AP/Wide World Photos

The Pentagon released this aerial photograph of the El Shifa pharmaceutical plant (outlined) in Sudan.

chose to neither condemn nor support the United States. U.N. secretary-general Kofi Annan, informed of the strikes by American diplomats only moments before they occurred, also maintained a cautious silence. Traditional U.S. allies were generally supportive of the military retaliation against state-sponsored terrorism. The strongest support was offered by Great Britain, but Israel, Germany, and Australia also spoke out in favor of the U.S. measures.

As a general rule, Muslim countries expressed the loudest outrage at the U.S. attacks. Muslim nations that most vehemently condemned the American actions included Iraq, Iran, Libya, and Pakistan. Muted protests came from other Muslim nations such as Turkey, Egypt, and Indonesia. Non-Muslim nations condemning the strikes included Russia and Cuba. Anti-American public demonstrations took place in many of these countries. U.S. flags were burned in protest in dozens of cities around the globe. Many critics of the unilateral American response, at home and abroad, charged that President Clinton had ordered the missile attacks only to draw attention away from the sex scandal in which he was personally becoming embroiled.

After the terrorist attacks on the Pentagon and the World Trade Center on September 11, 2001, suspicion again fell on bin Laden's organization. With the backing of the world's most powerful nations, the United States issued an ultimatum to Afghanistan's Taliban regime to surrender bin Laden and his followers. After the Taliban refused to comply, the United States led a massive missile and bomber assault on Taliban targets. This assault, in conjunction with a ground-based attack from the regime's Afghan opponents in the north, brought down the government in December. Many of bin Laden's subordinates were killed in the attacks, but the whereabouts of bin Laden himself remained unknown in early 2002.

Michael S. Casey

Former Ku Klux Klan Member Found Guilty of Murder in 1966 Firebombing

Samuel Bowers, the Imperial Wizard of the White Knights of the Ku Klux Klan in Mississippi in the 1960's, was convicted of murder and arson in the January, 1966, slaying of Vernon Dahmer, Sr., an African American civil rights activist in Forrest County, Mississippi.

What: Crime; Civil rights and liberties; Social reform
When: August 21, 1998
Where: Hattiesburg, Mississippi
Who:
SAMUEL BOWERS (1925-   ), former Imperial Wizard of the White Knights of the Ku Klux Klan in Mississippi
VERNON DAHMER, SR., president of the local chapter of the National Association for the Advancement of Colored People (NAACP) in Forrest County, Mississippi

Murder and Arson Conviction

On August 21, 1998, a jury in Hattiesburg, Mississippi, convicted former Ku Klux Klan leader Samuel H. Bowers of arson and murder for ordering a 1966 firebombing that resulted in the death of Vernon Dahmer, Sr. Dahmer, a merchant and farmer, was also president of the local chapter of the National Association for the Advancement of Colored People (NAACP) in Forrest County, Mississippi. He was apparently targeted by the Klan because he had allowed his store to be used by African Americans to pay the two-dollar poll tax required at that time for voter registration. Following the verdict, the seventy-three-year-old Bowers was sentenced to life in prison.

Background to the Case

The White Knights of the Ku Klux Klan was organized in the 1960's in Mississippi in response to civil rights activities in that state. Bowers, the owner of Sambo Amusement Company, from Laurel, Mississippi, was founder and leader of the organization. The White Knights of the Ku Klux Klan was the most violent Klan organization active in the South in the 1960's. Members were linked to a number of terrorist and illegal activities including cross burnings, whippings, church bombings, and murders. The most notorious crimes linked to the White Knights were the murders of three civil rights activists, James Chaney, Andrew Goodman, and Michael Schwerner in Neshoba County, Mississippi, in 1964. Bowers was convicted in federal court for his role in the crime and served six years in federal prison; however, the federal conviction was for the violation of the civil rights of the victims, not murder. He was never tried in Mississippi on state charges of murder.

The base of operation for Bowers's organization was Jones County, Mississippi, just north of Hattiesburg. In the winter of 1966, Bowers voiced frustration over the inability of Klan members in neighboring Forrest County to stop the voting registration efforts of Dahmer, the local NAACP leader. The Klan had put signs up on trees, and Dahmer had taken them down. Anonymous phone callers had threatened to kill Dahmer, and night riders had burned down a hay shed on his property. Finally, Bowers ordered a Code Four on Dahmer. This was the Klan code for assassination.

On the night of January 10, 1966, two carloads of Klansmen attacked Dahmer's home and store. They shot out the display window of the grocery and the plate-glass window of the house. They then drenched the two buildings with gasoline and tossed burning rags to set them on fire. As Dahmer shot several loads of buckshot at the night riders, his wife, three children, and an elderly relative escaped from the burning house. Dahmer, whose lungs were seared by the flames,

AP/Wide World Photos

Samuel H. Bowers, former Imperial Wizard of the Ku Klux Klan, leaves the Forrest County Courthouse in Hattiesburg, Mississippi, after his conviction.

died twelve hours after the attack. Bowers later bragged that his "boys" had carried out his instructions.

Bowers and twelve other men were indicted on murder and arson charges in state court; however, there were few convictions—three men were sentenced to life but none served more than ten years. Bowers was tried four times in the 1960's, including twice for murder; however, each resulted in a mistrial when all-white juries could not reach a verdict.

The New Trial

The Dahmer case received renewed interest following the 1994 conviction in Mississippi of Byron de la Beckwith for the 1963 murder of Medgar Evers, the state NAACP leader. The Dahmer family requested that the county and state reopen the case. The legislature responded by appropriating funds for a new investigation, and both candidates for district attorney promised to reexamine the case. A major break in the case occurred when Bob Stringer, a one-time protégé of Bowers, agreed to talk publicly about the case. When he was a teenager, Stringer worked for Bowers, typing manifestos and distributing leaflets for the Klan. His conscience was jarred after he saw the Dahmer family appeal for new information on television in 1994. Wanting to make amends, he became an informant over the next four years, gathering information that led to the arrest of Bowers and two of his associates, Charles R. Noble and Devours Nix. Bowers and Noble were charged with murder and arson, and Nix was charged with arson. In the trial that followed, the jury convicted Bowers of arson and murder. The judge then sentenced the seventy-three-year-old Bowers to life in prison.

Consequences

The conviction of Bowers made history for several reasons. First, the Forrest County circuit court jury consisted of six whites, five blacks, and one Asian. During the 1960's, African Americans were prevented from registering to vote in Mississippi; therefore, they did not serve on juries. Second, the fact that a Mississippi jury convicted a high-profile Klan leader for the murder of a civil rights activist illustrated how much the state of Mississippi had changed. Bowers had once told a fellow Klan member that no jury in Mississippi would convict a white man for killing a black man.

Mississippi state attorney general Mike Moore, whose office had helped Forrest County prosecutors revive the case, stated that the verdict "makes me feel very good, makes me feel proud of my state, proud of the Dahmers, proud of the entire judicial system." James Ingram, the state commissioner of the Mississippi Highway Patrol said, "The people of Mississippi have spoken with a strong conviction, righting a wrong of thirty-two years." Vernon Dahmer, Jr., who had served in the military for twenty years and had returned home said, "This is a new Mississippi. I came back home to be with my people, black and white. Mississippi has good people. I intend to die in Mississippi." Dahmer also noted that the verdict proved that the society his father died helping to create was now in Mississippi.

William V. Moore

North Korea Fires Missile Across Japan

Communist North Korea surprised the world, firing a missile across Japan and thereby demonstrating its newly achieved missile and nuclear technology and the weakness of the U.S.-Japan mutual defense system in East Asia.

What: International relations; Military capability
When: August 31, 1998
Where: North Korea, Japan, and the United States
Who:
Kim Il Sung (1912-1994), founder of the North Korean Workers' Communist Party (NKWP) and the Democratic People's Republic of Korea (DPRK)
Kim Jong Il (1942-   ), Kim Il Sung's son, NKWP chairman and DPRK president

Nuclear Crisis

On August 31, 1998, using a crude three-stage rocket, North Korea launched a missile that traveled across Japan and landed in the Pacific Ocean. The news shocked Japan because its national defense failed to respond or even warn its people. The undetected missile also worried the United States, which maintains a mutual security treaty with Japan and regularly deploys 100,000 American troops in East Asia (35,000 in South Korea and 65,000 in Japan and other Pacific islands). A few days later, North Korea announced that the firing had been a test of its new long-range missile, Taepo Dong 2, which was part of its nuclear program in connection with the development of its next generation of intercontinental ballistic missiles.

North Korea's nuclear programs, which began in the late 1980's, were part of an ongoing modernization of its military capability and represented a shift of its military focus from Cold War international concerns to regional ones. Having lost its nuclear protection after the collapse of the Soviet Union, the Democratic People's Republic of Korea (DPRK) prepared to deal with the United States by creating a nuclear balance in the Korean peninsula and its own nuclear deterrent. DPRK president Kim Il Sung felt compelled by U.S. tactical nuclear weapons deployed in South Korea. He wanted his own nuclear weapons so that he could break the nuclear monopoly established not only by the United States but also by communist nuclear powers such as China and Russia.

In 1990, Kim accelerated his nuclear programs by hiring former Soviet nuclear scientists and engineers and smuggling plutonium from Russia. The United States denounced North Korea's violation of the Nonproliferation of Nuclear Weapons Treaty and demanded an international on-site inspection in 1991. Kim refused to give up his nuclear programs and announced in 1993 that North Korea would withdraw from the nuclear weapons treaty. Although he promised he would never use atomic bombs on South Korea, he made no such pledges regarding their use in other countries.

North Korean Intentions

Since the end of the Cold War, famine, economic difficulties, political crisis, and the loss of powerful communist patrons have reduced North Korea's conventional military capabilities. After Kim died in 1994, his son, Kim Jong Il, became the new leader of the communist regime. Young Kim increased North Korea's nuclear capacity by developing ballistic missile technology. His unambiguous objective was to threaten South Korea, Japan, and the United States; therefore, he began with the short-range missile, Taepo Dong 1, and progressed to the long-range missile, Taepo Dong 2.

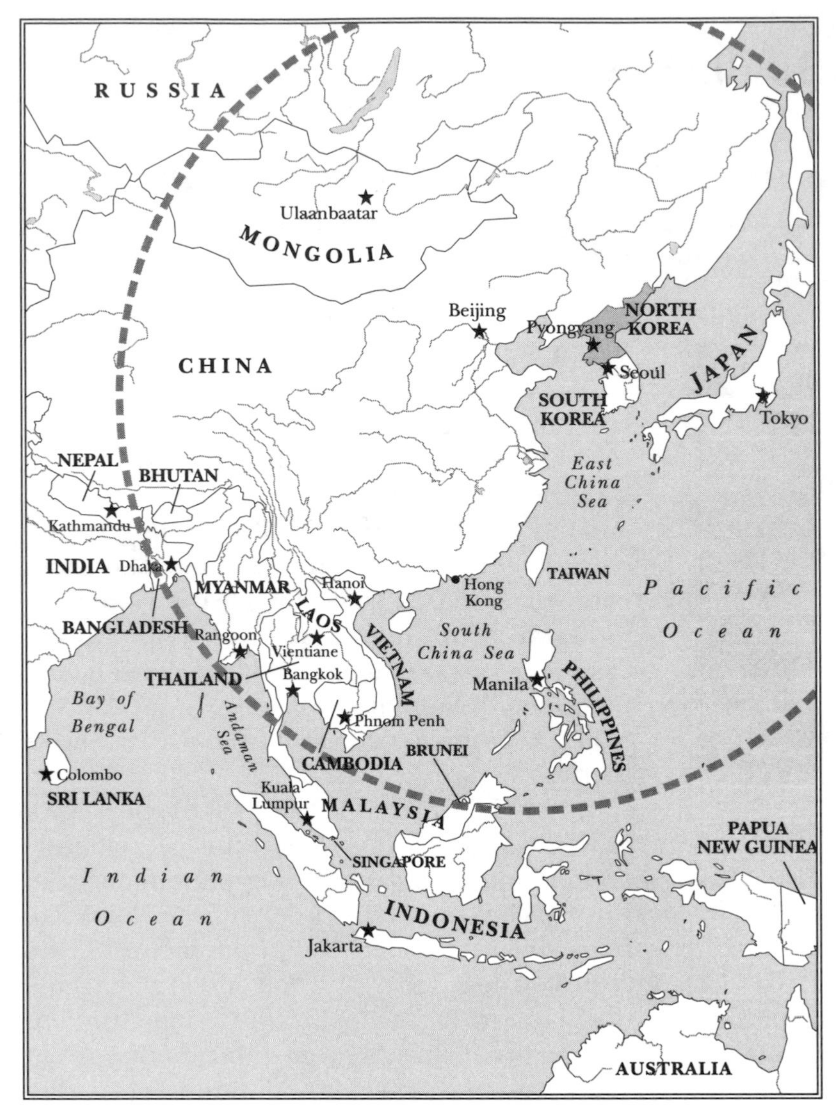

Approximate range of North Korea's Taepo Dong 2 missiles.

Kim's nuclear arms sales to the Middle East and other parts of the world seemed to have quickly solved North Korea's economic problems; however, his export of ballistic missiles to terrorists created a new threat for the United States. By developing the capability to manufacture and sell nuclear weapons, Kim reckoned that he could pressure the United States into providing economic assistance in exchange for his promise not to make additional bombs. At the Geneva negotiations in 1998-1999, when the U.S. delegation insisted that the international inspection of underground nuclear facilities had to be included in any agreement, the North Korean delegation made a counterproposal that they would agree with such an inspection only if the United States could provide a no-interest loan of $300 million or 300,000 tons of grain. Nuclear weapons negotiations became a powerful tool for Kim.

Consequences

The controversy surrounding North Korea's missile test frustrated the United Nations. Kim refused to give in to the pressure from the international community, and worldwide criticism of his irresponsible behavior did not appear to affect his determination to pursue his own course. The U.S. government was frustrated by North Korea's unwillingness to succumb to international laws or to the power politics of the world community. The missile crisis was not expected to be resolved easily. North Korea continued to harbor a deep resentment of the United States and Japan and maintained the fifth-largest standing army in the world, situated just north of the world's most fortified demarcation line at the thirty-eighth parallel, which has divided North and South Korea since the Korean War (1950-1953). Most experts in international politics believed that North Korea would not accept a full and unconditional nuclear inspection without considerable concessions from its adversaries, especially the United States.

After the missile test, the Japanese government expressed concerns about stringent limits on its participation in U.S.-North Korean negotiations as well as fears that U.S. military actions might jeopardize Japan's national defense in the case of a missile attack. The difficulties posed by this situation may force Japan and the United States to improve their mutual defense system in East Asia. North Korea's ballistic missile capability provided an unexpected but significant boost to the policy debate in the United States regarding the deployment of ballistic missile defense systems at home and overseas, including the National Missile Defense (NMD) and the Theater Missile Defense (TMD). The TMD is designed to provide a limited defense for the United States and its allies in East Asia to counter the primitive and small-scale arsenals of states of concern such as North Korea or China.

North Korea, China, and Russia challenged the proposed NMD and TMD systems, and some U.S. allies questioned their legality, feasibility, and desirability. For example, when Japan was asked to cooperate with the United States and to help finance the TMD system in East Asia, it questioned the combat readiness and effectiveness of the TMD, which seemed to many Japanese to be more of a psychological weapons system than an operational one. The consequences of the 1998 North Korean missile test for U.S. missile defense policy and the relationship between the United States and Japan may prove to be more profound than many people first thought they would be.

Xiaobing Li

United Nations Tribunal Convicts Rwandan Leader of Genocide

An international tribunal sentenced a former prime minister of Rwanda to life imprisonment for promoting genocide and committing crimes against humanity.

What: International law; Civil strife; Human rights; Ethnic conflict
When: September 4, 1998
Where: Rwanda
Who:
Jean Kambanda (1956-   ), prime minister of Rwanda from April to June 1994
Juvénal Habyarimana (1937-1994), president of Rwanda from 1973 to 1994 who died when his plane was shot down

International Criminal Tribunal for Rwanda

On September 4, 1998, the International Criminal Tribunal for Rwanda (ICTR) sentenced Jean Kambanda, a former prime minister of Rwanda, to life imprisonment for engaging in genocide, conspiring with others to engage in genocide, and for committing crimes against humanity. Before his sentence, Kambanda had pleaded guilty to these crimes so there was no formal trial.

Charges against Kambanda were based on claims that from April to July of 1994, somewhere between 500,000 and one million Rwandans were killed in what amounted to a civil war between the two dominant ethnic groups of Rwanda: the Hutus and the Tutsis. The vast majority of those who were killed were Tutsis, and the claim is that they were killed by Hutus who wanted to exterminate the entire Tutsi group. In his guilty plea, Kambanda confessed that he had encouraged the killing of Tutsis through radio broadcasts, by distributing weapons and ammunition, and by attending meetings and discussing plans for the killing of Tutsis.

Kambanda's sentence of life imprisonment was an important event for several reasons. It was the first decision of the ICTR. Prior to the sentence, the tribunal had been attacked for being poorly administered. After that sentence, the tribunal convicted eight other Rwandans and acquitted a ninth defendant.

In sentencing Kambanda to life imprisonment, the tribunal articulated several rules of international law applying to genocide and other crimes against humanity. Those rules have subsequently been developed and applied by other international tribunals.

Moreover, in punishing former head of a national government, the tribunal showed that top government leaders were no less subject to international law than rank-and-file soldiers. Since Kambanda's trial, leaders in such countries as Serbia, Croatia and Sierra Leone have been brought before similar international tribunals.

Background to Genocide

A densely populated, small, and landlocked country in east-central Africa, Rwanda is a creation of European colonial rule. It was originally colonized by Germany and then ruled by Belgium from the time of World War I until it and its neighbor Burundi became independent in 1962. Belgium administered both countries by relying on existing ruling cliques and reinforcing ethnic rivalries. However, when it granted independence to Rwanda, it left the crowded, impoverished country with the veneer of majority democratic rule.

Most members of Rwanda's ruling cliques during the colonial period belonged to the Tutsi group, when the majority of the people—as in Burundi—were Hutus. The Tutsis wanted to retain the power they had wielded under colonial rule. They controlled the civil service and the army. The Hutus saw in their own superior numbers the route to power. These were the lines of conflict, and the history of Rwanda between 1960 and 1994 is of Hutu riots and Tutsi use of the army to put those riots down, followed by political compromises. There were Hutus and Tutsis who found such compromises unacceptable. Thus, some Hutus sought significant changes in the distribution of governmental power, while some Tutsis not only resisted such changes, but mounted coups and invasions from outside of Rwanda to overthrow the governments of Rwanda.

Internal events in Rwanda were made even more chaotic by pervasive political instability in the general area of east-central Africa, known as the lake region, because of its great freshwater lakes. In addition to the former Belgian colonies of Rwanda, Congo (Kinshasa), and Burundi, the region also included the Sudan and Uganda, both of which were experiencing or had just experienced bitter civil wars. In early 1994, some unknown person shot down a plane carrying the Hutu president of Rwanda, Juvénal Habyarimana. Many Hutus suspected that the plane had been shot down by Tutsis seeking to regain control of the government. Some of them plotted to make sure that the government was not overthrown by the Patriotic Front, a group controlled primarily by Tutsis trained in Uganda who were fighting to seize control of the government. The chaos that followed witnessed unspeakable massacres—largely of Tutsis, but also of Hutus—that led to the charges for which Jean Kambanda pleaded guilty.

Because of the general political instability in Africa's Lake Region even before the massacres, United Nations peacekeeping soldiers under the command of French, Belgian, and Canadian officers had been stationed in Rwanda. As the massacres were developing, some of these officers asked for additional soldiers to be sent to Rwanda. This request, which had to be responded to by the United Nations Security Council was opposed by some members—most notably the United States.

AP/Wide World Photos

In 1998 U.N. secretary-general Kofi Annan visisted Rwanda and viewed the remains of victims of the genocide.

At the same time, the Security Council was busy dealing with a conflict between the Serbs and Muslims in Bosnia that was getting most of the attention in international news coverage.

Consequences

Human rights proponents compared the Rwanda atrocities to those to which Jews, Gypsies, and others had been subjected by Germany a generation earlier. They demanded that just as Nazi officials had been punished at Nuremberg following World War II, the United Nations ought to create a tribunal to try and punish perpetrators of human rights violations in the former Yugoslavia and in Rwanda. The U.N. Security Council created two tribunals to investigate, try, and punish persons accused of war crimes with regard to these conflicts. It was one of these tribunals, that for Rwanda, which accepted Kambanda's guilty plea, and imposed the sentence of life imprisonment on him.

Whether international tribunals such as the ICTR are appropriate institutions for punishing individuals alleged to have violated international law is a controversial matter. Many people argue that such trials should be undertaken by national courts. They fear that international tribunals may be controlled by powerful countries, so that only citizens of weak countries will be prosecuted. It seems clear, however, that the international community believes that tribunals like the ICTR serve a useful function in avoiding what some people call "impunity"; that is, that wrongdoers sometimes go unpunished for their acts.

Maxwell O. Chibundu

Scientists Report Record-Size Hole in Antarctic Ozone Layer

Despite efforts by scientists and the governments of 175 nations, the annual ozone hole covered a record 10.5 million square miles (27.3 million square kilometers).

What: Environment; International relations
When: September 19, 1998
Where: Stratosphere over Antarctica
Who:
JONATHAN SHANKLIN, senior scientific officer for the Meteorological and Ozone Monitoring Unit of the British Antarctic Survey

The Surprise of the Century

As disturbing as scientists found 1998's record-size ozone hole, they were far more alarmed by a 1985 paper published in the scholarly journal *Nature.* The British Antarctic Survey's team of scientists had noticed a regular springtime decline in ozone concentration since 1977, and the losses increased to 30 percent of the total by October, 1984. Three scientists from the team announced and defined the ozone hole for the scientific community. No mathematical models had predicted the magnitude of the changes their paper reported—not even for 50 to 100 years in the future. Michael H. Proffitt, the senior scientific officer for the World Meteorological Organization (WMO), stated, "That was the surprise of the century."

The ozone layer shields the earth from dangerous ultraviolet (particularly UV-B) radiation. Problems caused when too much UV-B radiation reaches Earth's surface include increased levels of skin cancer, premature aging of the skin, cataracts, and weakening of the immune system. Phytoplankton are highly sensitive to UV-B radiation, and many species of crops show impaired growth and reproduction when subjected to high levels.

Scientists determined that the damage to the ozone layer was caused by chlorofluorocarbons (CFCs) and other ozone-depleting substances. CFCs were used as coolants in refrigerators and air conditioners (Freon), as a propellant in aerosol cans, and to produce foam for bedding and packaging. Other ozone-depleting substances include halons (used in firefighting equipment), methyl bromide (used in pesticides), and methyl chloride (a solvent used to clean electronic circuit boards and in dry cleaning). Use of CFCs in aerosol cans was banned in the United States in 1978. Following the 1985 discovery of the Antarctic ozone hole, much more serious action was taken. In 1987, a treaty called the Montreal Protocol on Substances that Deplete the Ozone Layer was adopted. Along with later amendments, it set dates for phasing out a total of 95 ozone-depleting substances.

Beautiful, Lens-Shaped Clouds

A complex chemical process creates each year's Antarctic ozone hole. The ozone layer is located in the stratosphere between 6 miles (9.5 kilometers) and 18 miles (29 kilometers) above Earth's surface. Ozone is a compound consisting of three oxygen molecules. Measurements of total column ozone (the total amount above a point on Earth's surface) have been made from the ground using the Dobson spectrophotometer since the mid-1950's. In 1978, measurements were added from National Aeronautics and Space Administration (NASA) satellites using the total ozone mapping spectrometer (TOMS) system. Before ozone holes began appearing, the average ozone layer thickness was 300 to 350 Dobson units. An ozone hole (actually, a thinning) in the Antarctic is defined as an ozone column of 220 Dobson units or less. The Antarctic ozone hole is seasonal. Maximum depletions have usually been measured in late September, and the hole tends to disappear in late November or early December. During this period, monthly ozone column measurements will be 40 percent to 50 percent below pre-ozone hole (1970-1976) averages, with brief dips to 80 percent below. The 1998 record hole covered 10.5 million square miles (27.3 million square kilometers). The depth of the ozone depletion, 90 Dobson units, was the second lowest ever recorded. In addition, ozone loss was detected at the unusually high altitude of 79 miles (24 kilometers).

CFCs and other ozone-depleting substances are much heavier than air. It takes around two years for them to rise above Earth's surface and another three to five years to reach the ozone layer. In the winter in Antarctica, a strong westerly air circulation called the vortex allows the air inside it to become very cold, lower than minus 78 degrees Celsius (minus 108 degrees Fahrenheit). Only then can polar stratospheric clouds (PSCs) form. These beautiful, lens-shaped clouds—vibrant, iridescent blue and green, rimmed with pink—have a deadly side. When the sun returns in the Antarctic spring, they provide the surface on which a chemical reaction takes place. The CFC molecules break down, releasing chlorine. One chlorine atom can break apart more than 100,000 ozone molecules before it is finally removed from the stratosphere.

Shortly after the announcement of 1998's record-size ozone hole, scientists predicted that the ozone layer would remain seriously depleted for ten to twenty years but then slowly begin to heal. By the year 2050, it would return to its pre-1970 condition. The recovery would be slower if nations did not comply with the Montreal Protocol's phase-out schedules or if climate change worsened the environment.

Consequences

After the passage of the Montreal Protocol, and even after 1998's record-size ozone hole, the condition of the ozone layer remained complex and contradictory. By late 1999, CFC consumption had fallen 84 percent worldwide. However, in 2000, the ozone hole set a new record of 11.0 million square miles (28.4 million square kilometers). Shortly after this record was announced, Jonathan Shanklin, one of the three British scientists who wrote the *Nature* article announcing the ozone hole in 1985, predicted that in twenty years, the Arctic ozone hole could be as large as 2000's Antarctic hole. A hole of this size would extend over parts of Europe, North America, and Asia.

Ozone losses in the Arctic in the spring of 1995-1996, for the first time, were severe enough to be called ozone holes. Previously, ozone destruction in the Arctic had been less severe than in the Antarctic, partly because the northern stratosphere is not as cold. However, increases in greenhouse gases (such as carbon dioxide), which warm the surface of the earth, make the stratosphere cooler so that PSCs can form and begin the ozone-destroying chemical reaction.

Another factor in the continuing depletion of the ozone layer is the increase in the level of bromine in the stratosphere. Ozone-depleting substances form bromine when they are broken apart in the stratosphere. Bromine is fifty times more deadly to ozone than chlorine. The Montreal Protocol requires developing countries to phase out methyl bromide by 2005. Other disturbing indicators of accelerated ozone depletion have been reported. In September, 2000, the Antarctic ozone hole stretched over Punta Arenas, Chile—the first time it had covered a city. A mini ozone hole was detected over Hokkaido, Japan, in 1996 and over the Weddell Sea, east of the Antarctic peninsula, in July, 2000. All these reports indicate that there is no room—and, more important, no time—for complacency in addressing the ozone layer problem.

Glenn Ellen Starr Stilling

Iran Lifts Death Threat Against Author Rushdie

The Iranian government announced that it would end the death sentence against author Salman Rushdie, thereby reestablishing relations between Iran and Britain.

What: International relations; Religion
When: September 24, 1998
Where: New York City and London, England
Who:
SALMAN RUSHDIE (1947-), British author born in India
AYATOLLAH RUHOLLAH KHOMEINI (1902-1989), president of the Islamic Republic of Iran from 1979 to 1989
KAMAL KHARRAZI (1944-), foreign minister of the Islamic Republic of Iran
ROBIN COOK (1946-), foreign secretary of Great Britain

Death Threat Is Lifted

On September 24, 1998, Kamal Kharrazi, foreign minister of the Islamic Republic of Iran, announced that his government had "distanced itself" from the death sentence pronounced on British author Salman Rushdie by the revolutionary leader Ayatollah Ruhollah Khomeini on February 14, 1989. The death sentence, known as a *fatwa,* called on zealous Muslims to assassinate Rushdie because his book, *The Satanic Verses* (1989) blasphemed Islam. Kharrazi stated that under its new policy, the Iranian government would not take any action to threaten the life of Rushdie nor would it encourage anyone else to harm him.

Rushdie had been living in hiding under British police protection for almost a decade. When British foreign secretary Robin Cook assured him that the Iranians were very serious about the announcement, Rushdie was visibly relieved, "This looks like it's over. It means everything, it means freedom." He hoped that the fear in which he had lived would quickly dissolve and that his life would return to normal.

The Fatwa Continues

In reality, the proclamation was a carefully negotiated statement agreed on by the British and Iranian governments in order to improve their diplomatic relations. While disassociating the Iranian government from the death threat, the religious fatwa was not canceled. In fact, the two governments agreed that *The Satanic Verses* was a profane book but that its author had the right to free expression. Although the threat to Rushdie's life was greatly reduced, he continued to require police protection. Iqbal Sacranie of the United Kingdom Action Committee for Islamic Affairs stated that British Muslims never intended to harm Rushdie. However, the problem could not be resolved as long as *The Satanic Verses* remained in circulation.

On September 25, 1998, Rushdie stated that he was not sorry that he had written *The Satanic Verses,* which by then had sold more than one million copies in fifteen languages. He expressed sorrow for the demonstrations against his book that had taken many lives, particularly in India. British officials thwarted several attempts on Rushdie's life. Hitoshi Igarashi, the Japanese translator of the book, was stabbed to death in 1991. Ettore Caprioli, the Italian translator, and William Nygaard, the Norwegian publisher, were attacked and seriously wounded. Four bookstores owned by the book's publisher in England were bombed. Rushdie announced plans to write about the ordeal of living in hiding and seeing his associates threatened and killed.

Consequences

Within days of the Iranian government's announcement, the Iranian Islamic leader Ayatollah Hassan Saneii and his associates repeated the call for Muslims to kill Rushdie. On October 4, 1998, 150 of the 270 members of the Iranian parliament repeated their support of the fatwa. On October 12, 1998, the Iranian revolutionary Fifteenth Khordad Foundation increased its bounty on Rushdie's head to $2.8 million. The Association of Hezbollah Students at Tehran University also offered its own reward of $333,000.

In spite of the renewed fatwa by Iranian religious leaders, without governmental support, threats against Rushdie seemed to have lost much of their effect. In October, 1998, the Rushdie Defense Committee USA. and the International Rushdie Defense Committee, which campaigned on behalf of Rushdie, disbanded. However, on February 14, 1998, the tenth anniversary of the fatwa, Ayatollah Saneii proclaimed that the assassination of Rushdie would be carried out. As recently as February, 2001, both the elite Iranian Revolutionary Guards and the Islamic Propagation Organization issued statements repeating the fatwa against Rushdie and called on Muslims to cleanse the world of such "devils."

The Iranian government's decision to withdraw from the Rushdie affair was immediately followed by an agreement between Iran and Britain to improve their diplomatic relations by exchanging ambassadors. Cook announced that a new chapter had begun in the relationship between the United Kingdom and Iran. He also anticipated that relations would quickly improve between Iran and the European Union. In fact, the Iranian market for European goods was the primary motivation for the carefully negotiated agreement between Iran and Britain. Similarly, many U.S. policymakers were anxious to open negotiations for Iranian oil. In spite of the 1998 U.S. State Department annual report on terrorism clearly stating that there had been no change in Iran's violent conduct and concluding that Iran was the world's principal sponsor of terrorism, the administration of Bill Clinton continued to push to restore friendly relations with the Iranian government.

In spite of the renewed threats by Iranian Islamic leaders, Rushdie became able to move about more freely. Many of his lectures and appearances were frequently announced in advance. He was issued travel visas to several countries, including his home country of India, where he had been banned. He even appeared in a cameo role in the 2001 film *Bridget Jones' Diary*. Rushdie continued to write. His first novel after the 1998 announcement was *The Ground Beneath Her Feet* (1999), a nonpolitical love story involving an Indian rock-and-roll duo who travel to the United States in search of fame and glory. On a tour to promote this novel, Rushdie said that even though the danger was not completely gone, he and his third wife, Elizabeth West, were able to live a more relaxed life with their son Milan.

Gerald S. Argetsinger

AP/Wide World Photos

Salmon Rushdie, after Iran's announcement.

German Chancellor Kohl Loses Election to Schröder

In the election to the German lower house (Bundestag) on September 27, 1998, the voters for the first time in the history of the Federal Republic of Germany removed a sitting chancellor by replacing the Christian Democrat Helmut Kohl with the Social Democrat Gerhard Schröder.

What: National politics
When: September 27, 1998
Where: Germany
Who:
HELMUT KOHL (1930-), chancellor of the Federal Republic of Germany from 1982 to 1998
GERHARD SCHRÖDER (1944-), Social Democratic candidate and prime minister of the German state of Lower Saxony
JOSCHKA FISCHER (1948-), the most popular leader of the Green Party

German Politics

In 1990, the two German states, the Federal Republic of Germany (West Germany) and the Democratic Republic of Germany (East Germany), became a single nation. West Germany's Basic Law, or constitution, which was adopted in 1949 when West Germany was established, became the constitution of the united Germany. The Basic Law created a parliamentary government system. The chancellor, Germany's executive leader, can only be elected by a majority in the lower house (Bundesrat). Before 1990, four political parties held all seats in the lower house. The two largest parties, the Christian Democratic Party (together with the Bavarian Christian Social Party wing) and the Social Democratic Party (SPD) represented conservative and left-wing interests in Germany respectively. Liberals were organized in the small Free Democratic Party, and ecologists and libertarians came together in the small Green Party. After the incorporation of East Germany in 1990, the Party of Democratic Socialism, the heir to the old Communist Party, won seats in the lower house.

Since 1949, all governments in the Federal Republic of Germany were led either by a Christian Democratic (CDU/CSU) or a Social Democratic chancellor. However, between 1969 and 1998, the tiny Free Democratic Party played a key role in producing government majorities in the legislature by giving its support to either the SPD or the CDU. Between 1969 and 1982, the Free Democrats (FDP) joined a Social Democratic government, and in 1982, they switched their support to the Christian Democrats, allowing Helmut Kohl to become chancellor.

Kohl was born in Ludwigshafen on the Rhine River. He joined the Christian Democratic Party in 1946 and was elected minister-president of the German state of Rhineland-Palatinate in 1969. He held the office of chancellor (1982-1998) longer than any other chancellor of the Federal Republic of Germany. His most enduring accomplishment as chancellor was the reunification of the two German states in October, 1990. To pay for the cost of unification, Kohl had to raise taxes. In addition, the collapse of the eastern German economy led to widespread unemployment. By 1998, unemployment in Germany, particularly in the east, had reached record levels, and Kohl's popularity had declined dramatically.

The Election Contest

Although he was vulnerable in 1998, Kohl attempted to run for his fifth successive term as chancellor. More than four million people were unemployed, and 80 percent of the public polled cited that issue as Germany's major problem. In particular, the inhabitants of the five eastern states were frustrated by Kohl's apparent inability to solve the economic and social problems of unification. However, Kohl could enact few major reforms because the upper house (Bundesrat) was dominated by the Social Democrats, who blocked innovations.

The public clearly wanted a change after sixteen years of Kohl's leadership. SPD candidate Gerhard Schröder offered a younger and more dynamic image than Kohl. He also promised to move the socialists to the middle of the political spectrum. Although Schröder had been a member of the SPD's left wing during his early political career, he created a new moderate image for himself after being elected minister-president of the state of Lower Saxony in 1990. He made it clear that he was not hostile to a free market economy, and he told business leaders that he would support a reduction in the tax rate.

Schröder consciously copied the campaign style and moderate message of Tony Blair, the British prime minister, and of Bill Clinton, the president of the United States. The SPD adopted the left-centrist slogan "New Middle" to appeal to both traditional socialist voters (workers) and to middle class voters not committed to one party. Schröder also made a special effort to win support in eastern Germany. Media coverage during the campaign was much greater than in previous campaigns, and experts concluded that Schröder won the battle of the sound bites.

AP/Wide World Photos

Helmut Kohl.

Consequences

The election results clearly reflected the German public's feeling that it was time for a change. The CDU's share of the vote declined from 41.5 percent in 1994 to 35.2 percent, and the party experienced its worst election results since 1949. The decline was most dramatic in the east, where the party polled 10 percent fewer votes than in 1994. Many of these workers had turned to the SPD, which increased its votes from 36.4 percent in 1994 to 40.9 percent. The SPD was now the largest political party, a status it had not achieved since 1972.

For the first time in fourteen German national elections, a government was voted out of power. Previously, changes among the coalition partners had overturned governments. Like the CDU, the Free Democrats, who had been in coalition governments for twenty-nine years, were now out of power. The SPD would not accept the Free Democrats or the former communists (Democratic Socialists) as coalition partners. This made it possible for the Green Party to become a national government partner for the first time since it entered the lower house in 1983. On October 20, 1998, Schröder and the Greens agreed to form a coalition government. The former left-wing radical and most popular Green politician, Joschka Fischer, became foreign minister.

Following state elections in 1999, the SPD lost control of the upper house. Nevertheless, the following year, Schröder accomplished a major coup by overcoming criticism from the left wings of his coalition parties and obtaining the support of the upper house for his major tax reforms. Schröder benefited when a serious campaign finance scandal involving Kohl greatly weakened the CDU, which was forced to embark on a desperate search for a candidate for the next national election in 2002. By that date, however, to remain electable, Schröder needed to resolve allegations that Fischer participated in violent street demonstrations during the 1970's and had contacts with German terrorists and the Palestinian Liberation Organization.

Johnpeter Horst Grill

Vatican Grants Sainthood to Formerly Jewish Nun

Pope John Paul II declares Sister Teresa Benedicta of the Cross, better known as Edith Stein, a saint in the Roman Catholic Church. Stein, a former Jew, became a nun in the 1930's and was killed in the Nazi gas chambers at Auschwitz during World War II.

What: Religion; Gender issues
When: October 11, 1998
Where: Rome, Italy
Who:
EDITH STEIN (1892-1942), also known as Sister Benedicta of the Cross
JAMES RUDIN, director of Inter-religious Affairs for the American Jewish Committee
ABRAHAM H. FOXMAN (1940-), national director, Anti-Defamation League
LEON KLENICKI (1930-), rabbi and director of interfaith affairs, Anti-Defamation League
JOHN PAUL II (KAROL JÓSEF WOJTYŁA, 1920-), Roman Catholic pope from 1978

Stein's Life

Edith Stein was born in Breslau, Germany (in present-day Poland), to a devout Orthodox Jewish family. Despite her orthodox upbringing, Stein became an avowed atheist during her youth. She attended the University of Breslau and later the University of Goettingen, where she studied philosophy until World War I broke out, then she volunteered to help the Red Cross. At the war's end, she returned to her studies, earning her doctorate in philosophy from Freiburg University.

After the war, the conversion of a Jewish friend to Christianity inspired Stein to consider the possibility of conversion for herself. Through later readings, particularly the autobiography of Saint Teresa of Avila, a sixteenth century Spanish nun of Jewish ancestry, Stein became interested in Catholicism and, on January 1, 1922, was baptized into the Roman Catholic Church. From 1922 to 1932, she taught at a Catholic teacher's college and later at the German Institute for Scientific Pedagogy.

Stein's growing faith led her, in 1933, to enter a Carmelite convent in Cologne, Germany. She remained there until 1938, when superiors of her order decided to send her to Holland where they reasoned she would be safe from Nazi persecution because of her Jewish origins. By 1940, Hitler had begun his conquest of Europe, invading and taking over Holland, Belgium, and Denmark. Within two years, Stein would be just another casualty of Hitler's death camps.

Canonization

On October 11, 1998, Pope John Paul II declared Stein, a Jewish-born convert to Catholicism, a saint of the Roman Catholic Church. Sainthood, the final step in the canonization process, signifies that a person has died a martyr for the faith or has practiced an extraordinary and virtuous life, is truly in heaven, and merits recognition as a model of Christian imitation. Before sainthood, a candidate is beatified by the Church based on investigations of the person's holiness and, in many cases, miracles attributed to the person after death. Stein was beatified May 1, 1987, as a martyr to the Catholic faith.

As a Jewish-Catholic, Stein was arrested in Holland on August 2, 1942, when Nazis retaliated against an open letter from Catholic Dutch bishops protesting the Nazi treatment of Jews and asking protection for Jews who had converted to Christianity. Stein had been offered the opportunity to escape deportation to a concentration camp by renouncing the bishops' message, but she refused. Seven days later, on August 9, 1942,

AP/Wide World Photos

Edith Stein on a tapestry on the facade of St. Peter's Basilica, at the Vatican in October, 1998.

Stein died in the gas chambers of Auschwitz concentration camp.

The Catholic Church credited Stein with a 1987 miracle, the saving of the life of Teresia Benedicta, the two-year-old daughter of a married Melkite Catholic priest, who had named his daughter in honor of Stein's religious name—Teresa Benedicta of the Cross. The young Teresia had accidently swallowed an overdose of Tylenol, necessitating the need for a liver transplant. Those around her prayed for Stein's intercession to cure the girl, and Teresia miraculously recovered without surgery. Doctors found no medical explanation for the case. The Vatican certified the event in 1997 as the miracle needed for Stein's canonization.

The Catholic Church officially recognizes Stein as Saint Teresa Benedicta of the Cross. She chose the first name in honor of Saint Teresa of Avila. The remainder of the name comes from Benedicta, meaning "blessing," which refers to Stein's devotion to the blessing of the cross and the meaning it gave to her life.

Consequences

Stein's canonization set off a wave of controversy within Jewish circles. At issue was whether she was killed because she was a Catholic nun or a Jew. Many Jewish critics argue that Stein was killed because of her Jewish roots, not her Catholicism. Rabbi James Rudin, director of Interreligious Affairs for the American Jewish Committee, which has close ties with the Vatican on Catholic-Jewish dialogue, asked in response to Stein's canonization what symbol she conveyed. He asked, "Are they [the Catholic Church] saying that she is a martyr to the Catholic faith? Or is she a symbol of Jewish suffering during the Shoah, when conversion didn't mean a thing to the Nazis?" Abraham H. Foxman, national director of the Anti-Defamation League and a Holocaust survivor, and Rabbi Leon Klenicki, the league's director of interfaith affairs, in a joint statement, saw Stein's canonization as a distraction from the Vatican's failure to speak out against the Holocaust during World War II and as "a step toward Christianizing the Holocaust [while] diminishing Christian self-examination about this dark time in history."

Pope John Paul II endeavored to improve relations between Jews and Catholics. In 1979, he publicly prayed at Auschwitz and later became the first pope to visit a synagogue in Rome. John Paul also established diplomatic relations with the state of Israel and in 1998 issued a Vatican document calling Catholics to repent for the Church's failure to speak out against the Holocaust. However, such efforts fall short of what some Jews want, such as Shimon Samuels, European representative of the Los Angeles-based Simon Wiesenthal Center who says, "Turning Edith Stein into a saint will not atone for the silence of Pope Pius XII [Pontiff during World War II] nor for passive collaboration in the anti-Semitism that led to her murder."

Michael J. Garcia

Former Chilean Dictator Pinochet Is Arrested in London

Former Chilean president Augusto Pinochet, while in England receiving medical treatment, was arrested for violations of human rights committed while he had ruled Chile from 1973 to 1990.

What: Human rights; International law
When: October 16, 1998
Where: London Clinic, London, England
Who:
Augusto Pinochet Ugarte (1915-), dictator of Chile from 1973 to 1990
Baltasar Garzón (1956-), judge of the Spanish National Court, a special tribunal in Spain

The Arrest

On October 16, 1998, former Chilean dictator Augusto Pinochet Ugarte was arrested in his room at the London Clinic, a private London hospital. Pinochet had heeded the advice of his doctors and gone to London to undergo an operation to repair a herniated disc in his spinal column. The eighty-two-year-old general had diabetes and wore a pacemaker, making the operation potentially dangerous and recovery difficult and prolonged. He needed to spend several weeks on his back recuperating. He had traveled to England without incident several times to purchase arms for the Chilean military, but this time, during his recuperation, he was placed under arrest before he could return to Chile, where he had immunity from prosecution as a war criminal.

Legal Basis for the Arrest

General Pinochet, who had become the symbol of the excesses of military rule in Latin America, was accused of several crimes, including violations of human rights, while he ruled Chile. Consequently, various international human rights groups and exiled Chileans who had suffered or whose relatives had suffered during his rule supported attempts to bring him to justice.

The London Metropolitan Police told Pinochet that they were serving a Red Notice warrant issued by Interpol on behalf of the Spanish authorities who wanted to extradite him to Spain for questioning regarding crimes of murder, genocide, and terrorism. The Spanish authorities wanted the warrant served before Pinochet could return to Chile, where the chances of taking away his immunity were remote.

Judge Baltasar Garzón of the Spanish National Court issued the warrant based on the 1976 kidnapping and disappearance of Edgardo Henriquez, a Spanish citizen. The judge said he would send an additional seventy-six names of people of various nationalities who between 1976 and 1983 had been kidnapped, taken to Argentina, and disappeared. Judge Garzón also asked for the extradition of several Argentine junta members, but Argentina refused to honor the request.

Spanish authorities had sent the request for extradition to Scotland Yard through Interpol. The effort to have Pinochet extradited began in Paris and Geneva. The Geneva prosecutor, Bernard Bertosse, the families of three French citizens, and several human rights groups brought suits on behalf of people killed in Chile during the rule of Pinochet. The request was based on the European Convention on Terrorism, which requires members to help other member nations on terrorism-related matters. Several European nations had been investigating the so-called Condor Plan, a scheme in which military juntas in various Latin American nations were said to have cooperated in order to get rid of their enemies.

President Eduardo Frei Ruiz Tagle of Chile was opposed to the arrest of Pinochet, demanded his immediate release, and maintained that Pinochet had immunity as a former head of state and as senator for life, a position granted to him by the 1980 constitution Pinochet had forced on the nation.

The arrest of Pinochet presented the Labor government of Prime Minister Tony Blair of England with a quandary. While in opposition to the Conservative government, Blair had criticized the government for allowing two visits by Pinochet in 1991 and 1995 and had called these visits appalling. However, the Labor government did not want to disrupt diplomatic or economic relations with Chile. Prime Minister Blair and Foreign Secretary Robin Cook said that it was a judicial matter, and the government had no power to intervene. The decision on the request for extradition was in the hands of Home Secretary Jack Straw, who had ninety days to decide.

The Conservative prime minister of Spain, José Maria Aznar, opposed extradition but could do nothing to stop it. Foreign Minister Abel Matates insisted that the courts would decide the matter, but unidentified sources said that the Spanish government was working behind the scene against the request. Extradition efforts continued when an eleven-member panel of senior judges, the second highest court in Spain, ruled that Spain did have the right to ask for extradition. There is no appeal from this court.

In England, the High Court ruled that Pinochet did have immunity as head of state but ordered him held until an appeals ruling, expected the following week, was rendered. Pinochet was granted bail but was required to remain under police guard twenty-four hours a day in Grovelands Priory Hospital. During the first days of December, Pinochet was allowed to move to a rented private estate in Surrey while the case proceeded through the courts. After two weeks of consideration, the Law Lords ruled that Pinochet did have immunity, and on January 2, 2001, the home secretary refused to extradite Pinochet. Pinochet was not given approval to leave England until March 2. By that time, public opinion in Chile forced the Chilean government to state that Pinochet could face criminal proceedings in Chile for war crimes.

Consequences

Although Pinochet was not extradited, a precedent was established for holding dictators responsible for crimes committed while they were in office. Under the Geneva Convention signed after World War II, all nations theoretically are required to assist other members in matters of terrorism and deny a safe haven for those accused of war crimes. This part of the convention was meant to prevent people from avoiding prosecution for gross crimes by fleeing abroad. The extradition efforts aroused public opinion in Chile and around the world and forced the Chilean government to change its stance on prosecution of Pinochet, who was no longer assured of immunity from prosecution in Chile.

Robert D. Talbott

Former Astronaut Glenn Returns to Space

John H. Glenn, Jr., the first American to orbit Earth, pursued successful careers in business and national politics. While serving as a senator from Ohio, he fulfilled a long-standing desire to return to space.

What: Space and aviation; Government; Health; Technology
When: October 29 to November 7, 1998
Where: Earth orbit
Who:
John H. Glenn, Jr. (1921-   ), an original Mercury astronaut
Curt Brown (1956-   ), STS-95 commander
Daniel S. Goldin (1940-   ), National Aeronautics and Space Administration (NASA) administrator
Annie Glenn, wife of John H. Glenn, Jr.

Glenn's Second Flight

In 1997, while serving on the Special Committee on Aging, Senator John H. Glenn, Jr., a Democrat from Ohio, noted similarities between changes that astronauts' bodies undergo while weightless and physiological aspects of aging on Earth. The former Mercury astronaut campaigned for a space shuttle mission assignment, arguing that space-based medical research using an elderly subject could improve the quality of life of osteoporosis patients in particular and aging peoples' health in general. Initially, he received lukewarm responses at the National Aeronautics and Space Administration (NASA) and objections from his wife, Annie, and his two children. After much persistence, NASA administrator Daniel S. Goldin gave Glenn's proposal conditional approval. Glenn had to devise a research plan that passed scientific review and gained approval from the National Institutes of Health and Aging.

During a January 16, 1998, NASA headquarters press briefing, Goldin announced that former astronaut Glenn would fly aboard shuttle mission STS-95 later that year. Glenn entered preflight training, spending as much time at the Johnson Space Center as his Senate duties permitted. As launch approached, Glenn divested his Senate responsibilities and fully integrated himself with the STS-95 crew.

Enthusiastic spectators, media representatives, and famous people filled Kennedy Space Center on October 29, 1998. Even President Bill Clinton and his wife, Hillary, arrived to witness Glenn's return to space aboard space shuttle *Discovery.* Private aircraft circling the area intruded into restricted air space late in the countdown, forcing a delay in liftoff until they could retreat to safety.

Discovery lifted off at 2:19 P.M. under the command of Curt Brown, with payload specialist Glenn seated down on *Discovery*'s middeck. Eight and one-half minutes after liftoff, Glenn and his six fellow astronauts entered orbit. Thirty-six years had passed since Glenn first orbited Earth, and his heralded return to space came at the record-setting age of seventy-seven.

Glenn remained the focus of media interest throughout STS-95, but he stressed how much scientific research had been incorporated into this mission. *Discovery* carried two satellites. Petite Amateur Naval Satellite (PANSAT) was deployed into a separate orbit, supporting an educational program involving Naval Postgraduate School students. SPARTAN, a recoverable free-flying astrophysical observatory capable of conducting solar physics and ultraviolet astronomy research, was deployed early in the mission and retrieved two days later for analysis of recorded data back on Earth. Technology demonstration tests included work in support of upgrades to the Hubble Space Telescope and design of critical support systems for the upcoming International Space Station. By far, the majority of STS-95's eighty-three research programs involved life sciences

AP/Wide World Photos

Senator John Glenn (center) and Commander Curt Brown (front).

studies, many using Glenn as a test subject investigating subjects such as human adaptation to weightlessness, sleep behavior, and osteoporosis.

Numerous in-flight public affairs events were conducted during STS-95, all with Glenn as the center of attention. Most memorable of those included a taped appearance shown on NBC's *Tonight Show* with Jay Leno. Glenn, commander Brown, and the pilot traded jokes with Leno on late-night television.

STS-95 concluded on November 7 with a touchdown at Kennedy Space Center's shuttle landing facility runway at 12:03 P.M., having completed 135 Earth orbits over eight days, twenty-one hours, and forty-four minutes. Although Glenn initially displayed a wobbly gait and perceived heaviness in his legs after touchdown, he was able to disembark from *Discovery* under his own power.

The Early Days

The National Aeronautics and Space Act, passed by Congress and signed by President Dwight Eisenhower in 1958, created NASA. NASA's first major responsibility involved the race between the United States and the Soviet Union to send the first human being into space and return him to safely to Earth. To that end, Project Mercury was devised, and the first set of astronauts were selected in 1959 after a grueling selection process. Among the original seven Mercury astronauts was Glenn, who had previously achieved national recognition after a record-setting supersonic flight across the continental United States and several appearances on a popular game show in the late 1950's.

During the early days of the Space Age, the Soviets achieved the majority of space-firsts, including first satellite placed into orbit, first animal orbited, first satellite to reach lunar distance, first satellite image taken of the Moon's far side, and first human in space. Yuri Gagarin completed one full Earth orbit on April 12, 1961. At the time, Glenn was part of a three-astronaut team preparing for Project Mercury's first manned space flight, a suborbital launch to experience five minutes of weightlessness. That May 5, 1961, flight's prime pilot was Alan B. Shepard, Jr. NASA repeated that success with another suborbital mission flown by Virgil I. Grissom. Glenn then received assignment as prime pilot for Project Mercury's first manned orbital flight. However, following Grissom's brief voyage, the Soviets sent another cosmonaut, Gherman Titov, into orbit, this time for a full day.

After numerous weather and hardware delays, Glenn launched on February 20, 1962, atop an Atlas-D booster in his *Friendship 7* spacecraft. Glenn flew three Earth orbits and returned safely after some serious concern over his spacecraft's heat shield. Becoming the object of restored national pride, Glenn did not fly again during Project Mercury and did not receive even a backup assignment for the later Project Gemini flights. He left NASA in 1964, entering the business world. Glenn achieved considerable financial success, but his strong call to national service pushed the former Marine and astronaut to enter politics. Glenn won election in 1974, served four consecutive Senate terms, unsuccessfully ran for the Democratic presidential nomination in 1984, and retired in 1999.

Consequences

Whereas Glenn had first flown into space as a Cold War warrior, he returned to space in 1998 as a statesman advocating cooperative international scientific research. STS-95 was the final shuttle flight before initiation of construction of the International Space Station in December, 1998.

David G. Fisher

First Digital High-Definition Television Signals Are Broadcast

Through the use of technology developed in the computer industry, the long-awaited dream of high-resolution, cinema-quality television pictures becomes a commercial reality.

What: Communications; Technology; Entertainment
When: October 31, 1998
Where: New York City
Who:
William E. Kennard (1957-), chairman, Federal Communications Commission
Edward O. Fritts (1941-), president and chief executive officer of the National Association of Broadcasters

Television Goes Digital

On October 31, 1998, television broadcasting underwent one of its biggest changes since its pioneering days. For nearly five decades after the National Television Standards Committee (NTSC) established its standard for color television broadcasting in the United States, the basic structure of the television signal had remained relatively stable: 480 horizontal lines of resolution, scanned in two interlaced fields, with the color information sent separately from the brightness information so that color signals could still be viewed on black-and-white sets.

However, many people wanted a larger, sharper picture that would make television more like movies—high-definition television (HDTV). As early as the 1960's, television pioneer Philo T. Farnsworth demonstrated experimental systems with a thousand lines of resolution. However, these were not practical for the consumer market because the signal would require too large a portion of the radio spectrum to carry the necessary information—the bandwidth problem. Because the airwaves were already crowded, the television industry decided to maintain the older standard until a solution could be found.

That solution came from the computer industry, in the form of digital signals. A digital signal is a stream of 1's and 0's, which can be manipulated mathematically by a computer like any other digital data. This means that the signal can be compressed like a computer file by removing repetitions and replacing them with a set of instructions for reconstructing them by the receiver. This compressed signal takes up less space on the radio spectrum and is stronger and more reliable. In addition, digital television signals can share their bandwidth with audio and data transmissions.

In 1993, the International Organization for Standardization (ISO) approved standards for data compression of digital audio and video that would form the basis for an HDTV signal. In 1996, the Federal Communications Commission (FCC) adopted a standard for digital television broadcasts on the airwaves.

Getting Real

However, it was not sufficient to merely have standards for the signals. Before consumer HDTV broadcasts could become a reality, the necessary equipment would have to be produced. Without studio equipment or transmitters, television stations could not send HDTV broadcasts, and without the appropriate receivers in homes, there would be no audiences to view them. Therefore, the FCC developed a schedule for the conversion from NTSC to HDTV that would give stations and consumers time to make the switch in equipment.

During the transition period, each operating television station would be allowed a second frequency for HDTV broadcasts in addition to the existing frequency on which it would continue

NTSC broadcasts. The HDTV station would generally have the same call letters as the NTSC station but be distinguished by the letters DT instead of TV after it; for instance, Chicago's WLS-DT is the HDTV counterpart to WLS-TV broadcasting in NTSC.

According to the FCC schedule, the first HDTV broadcasts were scheduled to take place in cities with the largest populations, such as New York, Chicago, and Los Angeles. Later, additional stations would open in smaller cities, until every city in the United States was served by the year 2002. If the conversion to HDTV went as scheduled, NTSC broadcasts would cease by 2006, and stations would return one of their two frequencies to the FCC, which would then redistribute them to other uses, including public safety, police, and fire departments. However, the period of NTSC broadcasts could be extended if certain conditions were not met, for instance, if too few consumers had purchased HDTV equipment and many people would lose television service altogether if NTSC service were ended.

Although television stations were gearing up to broadcast HDTV, viewers still needed to purchase HDTV-capable receivers before HDTV broadcasts would become a reality. Much like the earliest color televisions in the 1950's and 1960's, the first HDTV systems were extremely expensive and intended primarily for wealthy consumers. Many of these were rear-projection systems with very large screens to create home theaters and cost between $5,000 to $10,000. Generally the tuner was sold as a separate unit.

Consequences

The short-term impact of HDTV has been somewhat mixed. Numerous additional stations in cities beyond the original ten have begun broadcasts in HTDV. However, they continue to have few viewers because HDTV-capable receivers remain expensive and difficult to obtain.

In spite of these problems, industry leaders remained optimistic about the long-term viability of HDTV. FCC chairman William E. Kennard allayed complaints regarding the continued high cost of HDTV receivers by comparing their prices to those of the original color televisions. After the prices were adjusted for inflation, each cost roughly the same. Even the first radios took a substantial bite from the consumer's paycheck. He also pointed out that prices were already dropping. In light of these factors, he argued that the government should not attempt to force growth, only to make sure that it occurs in an orderly fashion.

Similarly, Edward O. Fritts, president and chief executive officer of the National Association of Broadcasters, emphasized that the long-term interest of television broadcasting demanded a shift to digital technology. The old analog technology simply would not be able to continue to compete in a world in which such technologies as the Internet were steadily becoming more important. Although there had been some delays in various stations' transition to digital, on the whole, the television industry was actually staying ahead of the FCC's voluntary timetables—proof the free market was working for the benefit of the consumer.

In the long term, the development of HDTV is an opening toward the convergence of television and computers. Many experts predict that cable television and the Internet will ultimately merge into a seamless information appliance.

Leigh Husband Kimmel

DNA Tests Indicate That Jefferson Fathered Slave's Child

DNA tests strongly indicated that Eston Hemings, born in 1808, was the son of President Thomas Jefferson and the latter's slave Sally Hemings. It had been rumored since Jefferson's lifetime that he had fathered at least one child with Hemings, and the test results seemed to confirm this.

What: Genetics; Technology
When: November 5, 1998
Where: Charlottesville, Virginia
Who:
EUGENE FOSTER (1927-), retired pathologist formerly with University of Virginia
ANNETTE GORDON-REED, author of *Thomas Jefferson and Sally Hemings*
THOMAS JEFFERSON (1743-1826), third president of the United States from 1801 to 1809
SALLY HEMINGS (1773-1836), one of Jefferson's slaves

An Old Rumor Revived

On November 5, 1998, many Americans were surprised to learn that their third president, Thomas Jefferson, almost certainly fathered at least one child with Sally Hemings, one of his slaves. The report was based on deoxyribonucleic acid (DNA) tests performed by Eugene Foster, a pathologist retired from the University of Virginia. Foster's results were based on a comparison of DNA from descendants of Jefferson's uncle Field Jefferson and DNA from descendants of Hemings's son Eston, who was born in 1808 at Monticello, Jefferson's home. Foster concluded that the tests, in which nineteen genetic markers on the Y chromosome matched between the two groups studied, strongly pointed to Jefferson as Eston's father. According to Foster, "The chances are less than one in one hundred, maybe less than one in one thousand, that Eston is not a Jefferson descendant." Foster's study was released in the scientific journal *Nature.*

Rumors about Jefferson having sexual relations and children with his slave Hemings had begun circulating in 1802, during Jefferson's first term as president. The charge was repeated during his presidency by his political opponents. Jefferson never directly denied or confirmed the accusation. There were also rumors that Jefferson had fathered Hemings's son Madison, born in 1805 with red hair, but because Madison had no line of male descendants, DNA tests could not be done to help determine whether the rumors were true. Foster also had conducted DNA tests on descendants of Thomas Woodson, another son of Sally Hemings who had been linked to Jefferson by oral traditions, but Foster's results indicated that Jefferson was not the father of Woodson.

A Founding Father's Reputation at Stake

Americans reacted to the Jefferson DNA tests in various ways. Some refused to believe that one of the United States' greatest presidents and a founding father could have had sexual relations with a slave and, therefore, challenged the results of the DNA tests. Some argued that the Foster tests had examined only nineteen genetic markers on the Y chromosomes of the Jefferson and Hemings descendants, compared with more scientific tests that examine about two hundred genetic markers. Others suggested that Eston Hemings could have been fathered by Jefferson's brother Randolph, who lived within twenty miles of Monticello and visited there often. However, many African Americans, including Ray Farmer of Silver Spring, Maryland, were not surprised by the test results. Farmer told *USA Today,* "We've been saying it all along. . . . So Jefferson has black relatives? Well, all my relatives are black, and I like them just fine."

Historians came forward to explain the old rumors about Hemings and Jefferson, who developed a close relationship. Hemings was not originally Jefferson's slave but that of his wife, Martha, who inherited Hemings and was actually Hemings's half-sister. Hemings went to France with Jefferson's wife in 1787, joining Jefferson there. After Jefferson's return to the United States in 1789, Hemings remained with him as a house servant at Monticello. Much later, in 1873, Hemings's son Madison claimed that his mother and Jefferson had a love affair while in France and that Jefferson was his father. After Foster revealed the results of his DNA analysis, a debate began across the nation on television and radio talk-shows and especially on college campuses about whether, and to what extent, the DNA results would change Jefferson's status as a revered icon of U.S. history.

Consequences

The impact of the Jefferson-Hemings controversy went far beyond discussion of the DNA testing and the reliability of its results. The story added a historical dimension to the debate about President Bill Clinton's alleged relationship with Monica Lewinsky, which was sweeping the nation in 1998 and which ultimately led to Clinton's impeachment. Some used the Jefferson story to argue against Clinton's impeachment, arguing that if a figure as great as Jefferson had engaged in an affair with a black slave woman, Clinton's private behavior should not be linked with his professional capacity to do his job. Others maintained that any president ought to uphold a high standard of private morality, and that Jefferson's action, while not correct, took place in a very different time and place and had no relevance to President Clinton's case.

AP/Wide World Photos

Descendants of Sally Hemings and of Thomas Jefferson pose on the steps of Monticello in 1999.

Another result of the Jefferson-Hemings DNA test was a closer look by scholars at the nature and history of black-white relations in the United States. Some historians argued that white slave owners routinely had sexual relations and fathered children with their black female slaves, in order not only to satisfy their desires but also as a way of exercising control over their slaves. They emphasized Jefferson's action as one more piece of evidence of the evils of slavery in the United States. Moreover, the DNA controversy caused scholars and teachers to remind Americans that although white Americans had always feared miscegenation and even punished it severely, it had occurred during every phase of U.S. history. Finally, the controversy reminded Americans that the founding fathers were human and did have human weaknesses. According to Annette Gordon-Reed, author of *Thomas Jefferson and Sally Hemings* (1997), "What this proves is that Thomas Jefferson was a human being, not a marble icon."

Robert Harrison

Speaker of the House Gingrich Steps Down

Republican Newt Gingrich of Georgia resigned as Speaker of the U.S. House of Representatives three days after the Republicans unexpectedly lost five House seats in the 1998 midterm elections. His resignation came just hours after he was challenged for the leadership post.

What: Government; Law; Politics
When: November 6, 1998
Where: U. S. House of Representatives
Who:
NEWT GINGRICH (1943-), Speaker of the House of Representatives from 1995 to 1998
BILL CLINTON (1946-), president of the United States from 1993 to 2001
BOB LIVINGSTON (1943-), House Appropriations Committee chairman
J. DENNIS HASTERT (1942-), U.S. representative from Illinois

The Path to Resignation

Representative Newt Gingrich of Georgia spearheaded a Republican resurgence in the U.S. House of Representatives in the 1994 midterm elections. The Republicans gained fifty-two House seats for a 230-204 margin over the Democrats, giving them control of both houses of the U.S. Congress for the first time since the 1952 elections. Gingrich framed the Contract with America, a ten-point plan for smaller government, less regulation, term limits, tax cuts, family-oriented policies, strong national defense, and other conservative ideas. He replaced retired Bob Michel of Illinois as party leader and, on January 4, 1995, was elected Speaker of the House.

Gingrich's assertive, strategic vision of leadership initially transformed the speakership into a powerful public forum. Gingrich began crafting rules and reforms to place the most power in the speakership since 1910, hoping to supplant Democratic president Bill Clinton as the primary source of ideas about the direction of the nation. The House approved a balanced budget and welfare reform, giving the states responsibility for the poor.

Gingrich already had become a controversial, polarizing figure. In November, 1994, he had violated House rules by making a questionable tax arrangement on a book he wrote. Gingrich had signed a multimillion dollar book deal with Robert Murdoch's publishing company. In 1995, Democrats filed complaints with the House Ethics Committee that Gingrich had accepted free cable time to broadcast his class at Kennesaw State University and used his political action committee, GOPAC, to funnel campaign contributions and gifts to his own campaign. The Ethics Committee in December, 1995, appointed an outside counsel to investigate Gingrich's conduct.

President Clinton, meanwhile, outmaneuvered Gingrich during the deadlock in budget negotiations. Gingrich favored federal government shutdowns to prove his case for smaller government, but Republicans shouldered the blame for cutting school lunch programs and shutting down the government. Democrats labeled Gingrich an extremist willing to shut down the government to get his way and gained momentum in the November, 1996, elections. Republicans lost eight House seats, though they retained a 227-207 edge over the Democrats.

Gingrich faced further problems. The House Ethics Committee in December, 1996, concluded that Gingrich had violated House rules and misled the committee in its investigation of possible political use by GOPAC of tax-exempt donations. The House on January 7 narrowly defeated, 222-210, a motion to elect an interim speaker until completion of the committee investigation and reelected Gingrich, 216-205, over Democrat Dick Gephardt of Missouri. The Ethics Committee on January 17 reprimanded Gingrich and fined him

AP/Wide World Photos

Newt Gingrich announces his resignation.

$300,000 to cover approximate investigation costs. Four days later, the House overwhelmingly approved, 395-28, the reprimand and fine. Several Republican conservatives questioned the Speaker's leadership skills and complained that Gingrich had compromised too much with Clinton, but they failed in a July 10 attempt to remove him. Gingrich removed handpicked assistant Bill Paxon of New York from his leadership role but could not oust the others because they had been elected by the House caucus.

The Resignation

In 1998, Gingrich pushed for a House impeachment inquiry into Clinton's extramarital affair. House Republicans anticipated solidifying their impeachment moves by expanding their majority in the November, 1998, midterm elections. Democrats, however, picked up five House seats in those elections. Republican Representatives debated over who was to blame for the election results. The election marked only the second time since the Civil War that the party controlling the presidency had gained seats in midterm elections. The Republican agenda, however, seemed to run out of steam. Rank-and-file members claimed that Gingrich could no longer form the broad coalitions necessary to run the House. His public job approval rating had dropped to 38 percent, and he had become especially unpopular among women and minorities.

Gingrich, on November 6, announced his intention to run for reelection as Speaker of the House. Appropriations Committee chairman Bob Livingston the same day challenged Gingrich for the Speaker position. He had already rounded up commitments from more than one hundred members who vowed to support him. Livingston, who projected himself well on television and was an able money raiser, had demonstrated leadership abilities on the Appropriations Committee and had formed allies in different factions.

In a conference call with confidants that night, Gingrich announced that he would resign as the Speaker and leave his Sixth Congressional District seat at the end of his term. He declared, "The Republican conference needs to be unified, and it is time for me to move forward, where I believe I still have a significant role to play for our country and our party. I urge my colleagues to pick leaders who can better produce and discipline, who can work together and communicate effectively." Johnny Isakson replaced him as House representative from Georgia in February, 1999.

Consequences

The ideological, combative Gingrich had lost his ability to unify, produce, discipline, and communicate effectively with Republicans. Gingrich had lost the vision and message of the Republican resurgence because of his arrogance and abuse of power. The messenger changed, but the Republicans remained committed to the message of the Republican resurgence.

The Gingrich resignation made Livingston the leading contender for Speaker of the House. Republican representatives Bill Archer of Texas, Christopher Cox of California (also House Ways and Means Committee chairman), and James Talent of Missouri began lining up support for their candidacies. At a party caucus on November 18, 1998, Republicans unanimously supported Livingston as the next Speaker. Livingston on December 18, however, withdrew his candidacy and resigned his House seat because he had engaged in occasional extramarital affairs. Republicans on January 6, 1999, elected J. Dennis Hastert of Illinois as the new Speaker. Hastert, a conservative conciliator, was less ideological and combative than Gingrich.

David L. Porter

Brazil's Ailing Economy Gets $41.5 Billion in International Aid

Although President Fernando Henrique Cardoso saved the Brazilian economy from hyperinflation, the nation faced economic disaster in 1998 until it was rescued by an international aid package put together by the United States.

What: Economics; International relations
When: November 13, 1998
Where: Brazil
Who:
FERNANDO HENRIQUE CARDOSO (1931-), Brazilian minister of finance and president from 1995
ITAMAR FRANCO (1930-), vice president and later president of Brazil from 1992 to 1995
BILL CLINTON (1946-), president of the United States from 1993 to 2001

A Troubled Transition

Notorious throughout its history for political and financial turmoil, Brazil hit rock bottom in the early 1990's. After Brazil's twenty-one-year-long military dictatorship ended in 1985, the Brazilian people were faced with an annual inflation rate that reached more than 1,000 percent by 1990. By the early 1990's, they had witnessed one president elect die without ever taking office, another denied reelection because of his inability to overhaul the economy, and a third was impeached and forced to resign in 1992 for accepting bribes.

In 1992, Vice President Itamar Franco, a colorless but ambitious politician, assumed the presidency and, in 1993, appointed former federal senator and one-time sociologist Fernando Henrique Cardoso as minister of finance. Cardoso moved swiftly to dismantle the state-run sector of the economy, which he felt was chiefly responsible for the nation's economic quagmire. He instigated a number of important moves to revitalize Brazilian capitalism: selling off public companies to private parties, lowering tariffs on foreign-made products, and enhancing free trade with the neighboring countries of Argentina, Paraguay, and Uruguay. By these measures, Cardoso brought inflation down to a manageable level. He soon became more popular with Brazilian voters than Franco, and the leading candidate for president in 1994.

Cardoso easily won the election over his leftist rival and proceeded to integrate the Brazilian economy even further with the rest of the world, easing restrictions on foreign investment in key sectors of the economy such as communications and accepting harsh but inescapable terms from the International Monetary Fund (IMF), the World Bank, and the Inter-American Development Bank to cut the national budget in return for loans and financial aid for development projects. However, the president failed to appreciate how vulnerable the Brazilian economy had become to economic turmoil overseas. In August of 1998, Russia, reeling from stampeding inflation, announced it could not pay back loans due to the IMF. This set off a flight of capital not only from that country but also from other nations deemed unsafe by investors. Brazil fell into this category because Cardoso had overvalued the real, the Brazilian currency, keeping it nearly on par with the U.S. dollar.

Sensing Cardoso's wobbling position, former president and current governor Franco, who dreamed of succeeding Cardoso in 1998, announced that his state, Minas Gerais, the second most productive in the nation, would not pay taxes owed to the federal government, immediately touching off a financial panic: devaluation of the real by more than 40 percent against the dollar; the reappearance of inflation, which had been successfully contained for four years; and the resignation of three presidents of the Central Bank. Brazil seemed headed for an economic depression. Left unchecked, such a catastrophe would have spread to other Latin American countries, particularly Brazil's trade partners in the South American Common Market (MERSOCUR), and also endangered the economy of the United States because American companies had invested hundreds of millions of dollars in Brazil.

The IMF Bailout

U.S. president Bill Clinton, then in his second term, watched the developments in Brazil with great alarm. Clinton had praised Cardoso for his commitment to open markets and civilian government. The diplomatic giants of the Northern and Southern Hemispheres could not afford a falling out. A staggering financial crisis might force Cardoso to rethink the premises of his economic program and pour more money into the public sector just to ward off a recession, only to reignite inflation. U.S. plans to integrate the entire Western Hemisphere into one giant trade block would have to be postponed indefinitely.

In November of 1998, Clinton ordered the Treasury Department to arrange an aid package to Brazil that would include $18 billion from the IMF, $4.5 billion each from the World Bank and the Inter-American Development Bank, and billions more from European, Japanese, and other donors. The total package came to just over $41 billion, with the IMF promising another $9 billion in three years if Cardoso fulfilled the conditions attached to the bailout. These included spending cuts in social services such as health care, raising consumer interest rates to more than 100 percent to prevent a new round of inflation, and cutting the size of the federal government by firing thousands of civil employees. Cardoso, reelected in October of 1998, still had only a weak majority of supporters in the Brazilian congress and timidly accepted the agreement. The result for Brazilian workers was predictable

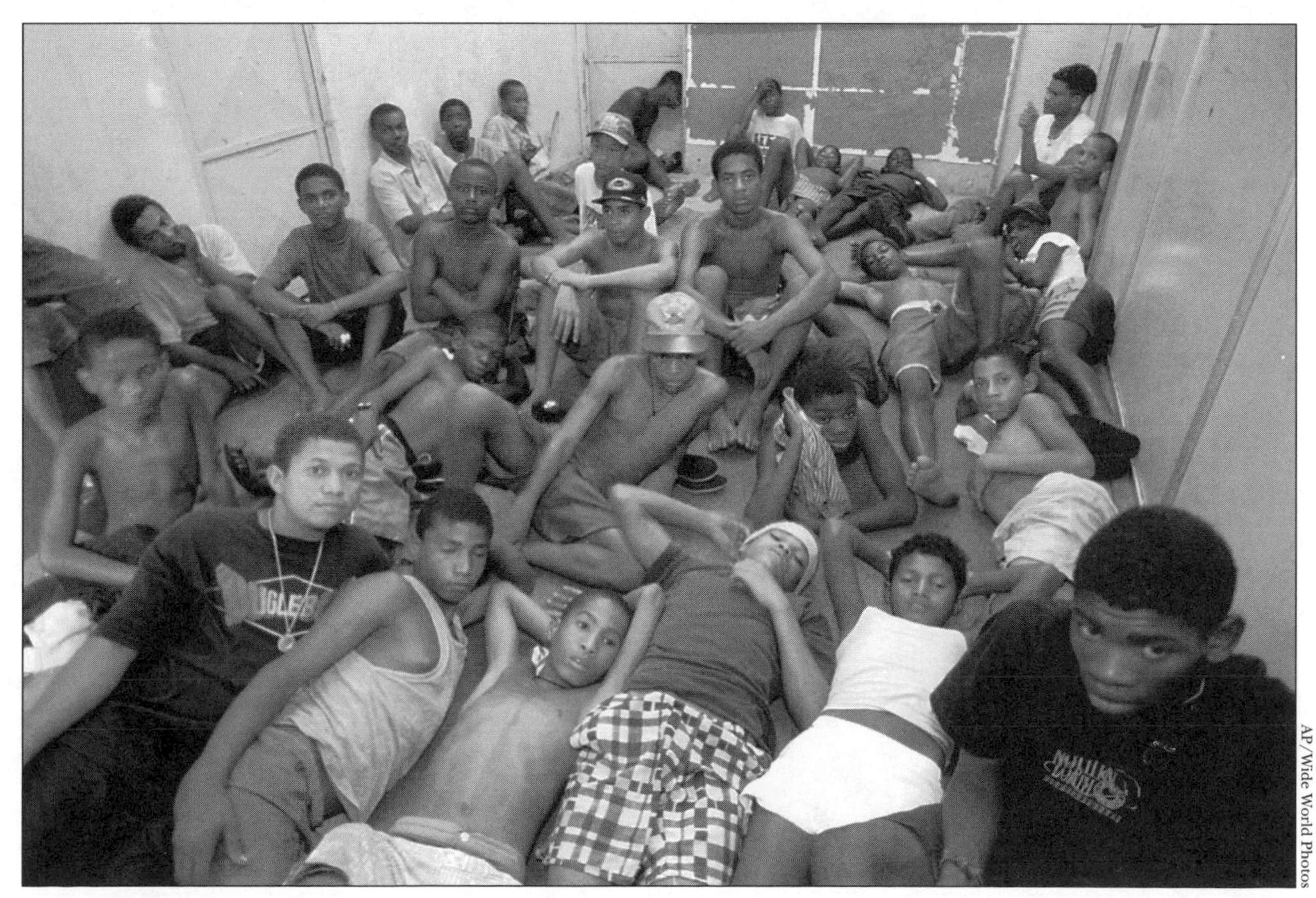

AP/Wide World Photos

Homeless children watch television at a social center in Rio de Janeiro, a week after the announcement of the international aid.

but nevertheless painful: Unemployment surged, especially in industrial cities, the income gap between rich and poor widened further in a country the World Bank had found to possess one of the world's most extreme and unjust distributions of wealth, and the president abandoned his bold plans to modernize Brazilian education. However, Cardoso and the IMF achieved their main goals—foreigners returned their money to Brazil, inflation was kept at double digits, and Brazil's real underwent no further serious devaluations.

Consequences

However, all of this was cold comfort for those without wealth in Brazil: the 60 million who lived below the official poverty line with no chance for employment except in odd jobs and the two-thirds of the population who never finished primary school. As long as the IMF insisted on giving priority to the private sector, there was little chance that Brazilian living standards would rise. In fact, Cardoso spent the first two years of his second term trying to enact cuts in public pensions and raising the retirement age. Both proposals were rejected by the Brazilian congress. By 2001, his low standing with the public ranked with that of many of his incompetent military and civilian predecessors.

Julio Cèsar Pino

GREAT EVENTS

1900-2001

CATEGORY INDEX

LIST OF CATEGORIES

AGRICULTURE

ANTHROPOLOGY

BIOLOGY

BUSINESS

CIVIL STRIFE

CIVIL WAR

COMMUNICATIONS

COMPUTER SCIENCE

CONSUMER ISSUES

COUPS

CRIME

DISASTERS

EARTH SCIENCE

ECONOMICS

EDUCATION

ENERGY

ENGINEERING

HEALTH

HUMAN RIGHTS

INTERNATIONAL LAW

INTERNATIONAL RELATIONS

LABOR

LAW

MATERIALS

MATHEMATICS

MEDICINE

NATIONAL POLITICS

PHOTOGRAPHY

PHYSICS

POLITICS

RELIGION

SOCIAL CHANGE

SOCIAL REFORM

SPACE AND AVIATION

SPORTS

TECHNOLOGY

TERRORISM

TRANSPORTATION

WAR

WEAPONS TECHNOLOGY